Fiber Reinforced Polymer Composites: Mechanical Properties and Applications

Fiber Reinforced Polymer Composites: Mechanical Properties and Applications

Editors

R.A. Ilyas
S.M. Sapuan
Emin Bayraktar
Shukur Abu Hassan
Nabil Hayeemasae
Khubab Shaker

MDPI • Basel • Beijing • Wuhan • Barcelona • Belgrade • Manchester • Tokyo • Cluj • Tianjin

Editors

R.A. Ilyas
Department of Chemical Engineering
Universiti Teknologi Malaysia
Skudai
Malaysia

S.M. Sapuan
Department of Mechanical and Manufacturing Engineering
Universiti Putra Malaysia
Serdang
Malaysia

Emin Bayraktar
School of Mechanical and Manufacturing Engineering
ISAE-SUPMECA
Paris
France

Shukur Abu Hassan
Centre for Advanced Composite Materials (CACM)
Universiti Teknologi Malaysia
Skudai
Malaysia

Nabil Hayeemasae
Department of Rubber Technology and Polymer Science
Prince of Songkla University
Pattani
Thailand

Khubab Shaker
Department of Materials
National Textile University
Faisalabad
Pakistan

Editorial Office
MDPI
St. Alban-Anlage 66
4052 Basel, Switzerland

This is a reprint of articles from the Special Issue published online in the open access journal *Polymers* (ISSN 2073-4360) (available at: www.mdpi.com/journal/polymers/special_issues/HK8P7953V9).

For citation purposes, cite each article independently as indicated on the article page online and as indicated below:

LastName, A.A.; LastName, B.B.; LastName, C.C. Article Title. *Journal Name* **Year**, *Volume Number*, Page Range.

ISBN 978-3-0365-8409-6 (Hbk)
ISBN 978-3-0365-8408-9 (PDF)

Contents

About the Editors

R.A. Ilyas

R.A. Ilyas is a senior lecturer in the Faculty of Chemical and Energy Engineering, Universiti Teknologi Malaysia, Malaysia. He is also a Fellow of the International Association of Advanced Materials (IAAM), Sweden, Fellow of the International Society for Development and Sustainability (ISDS), Japan, a member of Royal Society of Chemistry, UK, and the Institute of Chemical Engineers (IChemE), UK. R.A. Ilyas was the recipient of the MVP Doctor of Philosophy Gold Medal Award UPM in 2019 for Best Ph.D. Thesis and Top Student Award, INTROP, UPM. He was awarded with Outstanding Reviewer by Carbohydrate Polymers, Elsevier, United Kingdom, Top Cited Article 2020-2021 Journal Polymer Composite, Wiley, 2022, and Best Paper Award at various International Conferences. R.A. Ilyas was also listed and awarded Among World's Top 2% Scientist (Subject-Wise) Citation Impact during the Single Calendar Years of 2019, 2020, and 2021 by Stanford University, US, PERINTIS Publication Award 2021 and 2022 by Persatuan Saintis Muslim Malaysia, Emerging Scholar Award by Automotive and Autonomous Systems 2021, Belgium, Young Scientists Network - Academy of Sciences Malaysia (YSN-ASM) 2021, UTM Young Research Award 2021, UTM Publication Award 2021, and UTM Highly Cited Researcher Award 2021. In 2021, he won the Gold Award and Special Award (Kreso Glavac (The Republic of Croatia) at the Malaysia Technology Expo (MTE2022), Gold Award dan Special Award at International Borneo Innovation, Exhibition and Competition 2022 (IBIEC2022), and a Gold Award at New Academia Learning Innovation (NALI2022). His main research interests are (1) polymer engineering (biodegradable polymers, biopolymers, polymer composites, polymer-gels) and (2) material engineering (natural fiber-reinforced polymer composites, biocomposites, cellulose materials, nano-composites). To date, he has authored or co-authored more than 480 publications.

S.M. Sapuan

S.M. Sapuan is Professor (A grade) of composite materials in the Department of Mechanical and Manufacturing, Universiti Putra Malaysia (UPM), Head of Advanced Engineering Materials and Composite Research Centre, UPM, and Chief Executive Editor of Pertanika journals in the office of Deputy Vice Chancellor, UPM. He earned his BEng in Mechanical Engineering, University of Newcastle, Australia, MSc in Engineering Design, Loughborough University, UK, and PhD in Material Engineering, De Montfort University, UK. He is a Professional Engineer and Fellow of Society of Automotive Engineers, Academy of Science Malaysia, International Society for Development and Sustainability, Plastic and Rubber Institute Malaysia, Malaysian Scientific Association, International Biographical Association and Institute of Material Malaysia. He is an Honorary Member of Asian Polymer Association and Founding Chairman and Honorary Member of Society of Sugar Palm Development and Industry, Malaysia. He was World Class Visiting Professor, Andalas University Indonesia. He is Editor-in-Chief of the *Journal of Natural Fibre Polymer Composites*, co-Editor-in-Chief of *Functional Composites and Structures*, Associate Editor-in-Chief of *Defence Technology*, Elsevier, and Editorial Board Member of 30 journals. He has produced over 2000 publications, including over 900 journal papers, 55 books, and 220 chapters in book. He has delivered 60 plenary and keynote lectures, and 150 invited lectures. He has organized 31 journal Special Issues as Guest Editor, reviewed 1500 journal papers, and has 8 patents. He successfully supervised 98 PhD and 75 MSc students and 16 postdoctoral researchers. His h-index is 107 with 41710 citations (Google Scholar).

He received the ISESCO Science Award, Khwarizimi International Award, Kuala Lumpur Royal Rotary Gold Medal Research Award, National Book Award, Endeavour Research Promotion Award, TMU/IEEE India, Citation of Excellence Award, Emerald, UK, and Malaysia's Research Star Award.

Emin Bayraktar

Emin Bayraktar (Habil., Dr (PhD), DSc- Dr es Science) is an academic and research staff-member at Mechanical and Manufacturing engineering at SUPMECA/Paris, France. His research areas include manufacturing techniques of new materials (basically composites –hybrid), metal forming of thin sheets (design + test + FEM), static and dynamic behaviour and optimisation of materials (experimental and FEM –utilisation and design of composites-based metallic and non-metallic, powder metallurgy, energetic materials aeronautical applications)-based metallic, non-metallic, powder metallurgy and metallurgy of steels, welding, heat treatment, as well as the processing of new composites, sintering techniques, sinter –forging, thixoforming, etc. He has authored more than 200 publications in international journals and international conference proceedings, and has also authored more than 90 research reports (European = Steel Committee projects, Test + Simulation). He has already advised 32 PhDs and 120 MSc Theses, and 7 are ongoing. He is a Fellow of WAMME (World Academy of Science in Materials and Manufacturing Engineering), Editorial Board - Member of JAMME (International Journal of Achievement in Materials and Manufacturing Engineering), Advisory board member of AMPT-2009 (Advanced Materials Processing technologies), APCMP-2008, and APCMP-2010. He was a Visiting Professor at Nanyang Technology University, Singapore in 2012, Xi'an Northwestern Technical University, Aeronautical Engineering, in 2016, University of Campinas, UNICAMP-Brazil in 2013 until 2023. He is recipient of Silesian University Prix pour "FREDERIK STAUB Golden Medal-2009" by Academy of WAMME, "World Academy of Science"-Poland, materials science section, and recipient of the William Johnson International Gold Medal - 2014, AMPT academic association.

Shukur Abu Hassan

Shukur Abu Hassan, PhD (UTM,Malaysia), MSc (Warwick University, UK), BSc (UTM, Malaysia) is an Engineer, Associate Professor and the current Director of the Centre for Advanced Composite Materials (CACM), Universiti Teknologi Malaysia (UTM), Malaysia. In his teaching career of 33 years, he has worked as the Centre Director, Head of Composite Lab and Committee Member at the Faculty and University level. He is also involved as a Consultant and on the Editorial Advisory Board for various international journals. Assoc. Prof. Shukur research focuses on (1) composites structure durability, (2) recycling composites, (3) biomimetics design, (4) composites testing, (5) natural fibre composites, (6) hybrid composites, (7) composites bonding, and (8) polymer mortar composites. He is registered with the Board of Engineers Malaysia (BEM) and Member of Malaysian Board of Technologist (MBOT). To date, he has been a consultant to 16 projects amounting to RM 480,000.00 and Principal Investigator and member to 49 research grants worth RM 3.9 million. He has also supervised and co-supervised 16 postgraduates. He has also authored and co-authored 106 publications (published/accepted): 22 ISI Article/Journal, 9 Scopus Article/Journal, 10 Non-Indexed Article/Journal, 2 Indexed Conference Proceedings, 43 Non-Indexed Conference Proceedings, 7 Book Chapters, and 13 Classified/Technical/Expert Reports. He also has one patent granted, one patent filed, and seven registered industrial designs.

Nabil Hayeemasae

Nabil Hayeemasae is currently an Assistant Professor at the Department of Rubber Technology and Polymer Science, Faculty of Science and Technology, Prince of Songkla University, Pattani Campus. He obtained a Ph.D. in Polymer Engineering at the School of Materials and Mineral Resources Engineering, Universiti Sains Malaysia. He has conducted research in several research areas, particularly on the re-utilization of polymer and solid wastes into potential materials, rubber composites, and rubber recycling. So far, his research outputs have been selected to be published in over 80 papers internationally.

Khubab Shaker

Khubab Shaker currently serves as an Assistant Professor and Chairman of the Department of Materials at National Textile University, Pakistan. He is an active Textile Composite Materials Research Group member at NTU, with 73 publications (impact factor = 200), 6 books, and 34 conference presentations. He has more than 13 years of industrial and teaching experience in textiles and composite materials. His research areas include composites joining, natural fibers, biopolymers, green composites, and their applications as potential replacements for synthetic composite materials.

Preface to "Fiber Reinforced Polymer Composites: Mechanical Properties and Applications"

Composites are the most promising material available in the twenty-first century. Composites reinforced with synthetic or natural fibers are becoming extremely prevalent as the demand for lightweight materials with high strength for specific applications grows. Fiber-reinforced polymer composite has a high strength-to-weight ratio and excellent properties, such as high durability, stiffness, damping property, flexural strength, and resistance to corrosion, wear, impact, and fire. Various properties of composite materials have led to applications in construction, aerospace, automobile, biomedical, marine, and many other industries. Because their constituent elements and manufacturing techniques primarily determine the performance of composite materials, the functional properties of various fibers available worldwide, their classifications, and the manufacturing techniques used to fabricate the composite materials must be investigated.

R.A. Ilyas, S.M. Sapuan, Emin Bayraktar, Shukur Abu Hassan, Nabil Hayeemasae, and Khubab Shaker

Editors

polymers

MDPI

Editorial

Fibre-Reinforced Polymer Composites: Mechanical Properties and Applications

R. A. Ilyas [1,2,3,*], S. M. Sapuan [4], Emin Bayraktar [5], Shukur Abu Hassan [2], Nabil Hayeemasae [6], M. S. N. Atikah [7] and Khubab Shaker [8]

[1] School of Chemical and Energy Engineering, Faculty of Engineering, University Teknologi Malaysia (UTM), Johor Bahru 81310, Malaysia
[2] Centre for Advanced Composite Materials, University Teknologi Malaysia (UTM), Johor Bahru 81310, Malaysia
[3] Institute of Tropical Forest and Forest Products (INTROP), University Putra Malaysia (UPM), Serdang 43400, Malaysia
[4] Advanced Engineering Materials and Composites, Department of Mechanical and Manufacturing Engineering, Faculty of Engineering, University Putra Malaysia (UPM), Serdang 43400, Malaysia
[5] School of Mechanical and Manufacturing Engineering, ISAE-SUPMECA Institute of Mechanics of Paris, 93400 Saint-Ouen, France
[6] Department of Rubber Technology and Polymer Science, Faculty of Science and Technology, Prince of Songkla University, Pattani Campus, Pattani 94000, Thailand
[7] Department of Chemical and Environmental Engineering, Faculty of Engineering, University Putra Malaysia (UPM), Serdang 43400, Malaysia
[8] Department of Materials, School of Engineering and Technology, National Textile University, Punjab 37610, Pakistan
* Correspondence: ahmadilyas@utm.my

Citation: Ilyas, R.A.; Sapuan, S.M.; Bayraktar, E.; Hassan, S.A.; Hayeemasae, N.; Atikah, M.S.N.; Shaker, K. Fibre-Reinforced Polymer Composites: Mechanical Properties and Applications. *Polymers* **2022**, *14*, 3732. https://doi.org/10.3390/polym14183732

Received: 19 August 2022
Accepted: 26 August 2022
Published: 7 September 2022

"Fibre-Reinforced Polymer Composites: Mechanical Properties and Applications" is a newly open Special Issue of *Polymers*, which aims to publish original and review papers on new scientific and applied research and make boundless contributions to the finding and understanding of the reinforcing effects of various synthetic and natural fibres on the performance of biopolymer composites. This Special Issue also covers the fundamentals, characterisations, and applications of synthetic and natural fibre-reinforced biopolymer composites.

Rapid growth in the manufacturing industries has necessitated the improvement in materials in terms of density, stiffness, strength, and cost-effectiveness with increased sustainability. Composite materials have been developed as one of the materials with such improvements in these qualities that serve their promise in various applications. Composite materials are made up of two or more elements, one of which is present in the matrix phase (synthetic or biopolymer [1–6]) and the other in particle or fibre form. Composites have been discovered to be the most promising material available in the twenty-first century. Composites reinforced with synthetic or natural fibres are becoming extremely prevalent as the market grows in demand for lightweight materials with high strength for specific applications. The matrix, which serves primarily to hold the reinforcement together, is also regarded as resin, particularly in the case of polymers.

In order to create eco-friendly composites, natural fillers such as natural fibres [7–9], nanocrystalline cellulose, nanofibrillated cellulose [10,11], bacterial nanocellulose [12], and chitosan have been added to the polymer matrix. This increased material qualities while minimising the problem of residue formation. Many researchers have reported the benefits of cellulosic fibres, including the fact that they are abundant in nature, renewable, cost-effective, and non-toxic, as well as providing necessary bonding with the cement-based matrix for significant improvements in material properties such as flexural capacity, toughness, ductility, and impact resistance.

Several studies that investigated the fibre-loading effect on polymer composites found that it had a good relationship with tensile strength. Studies on the fibre loading effect that led to the tensile strength were observed [13,14]. It was demonstrated that the optimum fibre loading for kenaf/thermoplastic polyurethane composites was 30% [15]. Other studies regarding kenaf fibre and phenol-formaldehyde (KF/PF) composites reported that kenaf fibre loading up to 43% showed the best tensile strength for the composites [16].

A study on a kenaf-fibre-reinforced, corn-starch-based biocomposite film investigated the fibre loading effect for the tensile properties. It was found that at 6% of the kenaf fibre loading, the optimum tensile properties (17.74 MPa) were observed [17]. Studies on *Cymbopogan citratus* fibre with cassava starch showed higher tensile properties (19.27 MPa) with the use of 50% fibre loading [18]. Another study on arrowroot (*Maranta arundinacea*)-fibre-reinforced arrowroot starch biopolymer composites conducted by Tarique et al. [19] shows that the mechanical properties were enhanced up to 15.22 MPa with the optimum filler content was 10%. Previous studies on the correlation of fibre length to tensile strength were conducted by several researchers. Studies on wheat husk length reinforced rubber composites showed that the highest tensile strength achieved was the medium length of fibre (125–250μm), where the fibre was arranged longitudinally [20]. Jihua Zhu [21] conducted a study on glass-fibre-reinforced composites under acid–base and salt environments. The results indicated that the tensile strength of the GFRP decreased by 22%, 71%, and 87% after 56 d of exposure to 5% NaOH solutions at 20 °C, 50 °C, and 80 °C, respectively.

Not only do fibre-reinforced polymer composites have a high strength-to-weight ratio, but they also have excellent properties such as high durability, stiffness, damping property, flexural strength, and resistance to corrosion, wear, impact, and fire. Various properties of composites materials have led to applications in construction, aerospace, packaging [22–24], electronic, electrical, structural, energy storage [25], automotive [26], filter, coating, bone tissue engineering, and drug delivery [27] and many other industries. Because the performance of composite materials is primarily determined by their constituent elements and manufacturing techniques, the functional properties of various fibres available worldwide, their classifications, and the manufacturing techniques used to fabricate the composite materials must be investigated.

This Special Issue will also cover recent advances in composite processing, mechanical characterisation, and potential applications. Both synthetic and natural fibre-reinforced polymer composites are welcome. Moreover, we welcome approaches to this issue from several vital directions, such as the production of fibres, surface and interfacial characterisation of its properties, economic feasibility, challenges, and future perspectives in the field of polymer composites. As a result of this Special Issue, current and future literature data can be enriched.

Author Contributions: Conceptualization, R.A.I.; writing—original draft preparation, R.A.I., S.M.S., E.B., S.A.H., N.H., M.S.N.A. and K.S.; writing—review and editing, R.A.I., S.M.S., E.B., S.A.H., N.H., M.S.N.A. and K.S.; project administration, R.A.I., S.M.S., E.B., S.A.H., N.H., M.S.N.A. and K.S.; funding acquisition, R.A.I. All authors have read and agreed to the published version of the manuscript.

Funding: This research was funded by Universiti Teknologi Malaysia grant number PY/2022/02318—Q.J130000.3851.21H99, and Ministry of Higher Education Malaysia (MOHE) grant number JPT (BPKI) 1000/016/018/25 (57).

Institutional Review Board Statement: Not applicable.

Informed Consent Statement: Not applicable.

Data Availability Statement: Not applicable.

Acknowledgments: The authors would like express gratitude for the financial support received from the Universiti Teknologi Malaysia, the project "The impact of Malaysian bamboos' chemical and fibre characteristics on their pulp and paper properties, grant number PY/2022/02318—Q.J130000.3851.21H99". The research has been carried out under the program Research Excellence Consortium (JPT (BPKI) 1000/016/018/25 (57)) provided by the Ministry of Higher Education Malaysia (MOHE).

Conflicts of Interest: The authors declare no conflict of interest.

References

1. Ilyas, R.A.; Zuhri, M.Y.M.; Aisyah, H.A.; Asyraf, M.R.M.; Hassan, S.A.; Zainudin, E.S.; Sapuan, S.M.; Sharma, S.; Bangar, S.P.; Jumaidin, R.; et al. Natural Fiber-Reinforced Polylactic Acid, Polylactic Acid Blends and Their Composites for Advanced Applications. *Polymers* **2022**, *14*, 202. [CrossRef] [PubMed]
2. Norfarhana, A.; Ilyas, R.; Ngadi, N. A review of nanocellulose adsorptive membrane as multifunctional wastewater treatment. *Carbohydr. Polym.* **2022**, *291*, 119563. [CrossRef] [PubMed]
3. Ilyas, R.A.; Zuhri, M.Y.M.; Norrrahim, M.N.F.; Misenan, M.S.M.; Jenol, M.A.; Samsudin, S.A.; Nurazzi, N.M.; Asyraf, M.R.M.; Supian, A.B.M.; Bangar, S.P.; et al. Natural Fiber-Reinforced Polycaprolactone Green and Hybrid Biocomposites for Various Advanced Applications. *Polymers* **2022**, *14*, 182. [CrossRef]
4. Ilyas, R.A.; Aisyah, H.A.; Nordin, A.H.; Ngadi, N.; Zuhri, M.Y.M.; Asyraf, M.R.M.; Sapuan, S.M.; Zainudin, E.S.; Sharma, S.; Abral, H.; et al. Natural-Fiber-Reinforced Chitosan, Chitosan Blends and Their Nanocomposites for Various Advanced Applications. *Polymers* **2022**, *14*, 874. [CrossRef] [PubMed]
5. Haris, N.I.N.; Hassan, M.Z.; Ilyas, R.; Suhot, M.A.; Sapuan, S.; Dolah, R.; Mohammad, R.; Asyraf, M. Dynamic mechanical properties of natural fiber reinforced hybrid polymer composites: A review. *J. Mater. Res. Technol.* **2022**, *19*, 167–182. [CrossRef]
6. Ilyas, R.; Sapuan, S.; Asyraf, M.; Dayana, D.; Amelia, J.; Rani, M.; Norrrahim, M.; Nurazzi, N.; Aisyah, H.; Sharma, S.; et al. Polymer Composites Filled with Metal Derivatives: A Review of Flame Retardants. *Polymers* **2021**, *13*, 1701. [CrossRef] [PubMed]
7. Ilyas, R.; Sapuan, S.; Atikah, M.; Asyraf, M.; Rafiqah, S.A.; Aisyah, H.; Nurazzi, N.M.; Norrrahim, M. Effect of hydrolysis time on the morphological, physical, chemical, and thermal behavior of sugar palm nanocrystalline cellulose (*Arenga pinnata (Wurmb.) Merr*). *Text. Res. J.* **2021**, *91*, 152–167. [CrossRef]
8. Abral, H.; Ariksa, J.; Mahardika, M.; Handayani, D.; Aminah, I.; Sandrawati, N.; Pratama, A.B.; Fajri, N.; Sapuan, S.; Ilyas, R. Transparent and antimicrobial cellulose film from ginger nanofiber. *Food Hydrocoll.* **2020**, *98*, 105266. [CrossRef]
9. Sabaruddin, F.A.; Tahir, P.; Sapuan, S.M.; Ilyas, R.A.; Lee, S.H.; Abdan, K.; Mazlan, N.; Roseley, A.S.M.; Hps, A.K. The Effects of Unbleached and Bleached Nanocellulose on the Thermal and Flammability of Polypropylene-Reinforced Kenaf Core Hybrid Polymer Bionanocomposites. *Polymers* **2020**, *13*, 116. [CrossRef] [PubMed]
10. Ilyas, R.; Sapuan, S.; Ishak, M.; Zainudin, E. Sugar palm nanofibrillated cellulose (Arenga pinnata (Wurmb.) Merr): Effect of cycles on their yield, physic-chemical, morphological and thermal behavior. *Int. J. Biol. Macromol.* **2019**, *123*, 379–388. [CrossRef]
11. Syafri, E.; Jamaluddin; Sari, N.H.; Mahardika, M.; Amanda, P.; Ilyas, R.A. Isolation and characterization of cellulose nanofibers from Agave gigantea by chemical-mechanical treatment. *Int. J. Biol. Macromol.* **2022**, *200*, 25–33. [CrossRef] [PubMed]
12. Abral, H.; Chairani, M.K.; Rizki, M.D.; Mahardika, M.; Handayani, D.; Sugiarti, E.; Muslimin, A.N.; Sapuan, S.; Ilyas, R. Characterization of compressed bacterial cellulose nanopaper film after exposure to dry and humid conditions. *J. Mater. Res. Technol.* **2021**, *11*, 896–904. [CrossRef]
13. Bahrain, S.H.K.; Rahim, N.N.C.A.; Mahmud, J.; Mohammed, M.N.; Sapuan, S.M.; Ilyas, R.A.; Alkhatib, S.E.; Asyraf, M.R.M. Hyperelastic Properties of Bamboo Cellulosic Fibre–Reinforced Silicone Rubber Biocomposites via Compression Test. *Int. J. Mol. Sci.* **2022**, *23*, 6338. [CrossRef] [PubMed]
14. Bahrain, S.H.K.; Masdek, N.R.N.; Mahmud, J.; Mohammed, M.N.; Sapuan, S.M.; Ilyas, R.A.; Mohamed, A.; Shamseldin, M.A.; Abdelrahman, A.; Asyraf, M.R.M. Morphological, Physical, and Mechanical Properties of Sugar-Palm (*Arenga pinnata* (*Wurmb*) *Merr.*)-Reinforced Silicone Rubber Biocomposites. *Materials* **2022**, *15*, 4062. [CrossRef] [PubMed]
15. El-Shekeil, Y.A.; Sapuan, S.M.; Abdan, K.; Zainudin, E.S. Influence of fiber content on the mechanical and thermal properties of Kenaf fiber reinforced thermoplastic polyurethane composites. *Mater. Des.* **2012**, *40*, 299–303. [CrossRef]
16. Özturk, S. Effect of Fiber Loading on the Mechanical Properties of Kenaf and Fiberfrax Fiber-reinforced Phenol-Formaldehyde Composites. *J. Compos. Mater.* **2010**, *44*, 2265–2288. [CrossRef]
17. Hazrol, M.D.; Sapuan, S.M.; Zainudin, E.S.; Wahab, N.I.A.; Ilyas, R.A. Effect of Kenaf Fibre as Reinforcing Fillers in Corn Starch-Based Biocomposite Film. *Polymers* **2022**, *14*, 1590. [CrossRef] [PubMed]
18. Kamaruddin, Z.H.; Jumaidin, R.; Ilyas, R.A.; Selamat, M.Z.; Alamjuri, R.H.; Yusof, F.A. Biocomposite of Cassava Starch-Cymbopogan Citratus Fibre: Mechanical, Thermal and Biodegradation Properties. *Polymers* **2022**, *14*, 514. [CrossRef]
19. Tarique, J.; Zainudin, E.S.; Sapuan, S.M.; Ilyas, R.A.; Khalina, A. Physical, Mechanical, and Morphological Performances of Arrowroot (*Maranta arundinacea*) Fiber Reinforced Arrowroot Starch Biopolymer Composites. *Polymers* **2022**, *14*, 388. [CrossRef]
20. Sobhy, M.S.; Tammam, M.T. The Influence of Fiber Length and Concentration on the Physical Properties of Wheat Husk Fibers Rubber Composites. *Int. J. Polym. Sci.* **2010**, *2010*, 1–8. [CrossRef]
21. Zhu, J.; Deng, Y.; Chen, P.; Wang, G.; Min, H.; Fang, W. Prediction of Long-Term Tensile Properties of Glass Fiber Reinforced Composites under Acid-Base and Salt Environments. *Polymers* **2022**, *14*, 3031. [CrossRef]
22. Rozilah, A.; Jaafar, C.N.A.; Sapuan, S.M.; Zainol, I.; Ilyas, R.A. The Effects of Silver Nanoparticles Compositions on the Mechanical, Physiochemical, Antibacterial, and Morphology Properties of Sugar Palm Starch Biocomposites for Antibacterial Coating. *Polymers* **2020**, *12*, 2605. [CrossRef] [PubMed]
23. Alias, A.H.; Norizan, M.N.; Sabaruddin, F.A.; Asyraf, M.R.M.; Norrrahim, M.N.F.; Ilyas, A.R.; Kuzmin, A.M.; Rayung, M.; Shazleen, S.S.; Nazrin, A.; et al. Hybridization of MMT/Lignocellulosic Fiber Reinforced Polymer Nanocomposites for Structural Applications: A Review. *Coatings* **2021**, *11*, 1355. [CrossRef]

24. Nazrin, A.; Sapuan, S.M.; Zuhri, M.Y.M.; Tawakkal, I.S.M.A.; Ilyas, R.A. Flammability and physical stability of sugar palm crystalline nanocellulose reinforced thermoplastic sugar palm starch/poly(lactic acid) blend bionanocomposites. *Nanotechnol. Rev.* **2022**, *11*, 86–95. [CrossRef]
25. Nurazzi, N.M.; Sabaruddin, F.A.; Harussani, M.M.; Kamarudin, S.H.; Rayung, M.; Asyraf, M.R.M.; Aisyah, H.A.; Norrrahim, M.N.F.; Ilyas, R.A.; Abdullah, N.; et al. Mechanical Performance and Applications of CNTs Reinforced Polymer Composites—A Review. *Nanomaterials* **2021**, *11*, 2186. [CrossRef]
26. Nurazzi, N.M.; Asyraf, M.R.M.; Rayung, M.; Norrrahim, M.N.F.; Shazleen, S.S.; Rani, M.S.A.; Shafi, A.R.; Aisyah, H.A.; Radzi, M.H.M.; Sabaruddin, F.A.; et al. Thermogravimetric Analysis Properties of Cellulosic Natural Fiber Polymer Composites: A Review on Influence of Chemical Treatments. *Polymers* **2021**, *13*, 2710. [CrossRef] [PubMed]
27. Sharma, S.; Sudhakara, P.; Singh, J.; Ilyas, R.A.; Asyraf, M.R.M.; Razman, M.R. Critical Review of Biodegradable and Bioactive Polymer Composites for Bone Tissue Engineering and Drug Delivery Applications. *Polymers* **2021**, *13*, 2623. [CrossRef]

Short Biography of Authors

R. A. Ilyas is a senior lecturer at the School of Chemical and Energy Engineering, Faculty of Engineering, Universiti Teknologi Malaysia (UTM), Malaysia. He received his Diploma in Forestry at Universiti Putra Malaysia, Bintulu Campus (UPMKB), Sarawak, Malaysia, from May 2009 to April 2012. In 2012, he was awarded the Public Service Department (JPA) scholarship to pursue his bachelor's degree (BSc) in Chemical Engineering at Universiti Putra Malaysia (UPM). Upon completing his BSc. programme in 2016, he was again awarded the Graduate Research Fellowship (GRF) by the Universiti Putra Malaysia (UPM) to undertake a PhD degree in the field of Biocomposite Technology and Design at Institute of Tropical Forestry and Forest Products (INTROP) UPM. R.A. Ilyas was the recipient of MVP Doctor of Philosophy Gold Medal Award UPM 2019, for Best Ph.D. Thesis and Top Student Award, INTROP, UPM. He was awarded Outstanding Reviewer by Carbohydrate Polymers, Elsevier United Kingdom, Best Paper Award (11th AUN/SEED-Net Regional Conference on Energy Engineering), and National Book Award 2018, Best Paper Award (Seminar Enau Kebangsaan 2019, Persatuan Pembangunan dan Industri Enau Malaysia) and Top Cited Article 2020-2021 Journal Polymer Composite, Wiley, 2022. R.A. Ilyas also was listed and awarded among the World's Top 2% Scientists (Subject-Wise) citation impact during the single calendar year 2019 and 2020 by Stanford University, US, PERINTIS Publication Award 2021 and 2022 by Persatuan Saintis Muslim Malaysia, Emerging Scholar Award by Automotive and Autonomous Systems 2021, Belgium, Young Scientists Network—Academy of Sciences Malaysia (YSN-ASM) 2021, UTM Young Research Award 2021, UTM Publication Award 2021, and UTM Highly Cited Researcher Award 2021. His main research interests are: (1) polymer engineering (biodegradable polymers, biopolymers, polymer composites, polymer gels) and (2) material engineering (natural fibre-reinforced polymer composites, biocomposites, cellulose materials, nano-composites). To date, he has authored or co-authored more than 404 publications (published/accepted): 164 Journals Indexed in JCR/Scopus, 2 non-index Journal, 15 books, 104 book chapters, 78 conference proceedings/seminars, 4 research bulletins, 10 conference papers (abstract published in book of abstract), 17 Guest Editor of Journal Special Issues and 10 Editor/Co-Editor of Conference/Seminar Proceedings on green materials related subjects.

Mohd Sapuan Salit is an "A" Grade Professor of composite materials at Department of Mechanical and Manufacturing, Universiti Putra Malaysia (UPM) and a Head of Laboratory of Biocomposite Technology, INTROP, UPM. He has a BEng in Mechanical Engineering from University of Newcastle, Australia, an MSc in Engineering Design from Loughborough University, UK and PhD in Material Engineering from De Montfort University, UK. He is a Professional Engineer, a Society of Automotive Engineers Fellow, an Academy of Science Malaysia Fellow, a Plastic and Rubber and Institute Malaysia Fellow, a Malaysian Scientific Association Fellow, an International Biographical Association Fellow and an Institute of Material Malaysia Fellow. He is an Honorary Member and immediate past Vice President of Asian Polymer Association based in IIT Delhi and Founding Chairman and Honorary Member of Society of Sugar Palm Development and Industry, Malaysia. He is the co-editor-in-chief of Functional Composites and Structures, and member of editorial boards of more than two dozen journals. To date he has produced more than 1800 publications including over 860 journal papers, 50 books, and 175 chapters in book. He has delivered over 50 plenary and keynote lectures, and over 150 invited lectures. He organized 30 journal special issues as a guest editor, presented over 650 technical articles in conferences and seminars, reviewed over 1300 journal papers and has 8 patents. He successfully supervised 91 PhD and 70 MSc students and 15 postdoctoral researchers. His current h-index is 93 and the number of citations is 31,647 (Google Scholar). He received nine Outstanding Researcher Awards from UPM, ISESCO Science Award (Gold Medal), Plastic and Rubber Institute Malaysia Fellowship Award and Forest Research Institute Malaysia First Prize Publication Award. He also received Khwarizimi International Award, SEARCA Regional Professorial Chair award, Kuala Lumpur Royal Rotary Gold Medal Research Award and two National Book Awards. He received the Endeavour Research Promotion Award by TMU/IEEE India, Citation of Excellence Award, Emerald, UK, Malaysia's Research Star Award, Elsevier/Ministry of Education Malaysia, Publons Peer Review Award, Publons, USA, Professor of Eminence Award from Aligarh Muslim University, India, Top Research Scientists' Malaysia Award, Academy of Science Malaysia, Gold in Invention and Innovation Awards, Malaysia Technology Expo and PERINTIS Publication Award, PERINTIS, Malaysia. He was listed among the World Top 2% Scientists by Stanford University, USA. He is the finalist of IET Achievements Award, IET, UK and 2021 SAE Subir Chowdhury Medal of Quality Leadership, SAE, USA.

Emin Bayraktar (Prof. Emeritus, Habil., Dr (Ph.D.), DSc—Doctor of Science) is an academic and research staff member in Mechanical and Manufacturing engineering at SUPMECA/Paris, France. His research areas include manufacturing techniques of new materials (basic composites—hybrid), metal Forming of thin sheets (Design + test + FEM), static and dynamic behavior and optimization of materials (experimental and FEM—utilization and design of composite-based metallic and non-metallic, powder metallurgy, and energetic material aeronautical applications), metallic-based and non-metallic materials, powder metallurgy and metallurgy of steels, welding, and heat treatment, as well as the processing of new composites, sintering techniques, sinter–forging, thixoforming, etc. He has authored more than 200 publications in the International Journals and International Conference Proceedings, and has also authored more than 90 research reports (European = Steel Committee projects, Test + Simulation). He already advised 32 Ph.D. and 120 MSc theses, and is currently advising 7. He is a Fellow of WAMME (World Academy of Science in Materials and Manufacturing Engineering), Editorial Board—Member of JAMME (*International Journal of Achievement in Materials and Manufacturing Engineering*), Advisory board member of AMPT—2009 (*Advanced Materials Processing technologies*), APCMP—2008 and APCMP—2010. He was Visiting Professor at Nanyang Technology University, Singapore in 2012, Xi'an Northwestern Technical University, Aeronautical Engineering, in 2016, University of Campinas, UNICAMP-Brazil in 2013 until 2023. He is a recipient of the Silesian University Prix pour "FREDERIK STAUB Golden Medal-2009" by the Academy of WAMME, "World Academy of Science"—Poland, materials science section, and a recipient of the William Johnson International Gold Medal—2014, AMPT academic association.

Shukur Abu Hassan, PhD (UTM, Malaysia), MSc (Warwick University, UK), BSc (UTM, Malaysia) is an Engineer, associate professor and current Director of Centre for Advanced Composite Materials (CACM), Universiti Teknologi Malaysia (UTM), Malaysia. In his teaching career of 33 years, he has been working as Centre Director, Head of Composite Lab and Committee Member at the Faculty and University level. He is also involved as Consultant and on the Editorial Advisory Board for various international journals. Assoc. Prof. Shukur research focuses on (1) composites structure durability, (2) recycling composites, (3) biomimetics design, (4) composites testing, (5) natural fibre composites, (6) hybrid composites, (7) composites bonding and (8) polymer mortar composites. He is registered with Board of Engineers Malaysia (BEM) and Member of Malaysian Board of Technologist (MBOT). To date he has been a consultant to 16 projects amounting to RM 480,000. 00 and Principal Investigator and member to 49 research grants worth RM 3.9 million. He has also supervised and co-supervised 16 postgraduates. He has also authored and co-authored 106 publications (published/accepted); 22 ISI Article/Journal, 9 Scopus Article/Journal, 10 Non-Indexed Article/Journal, 2 Indexed Conference Proceedings, 43 Non-Indexed Conference Proceedings, 7 Book Chapters, 13 Classified/Technical /Expert Report. He also has 1 patent granted, 1 patent filed and 7 registered industrial design.

Nabil Hayeemasae is currently an Assistant Professor at the Department of Rubber Technology and Polymer Science, Faculty of Science and Technology, Prince of Songkla University, Pattani Campus. He obtained a Ph.D. in Polymer Engineering at the School of Materials and Mineral Resources Engineering, Universiti Sains Malaysia. He has been conducting several research areas, particularly on the re-utilization of polymer and solid wastes into potential materials, rubber composites, and rubber recycling. So far, his research outputs have been selected to be published in over 80 papers internationally.

M.S.N. Atikah is a PhD candidate in the field of Chemical Engineering at Universiti Putra Malaysia. She graduated with a Bachelor of Engineering (Chemical) and a Master of Science (Chemical) from the same university. She has authored and co-authored 54 publications from 2016-2022 and has an h-index of 20. She was awarded scholarships for her bachelor's degree and doctor of philosophy programs from Jabatan Perkhidmatan Awam (JPA), Malaysia.

Khubab Shaker is currently serving as Assistant Professor and Chairman, Department of Materials at National Textile University, Pakistan. He is an active member of the Textile Composite Materials Research Group at NTU, with 63 publications (impact factor=166), 3 books, and 32 conference presentations. He has more than 12 years of industrial and teaching experience in textiles and composite materials. His areas of research include composites joining, natural fibres, biopolymers, green composites, and their applications as a potential replacement for synthetic composite materials.

polymers MDPI

Article

Effect of *Cymbopogan citratus* Fibre on Physical and Impact Properties of Thermoplastic Cassava Starch/Palm Wax Composites

Zatil Hafila Kamaruddin [1,2,*], Ridhwan Jumaidin [3,*], Zatil Hazrati Kamaruddin [2], Muhammad Rizal Muhammad Asyraf [4,5,*], Muhammad Rizal Razman [6] and Tabrej Khan [7]

1 Fakulti Kejuruteraan Mekanikal, Universiti Teknikal Malaysia Melaka, Hang Tuah Jaya, Durian Tunggal 76100, Melaka, Malaysia
2 German-Malaysian Institute, Jalan Ilmiah Taman Universiti, Kajang 43000, Selangor, Malaysia
3 Fakulti Teknologi Kejuruteraan Mekanikal dan Pembuatan, Universiti Teknikal Malaysia Melaka, Hang Tuah Jaya, Durian Tunggal 76100, Melaka, Malaysia
4 Engineering Design Research Group (EDRG), Faculty of Mechanical Engineering, Universiti Teknologi Malaysia, Johor Bahru 81310, Johor, Malaysia
5 Centre for Advanced Composite Materials (CACM), Universiti Teknologi Malaysia, Johor Bahru 81310, Johor, Malaysia
6 Research Centre for Sustainability Science and Governance (SGK), Institute for Environment and Development (LESTARI), Universiti Kebangsaan Malaysia, UKM Bangi 43600, Selangor, Malaysia
7 Department of Engineering Management, College of Engineering, Prince Sultan University, Riyadh 11586, Saudi Arabia
* Correspondence: zatilhafila@gmail.com (Z.H.K.); ridhwan@utem.edu.my (R.J.); muhammadasyraf.mr@utm.my (M.R.M.A.)

Abstract: *Cymbopogan citratus* fibre (CCF) is an agricultural waste plant derived from a natural cellulosic source of fibre that can be used in various bio-material applications. This paper beneficially prepared thermoplastic cassava starch/palm wax blends incorporated with *Cymbopogan citratus* fibre (TCPS/PW/CCF) bio-composites at different CCF concentrations of 0, 10, 20, 30, 40, 50 and 60 wt%. In contrast, palm wax loading remained constant at 5 wt% concentration using the hot moulding compression method. TCPS/PW/CCF bio-composites were characterised in the present paper via their physical and impact properties. The addition of CCF significantly improved impact strength by 50.65% until 50 wt% CCF loading. Furthermore, it was observed that the inclusion of CCF resulted in a little decrement in biocomposite solubility compared to neat TPCS/PW biocomposite from 28.68% to 16.76%. Water absorption showed higher water resistance in the composites incorporating 60 wt.% fibre loading. The TPCS/PW/CCF biocomposites with different fibre contents had 11.04–5.65% moisture content, which was lower than the control biocomposite. The thickness of all samples decreased gradually with increasing fibre content. Overall, these findings provide evidence that CCF waste can be utilised as a high-quality filler in biocomposites due to its diverse characteristics, including improving the properties of biocomposites and strengthening their structural integrity.

Keywords: starch; *Cymbopogan citratus* fibre (CCF); thermoplastic cassava starch

Citation: Kamaruddin, Z.H.; Jumaidin, R.; Kamaruddin, Z.H.; Asyraf, M.R.M.; Razman, M.R.; Khan, T. Effect of *Cymbopogan citratus* Fibre on Physical and Impact Properties of Thermoplastic Cassava Starch/Palm Wax Composites. *Polymers* **2023**, *15*, 2364. https://doi.org/10.3390/polym15102364

Academic Editor: Alexey Iordanskii

Received: 15 February 2023
Revised: 5 May 2023
Accepted: 14 May 2023
Published: 18 May 2023

1. Introduction

Over the past two decades, demand for agriculturally based products has grown due to environmental concerns and the awareness that petroleum resources are limited [1,2]. Polymers derived from biological sources have often been raised as the most promising solutions since they are renewable and biodegradable [3–6]. The development of organic, renewable, and biodegradable materials has been driven by the escalating amount of plastic trash that threatens the environment [7]. Sustainable development and the absence of damaging carbon emissions during processing and after destruction are desirable properties of prospective materials [8–10]. Excess plastic garbage is too much for recycling programmes because it is difficult to distinguish between polymers and eco-friendly plastics [11,12]. It is urgent to provide eco-friendly packaging items at comparable prices to preserve our planet, since synthetic plastic packaging pollution has exacerbated the issue of environmental destruction. In order to solve this recurring issue brought on by non-biodegradable plastics, natural biopolymers such as starch-based composites are being investigated as potential alternatives to conventional plastics [13–15].

Similar traits to conventional synthetic fibre-reinforced composites can be seen in natural fibre-reinforced composites. Due to considerable environmental concerns, they have been considered an alternative to organic and inorganic fillers and fibres [16,17]. These natural-source polymers can be found in various plant parts, such as grass [18], stalks [19], and leaves [20]. Hemicellulose, cellulose, and lignin are all examples of natural fibres [21]. Cellulose and hemicellulose components are all hydrophilic, whereas lignin is slightly hydrophobic [22,23]. Natural fibres are desirable filler materials for polymer composites because of their low manufacturing costs, particular acceptable characteristics, energy consumption, and environmentally friendly nature [24–26]. Natural fibres were incorporated to reinforce starch-based polymers, significantly impacting the materials' physical and impact qualities. These were primarily attributable to cellulose and starch's close structural resemblance and connection [27,28].

Starch, a form of carbohydrate, is one of the natural polymers that has received attention for use in the production of biodegradable polymers. Starch is generally composed of linear and branched chains of glucose molecules known as amylose and amylopectin, respectively [29]. Starch is combined with a plasticiser and fibre reinforcement and then subjected to high-temperature compression moulding to produce thermoplastic starch. Native starch can be transformed into a thermoplastic material by adding plasticising agents such as sorbitol, xylitol and glycerol. Incorporating plasticisers results in reduced intermolecular forces, increased mobility of polymer chains, and decreased glass transition temperature [30]. In our prior studies, we discovered that combining thermoplastic starch and palm wax had better properties than biopolymer blends [31,32]. As well, palm wax is considered an excellent matrix because it may reduce starch's hydrophobicity, improving processability.

The use of starch in both food and non-food products has been widespread. Since starch-based composites have poor mechanical efficiency, particularly impact strength, improving their qualities is a big challenge [33,34]. Numerous blending and compositing techniques have been devised to enhance these mechanical properties, such as mixing with other polymers or strengthening natural fillers [35–37]. Various additional agents support the interaction between starch and other components. In order to build polymer blend composites and biocomposites, filler reinforcement added to starch-based polymers has recently received more attention [38,39]. It was established that using natural fillers to reinforce materials with specific capabilities and add new features is a successful strategy. Although pure thermoplastic starch has many benefits, it also has some drawbacks that restrict the range of its potential uses. These drawbacks include limited water barriers, mechanical strength, and long-term stability [40–42]. Thus, modifications are typically needed to make thermoplastic starch useful in actual applications. The utilization of natural fibers as reinforcement for thermoplastic starch is an intriguing approach that could be utilized to solve these shortcomings. The mechanical properties of natural fibers, when combined with thermoplastic starch, are obviously improved due to the chemical similarity of starch and plant fibers [43]. In addition, thermoplastic starch can be used in combination with other natural polymers to reduce the downsides of this biopolymer while maintaining the material's biodegradability [15]. Results showed that adding palm wax significantly improved the mechanical characteristics of a cassava starch/palm wax thermoplastic blend [31] and improved the biopolymer's performance. Additionally, it contributed to the biopolymer's distinctiveness as an edible food coating and food packaging material.

Lemongrass, also called *Cymbopogan citratus,* is one of the Poaceae family's aromatic plants. These natural crops are grown in the tropical and semitropical areas of Africa, Asia, South America, and India [32]. In Malaysia, *Cymbopogon citratus* fibre (CCF) is becoming more well-liked as a natural resource that can help create ecologically friendly resources. The leaves of *Cymbopogan citratus* grow directly from the earth and have short rhizomes. All parts of the *Cymbopogan citratus* plant, including the leaves, are abundantly produced since the plant is simple to grow. The leaves have a glabrous surface and a greenish interior and can reach a length and width of about 50 and 1.5 cm, respectively [44]. The leaves of the *Cymbopogon citratus* plant are thrown away as trash in Malaysia, where the stalks are most frequently utilised as a constituent to flavour cuisine [44]. Due to the significant volume of garbage created, various trash reduction attempts have been conducted by preventing disposal and reusing waste. CCF has recently been employed as an adsorbent from aqueous solutions to remove methylene blue dye, as a raw material for making paper and pulp, and broadly in medicinal activities [45,46]. In addition, CCF is one of the substitute materials used to manufacture polymer composites as a reinforcing agent. The fibre of *Cymbopogan citratus* contains a significant quantity of cellulose, making it a source of cellulose fibre, according to research [47].

Although research has been reported on using Cymbopogan species from China in synthetic polymers, none was discovered on the characterisation of Cymbopogan citratus fibre from Malaysia in a thermoplastic cassava starch/palm wax matrix. As a result, the primary objective of this research topic is to investigate the effects of Malaysian-sourced Cymbopogan citratus fibre on the impact resistance and physical properties of thermoplastic cassava starch/palm wax composites. This study aims to produce decomposable materials that reduce pollution and are more eco-friendly.

2. Materials and Methodology

2.1. Materials

CCF, with contents of cellulose (37.56%), hemicellulose (29.29%), lignin (11.14%), and ash (4.28%), was collected from a rural area located in Beranang, Selangor (West Malaysia). The extraction of CCF was performed according to our previous research [48] via the water-retting process. Antik Sempurna Sdn. Bhd., based in Selangor, Malaysia, provided the food-grade cassava starch used in this investigation. The palm wax in the commercial grade was provided by Green & Natural Industries Sdn. Bhd and the glycerol (99.5% purity) was purchased from QRec Chemicals Sdn, Selangor, Malaysia.

2.2. Sample Preparation of Biocomposites

The thermoplastic cassava starch (TPCS) preparation was made by combining 5% palm wax with 65% starch and 30% glycerol. The mixture was then blended at room temperature for 5 min at 1200 rpm using a Panasonic Dry Mixer MX-GM1011. The materials were subjected to thermo-pressing at 150 °C for 30 min using a Malaysian-made Technopress-40HC-B Plastic Hydraulic Moulding Press under a weight of 10 tonnes to produce plates with a thickness of about 3 mm. Similar techniques were applied to create TPCS/PW/CCF composites. The matrix's characteristics were altered by introducing various CCF ratios ranging from 0 to 60 wt.%. The specimens were immediately put in a desiccator with silica gel to avoid unforeseen water absorption before conditioning.

2.3. Density

The ASTM D1895 [49] standard was followed to determine the biocomposite material density of the samples (10 mm × 10 mm × 3 mm), which were fabricated and dried for 24 h at 105 °C in the oven. The specimens were then positioned in a desiccator with silica gel in granulated form, and their weights and volumes were calculated using an electronic densimeter and weighing balance. Equation (1) was used to determine the density value.

$$\text{Density } (\text{g/cm}^3) = \frac{\text{Mass (g)}}{\text{Volume } (\text{cm}^3)} \quad (1)$$

2.4. Moisture Content of Composite Materials

Moisture content analysis was performed for five samples that were prepared and heated in the oven for 24 h at 105 °C. The initial weight (W_i, grammes) and the final weight (W_f, grammes) after heating are needed to calculate the moisture content [50]. Equation (2) was utilised to compute the moisture content.

$$\text{Moisture content }(\%) = \frac{W_i - W_f}{W_i} \times 100 \quad (2)$$

2.5. Water Absorption

Five samples (10 mm × 10 mm × 3 mm) were dried for 24 h at 105 ± 2 °C in an air-circulating oven to remove existing moisture. The test samples were fully submerged in water for 2 h at ambient temperature (23 ± 1 °C). The initial weight (W_i) and final immersion weight (W_f) were calculated for determining the water absorption rate using Equation (3).

$$\text{Water absorption }(\%) = \frac{W_i - W_f}{W_i} \times 100 \quad (3)$$

2.6. Thickness Swelling

Analysis of the samples' swelling was conducted based on the amended procedure by Jawaid et al. [51]. Five samples with measurements of 10 mm × 10 mm × 3 mm were prepared for thickness swelling calculation. Specimens were then dried in a dry laboratory oven at 105 °C for 24 h. Each sample was measured for thickness before testing (T_i). After that, each sample was submerged in 30 mL of distilled water for 2 h at room temperature (23 ± 1 °C). During the immersion process, the starting and final thicknesses, T_i and T_f, were measured using a Mitutoyo brand digital vernier calliper with a precision of 0.01 cm. This allowed for more accurate reading. Equation (4) summarizes the steps used to calculate the samples' thickness swelling ratio:

$$\text{Thickness swelling }(\%) = \frac{T_i - T_f}{T_i} \times 100 \quad (4)$$

2.7. Impact Testing of Composites Material

Analysis via Izod impact tests was performed using ASTM D256 [52] with 50 ± 5% relative humidity and temperature of 23 ± 1 °C, and five replicates of each sample with dimensions of 13 mm (W) × 3 mm (T) × 60 mm (L) were set up. Victor Equipment Resources Sdn. Bhd. (Subang Jaya, Malaysia) provided testing using a digital pendulum impact tester. Before testing, all samples were preconditioned for two days and then processed at 53% RH. The results of the impact properties were obtained by taking the average of the data using Equation (5).

$$\text{Impact strength} = \text{Impact energy (J)/area (mm}^2) \quad (5)$$

2.8. Water Solubility of Composite Materials

The procedure described by Zhang et al. [53] was used to measure the samples' water solubility with a few minor adjustments. Five samples (10 mm × 10 mm × 3 mm) were cut and dried in a dry laboratory oven for 24 h at 105 °C 2 and the initial dry matter of each sample was noted as W_i. Each sample was immersed in 30 mL of distilled water while stirring vigorously. The undissolved portions of the sample were then taken out of the laboratory cup after a 24-h immersion, and any remaining water on the sample's surface was wiped away using filter paper. The samples were then once again dried for

24 h at 105 °C to produce the final dried sample, indicated as W_f. Equation (6) was used to calculate the samples' water solubility.

$$\text{Water solubility } (\%) = \frac{W_i - W_f}{W_i} \times 100 \quad (6)$$

2.9. Statistical Analysis of Composites Material

Analysis of variance (ANOVA) was used for statistical analysis of the experimental results in SPSS. Duncan's test was used to compare the means with a significance level of 0.05 ($p < 0.05$).

3. Results and Discussions

3.1. Density

Biocomposites of TPCS/PW/CCF were produced by varying the CCF loading from 0% to 60% of fibre. There was a slightly significant difference in TPCS/PW/CCF biocomposite density. However, when fibre was introduced, it was discovered that the density value of biocomposite reduced as the fibre loading increased, as illustrated in Figure 1, ranging from 1.31 ± 0.01 to 1.22 ± 0.01 g/cm^3. This finding might be related to the low density of the added CCF, which is known to be 0.25 ± 0.002 g/cm^3, relative to the TPCS/PW control sample (CCF0), which had a density value of 1.31 ± 0.01 g/cm^3. This affected the overall density of the bio-composite that was produced. Considering the density value of TPCS/PW/CCF, it can be noticed that the density of samples decreased by 6.87%. This specifies that the decreasing trend in density value might be associated with the reduction in the proportion of biocomposite mass when CCF is added, as the percentage of the TPCS/PW matrix is reduced while the volume remains constant [28]. As a result, reduction in biocomposite mass had a corresponding effect on the density values of the samples. This finding agreed with a previous study on developing Dioscorea hispida fibre-reinforced Dioscorea hispida starch, where the biocomposites' density values decreased as fibre loading was added [29]. Meanwhile, a similar result was reported when developing wood apple shell-reinforced epoxy composites. The density of the composites decreasing as the amount of filler used increased might be due to the lighter density of the filler material [54]. In terms of practicality, the decrease in bio-composite density when fibre loading is increased makes a substantial contribution by providing lightweight materials [55].

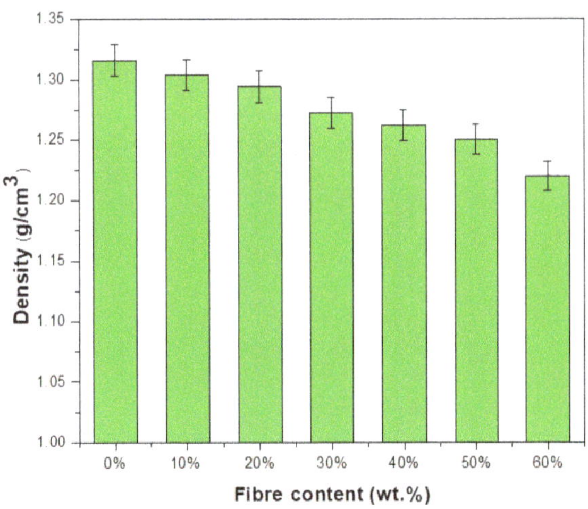

Figure 1. Density of TPCS/PW/CCF biocomposites.

3.2. Moisture Content

Thermoplastic starch exhibits a high moisture content, which is caused by the inherent hydrophilic nature of native cassava starch, and results in poor moisture barrier capabilities. In spite of this, thermoplastic starch's resistance to water can be increased by the inclusion of components that are either hydrophilic or hydrophobic. This study investigated the effect of Cymbopogan citratus fibre on thermoplastic cassava starch/palm wax composites' moisture content. The moisture content of the thermoplastic starch/palm wax composite without *Cymbopogan citratus* fibre was the highest, with absorption rates of 11.04%, which might be due to the strong hydrogen interactions between the hydroxyl groups in starch and the free water molecules [56]. The TPCS/PW/CCF biocomposites with different fibre contents had moisture contents of 11.04–5.65%, which was lower than that of the control biocomposite sample (11.04%), as illustrated in Figure 2. However, the increase in CCF content from 10 to 60 wt.% resulted in a slight reduction in the moisture content of the TPCS/PW/CCF samples. This could be explained by the strong hydrogen bonds between the Cymbopogan citratus fibre and the starch matrix, reducing the number of accessible hydroxyl groups.

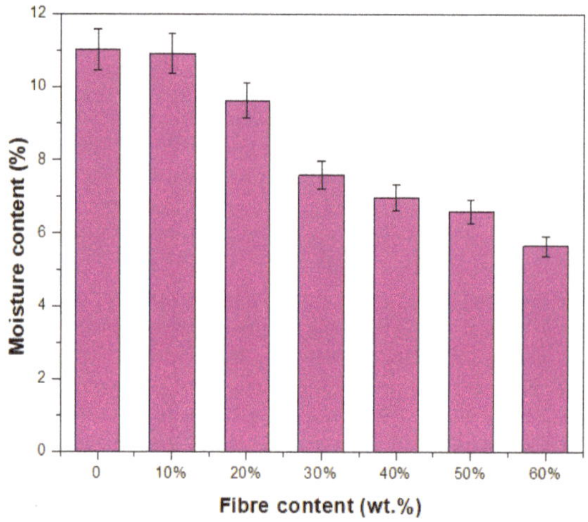

Figure 2. Moisture content of TPCS/PW/CCF biocomposites.

Consequently, the biocomposites have increased resistance to moisture. This result agrees with the previous study's finding on the effect on moisture content of cassava starch foam with the addition of natural fibre and chitosan [57]. Likewise, Soykeabkaew et al. [58] reported that similar results occurwhen jute and flax are added to tapioca starch foam, decreasing the amount of moisture content. This could be related to the hydrophobic nature of CCF, especially when compared to the CCF 0% sample. Although all-natural fibre is classified as hydrophilic, it is not as hygroscopic as TPCS. This might be due to the presence of lignin and wax in CCF composition, which makes it comparatively hydrophobic and gives it better water-resistant properties than TPCS [59]. In contrast, palm wax is considered a superior matrix since it can reduce starch hydrophobicity, which improves processability [31,60].

3.3. Water Absorption of Composites

Water absorption is crucial for many applications of TPS products. The results of the water absorption tests performed on TPCS/PPW/CCF biocomposites are shown in Figure 3. In general, water absorption followed the same trend as moisture content, which exhibited decreasing values when fibre content is increased. In this study, the neat

TPCS/PW biocomposite showed the highest water absorption at 33.15%. However, the absorption rate appeared to be reduced with CCF reinforcement, which reached 10.90% with 60 wt.% loading.

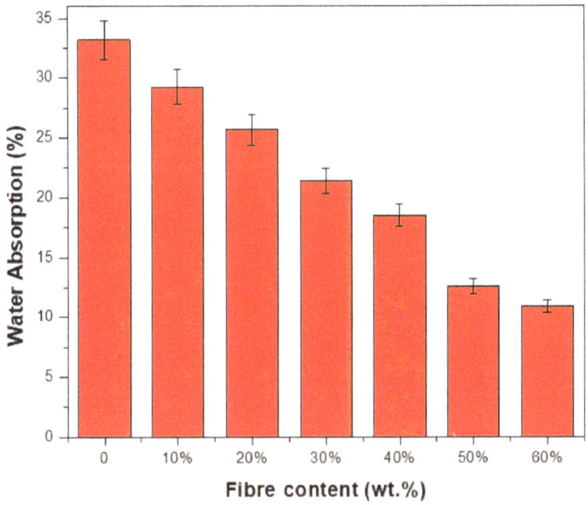

Figure 3. TPCS/PW/CCF biocomposites' water absorption.

The ability of fibres to produce strong interfacial bonding, which prevents water from penetrating through the matrix, may play a role in the decrease in absorption rate in the presence of CCF [61]. In addition, the presence of lignin as the main component, in conjunction with wax and fatty substances, in fibre enhances the ability of composites to withstand the effects of water [61,62]. A prior study also reported that the gradual loading of sugar palm fibre into TPS-based hybrid composites resulted in a considerable reduction in the water absorption rate [63]. Meanwhile, a previous study by Sarifuddin et al. [64] on developing thermoplastic sago starch reinforced kenaf core fibre biocomposites reported low water absorption when adding kenaf core fibre into the samples. This was associated with the fibre having a less hydrophilic nature than the starch. The higher the fibre loading, the lower the amount of absorbed water. The strong hydrogen bonding between the matrix phase and the fibre phase can be attributed to the overall lower water absorption of the material. However, it is clear that the CCF-reinforced starch biocomposite is less water-resistant than the control CCF samples.

3.4. Thickness Swelling

A material's dimensional stability can be evaluated by determining the degree to which it swells or contracts as a direct result of the movement of moisture in the material. The thickness swelling ratio of TPCS/PW/CCF composites was evaluated to determine changes in TPCS/PW dimensional stability following the introduction of CCF. Figure 4 illustrates the thickness swelling percentage for TPCS/PW with the inclusion of CCF after immersion for 2 h. The thickness of all samples decreased gradually with increasing fibre content. In comparison to the TPCS/PW control sample, the sample containing 60% CCF content exhibited the lowest water uptake and thickness swelling after 2 h of immersion, decreasing by 54.24%. This finding can be attributed to the presence of fibre in the composites, which possesses a more rigid structure than starch, providing higher dimensional stability to the composites [65]. This study's findings agree with those of a prior investigation into composites made of cassava starch and green coir fibres [66]. Additionally, similar findings were reported when developing a cassava/sugar palm fibre-reinforced cassava starch hybrid. In this study, the incorporation of sugar palm fibre made the sample swell less and influenced the interaction between the polymer chains, which could reduce the polymer's ability to swell. In each sample, increasing the fibre concentration resulted in a decrease in swelling behaviour. The sample that contained a low fibre loading was more swollen than the ones that contained a high fibre loading [67].

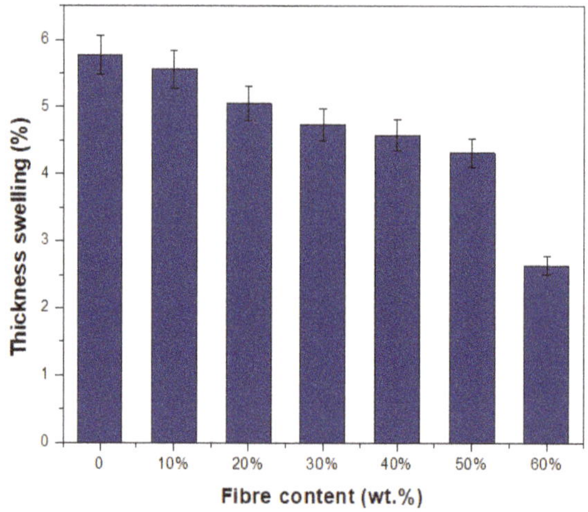

Figure 4. Thickness swelling of TPCS/PW/CCF biocomposites.

3.5. Impact Testing of Composites

The impact strength of the biocomposites was also determined to measure the material's impact resistance, indicating the capability to endure an unexpectedly applied load. Figure 5 demonstrates the TPCS/PW/CCF's impact strength with different fibre contents. Analysis of variance (ANOVA) of the impact characteristics is presented in Table 1. This indicates that there is a statistically significant difference ($p < 0.05$) between the mean impact strength from one level of composites to another. In general, the findings of the impact test results demonstrate the highest impact strength at 50 wt.% fibre loading. Significant improvement ($p < 0.05$) was evidenced in the impact properties of the composite when incorporated with *Cymbopogan citratus* fibre. Failure of a fibre composite to withstand impact loading can be attributed to several other factors, including fibre pull-out, matrix fracture, fibre breakage, and debonding of the fibres from the matrix [68]. It could be observed that the TPCS/PW/CCF biocomposite had significantly increased impact strength by 50.65% ($p < 0.05$) with increasing fibre content from (0 to 50 wt.%). This finding might be associated with the strong interfacial bonding at fibre–matrix interfaces, which improved energy absorption during impact loading [69]. However, impact resistance declined after increasing reinforcement from 50 to 60 wt.%. The reduction in impact strength at higher fibre contents might be attributed to the higher rigidity of the materials, which led to more brittle properties and, as a result, a decreased capacity to absorb the impact energy [31]. In our prior studies, we discovered that the results revealed that elongation at break values increased from 40 to 50 wt.% as a result of the addition of *Cymbopogan citratus* fibre loading; however, an increase in the amount of *Cymbopogan citratus* fibre up to 60 wt.% caused a decrease in elongation at break values. Overall, the inclusion of *Cymbopogan citratus* fibre content resulted in a decrease in the molecular mobility of the TPCS matrix, which produced biocomposite materials with a greater degree of rigidity [48]. A similar result was observed in a study on the development of date palm-reinforced epoxy composites. A significant improvement in impact strength was found at 50% date palm fibre reinforced epoxy composites in comparison to loadings of 40% and 60%, which can be primarily attributed to better adhesion of the date palm fibre with the epoxy matrix in order to overcome the high impact stress/load. Furthermore, the increase in impact strength might be associated with enhanced stress capability, which would limit the contribution of fibre-related mechanisms such as fibre pull out [70].

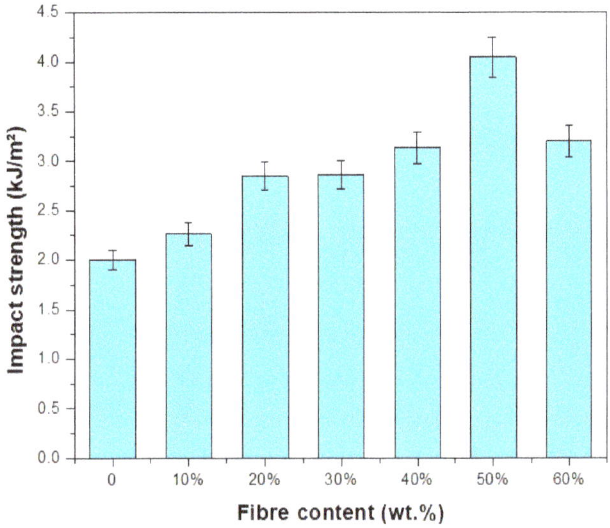

Figure 5. Impact strength of TPCS/PW/CCF biocomposites.

Table 1. Analysis of variance (ANOVA) summary of TPCS/PW/CCF composites.

Variables	df	Impact Strength
Mixture	6	0.00 *

Note: * Significantly different at $p < 0.05$.

3.6. Water Solubility of Composites

The solubility of samples in water is one of the essential properties worthy of consideration in many applications. Applications that require moisture and water loss protection must have low water solubility. Figure 6 shows the TPCS/PW/CCF biocomposites' water solubility. This solubility demonstrates the effect of water immersion with continuous stirring on the composite samples. The inclusion of CCF resulted in a little decrement in biocomposite solubility compared to neat TPCS/PW biocomposite, from 28.68% to 16.76%. This phenomenon could be attributed to the role of the fibre, which produces a network that tightly keeps composites together and prevents the dissolution of composites by lowering the solubility of the samples [55]. Additionally, the solubility of samples reduced with increasing fibre content, which might be due to fibre's capacity to resist water diffusion and enhance composite integrity by limiting water penetration, thus reducing sample solubility [63]. This observation agrees with previous findings on the development of a cornhusk/sugar palm fibre reinforced corn starch-based hybrid, which reported that adding sugar palm fibre decreased the solubility of the samples [61]. The result supports this study's water absorption, moisture content, and thickness swelling findings.

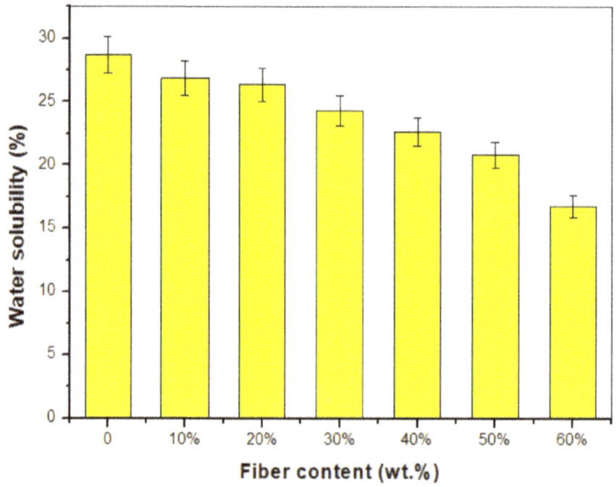

Figure 6. Water solubility of TPCS/PW/CCF biocomposites.

4. Conclusions

Utilising hot pressing, a novel biocomposite made of thermoplastic cassava starch and varying concentrations of *Cymbopogan citratus* fibre (CCF) was created. Its water barrier behaviour and impact qualities were studied. The experimental results revealed that the inclusion of CCF resulted in a little decrement in biocomposite solubility compared to neat TPCS/PW biocomposite, from 28.68% to 16.76%, and a reduction in the absorption rate, which reached 10.90% with 60 wt.% loading. TPCS/PW/CCF biocomposites with different fibre contents had moisture contents of 11.04–5.65%, which was lower than the control biocomposite sample, and a slightly significant difference in biocomposite density. TPCS-reinforced CCF composites show promise as an alternative to non-environmentally friendly polymers and composites, with better water barrier and impact performance.

Author Contributions: Conceptualization, Z.H.K. (Zatil Hafila Kamaruddin) and R.J.; methodology, Z.H.K. (Zatil Hafila Kamaruddin); formal analysis, Z.H.K. (Zatil Hazrati Kamaruddin) and R.J.; investigation, Z.H.K. (Zatil Hazrati Kamaruddin); resources, M.R.M.A. and M.R.R.; data curation, Z.H.K. (Zatil Hazrati Kamaruddin) and M.R.M.A.; writing—original draft preparation, Z.H.K. (Zatil Hazrati Kamaruddin); writing—review and editing, Z.H.K. (Zatil Hazrati Kamaruddin); visualization, Z.H.K. (Zatil Hafila Kamaruddin) and M.R.M.A.; supervision, R.J.; project administration, Z.H.K. (Zatil Hafila Kamaruddin); funding acquisition, M.R.M.A. and T.K. All authors have read and agreed to the published version of the manuscript.

Funding: The financial support was received from the Universiti Teknikal Malaysia Melaka and the Ministry of Higher Education Malaysia for financial aid through research grant PJP/2021/FTKMP/S01815 and to publish these project findings. Additionally, the paper has received funding from Universiti Teknologi Malaysia through the project "Characterizations of Hybrid Kenaf Fibre/Fibreglass Meshes Reinforced Thermoplastic ABS Composites for Future Use in Aircraft Radome Applications" under grant number PY/2022/03758—Q.J130000.3824.31J25.

Institutional Review Board Statement: Not applicable.

Data Availability Statement: The data is available within the article.

Acknowledgments: The authors would like to acknowledge the support of Universiti Teknikal Malaysia Melaka and the Ministry of Higher Education Malaysia for financial aid through research grant PJP/2021/FTKMP/S01815 and to publish these project findings. Additionally, the authors would like to express gratitude for the financial support from the Universiti Teknologi Malaysia through the project "Characterizations of Hybrid Kenaf Fibre/Fibreglass Meshes Reinforced Thermoplastic ABS Composites for Future Use in Aircraft Radome Applications" under grant number PY/2022/03758—Q.J130000.3824.31J25. In addition, thank you to the German Malaysian Institute for providing a scholarship award to the principal author of this project.

Conflicts of Interest: The authors declare no conflict of interest.

References

1. Ali, S.S.S.; Razman, M.R.; Awang, A.; Asyraf, M.R.M.; Ishak, M.R.; Ilyas, R.A.; Lawrence, R.J. Critical Determinants of Household Electricity Consumption in a Rapidly Growing City. *Sustainability* **2021**, *13*, 4441. [CrossRef]
2. Roslan, Z.B.; Ramli, Z.; Razman, M.R.; Asyraf, M.R.M.; Ishak, M.R.; Ilyas, R.A.; Nurazzi, N.M. Reflections on Local Community Identity by Evaluating Heritage Sustainability Protection in Jugra, Selangor, Malaysia. *Sustainability* **2021**, *13*, 8705. [CrossRef]
3. Rafidah, M.; Asyraf, M.R.M.; Nurazzi, N.M.; Hassan, S.A.; Ilyas, R.A.; Khan, T.; Saad, W.A.A.; Rashedi, A.; Sharma, S.; Hussein, E.K. Unlocking the Potential of Lignocellulosic Biomass in Road Construction: A Brief Review of OPF. *Mater. Today Proc.* **2023**. [CrossRef]
4. Asyraf, M.R.M.; Khan, T.; Syamsir, A.; Supian, A.B.M. Synthetic and Natural Fiber-Reinforced Polymer Matrix Composites for Advanced Applications. *Materials* **2022**, *15*, 6030. [CrossRef]
5. Norrrahim, M.N.F.; Knight, V.F.; Nurazzi, N.M.; Jenol, M.A.; Misenan, M.S.M.; Janudin, N.; Kasim, N.A.M.; Shukor, M.F.A.; Ilyas, R.A.; Asyraf, M.R.M.; et al. The Frontiers of Functionalized Nanocellulose-Based Composites and Their Application as Chemical Sensors. *Polymers* **2022**, *14*, 4461. [CrossRef] [PubMed]
6. Madenci, E.; Özkılıç, Y.O.; Aksoylu, C.; Asyraf, M.R.M.; Syamsir, A.; Supian, A.B.M.; Mamaev, N. Buckling Analysis of CNT-Reinforced Polymer Composite Beam Using Experimental and Analytical Methods. *Materials* **2023**, *16*, 614. [CrossRef] [PubMed]
7. Huang, M.R.; Li, S.; Li, X.G. Longan Shell as Novel Biomacromolecular Sorbent for Highly Selective Removal of Lead and Mercury Ions. *J. Phys. Chem. B* **2010**, *114*, 3534–3542. [CrossRef] [PubMed]
8. Ilyas, R.A.; Sapuan, S.M.; Harussani, M.M.; Hakimi, M.Y.A.Y.; Haziq, M.Z.M.; Atikah, M.S.N.; Asyraf, M.R.M.; Ishak, M.R.; Razman, M.R.; Nurazzi, N.M.; et al. Polylactic Acid (PLA) Biocomposite: Processing, Additive Manufacturing and Advanced Applications. *Polymers* **2021**, *13*, 1326. [CrossRef]
9. Nurazzi, N.M.; Asyraf, M.R.M.; Fatimah Athiyah, S.; Shazleen, S.S.; Rafiqah, S.A.; Harussani, M.M.; Kamarudin, S.H.; Razman, M.R.; Rahmah, M.; Zainudin, E.S.; et al. A Review on Mechanical Performance of Hybrid Natural Fiber Polymer Composites for Structural Applications. *Polymers* **2021**, *13*, 2170. [CrossRef]
10. YHazrati, K.Z.; Sapuan, S.M.; Zuhri, M.Y.M.; Jumaidin, R. Recent and Potential Developments in Dioscorea Hispida Biopolymer Composites: A Review. In Proceedings of the 7th Postgraduate Seminar on Natural Fibre Reinforced Polymer Composites 2020, Serdang, Malaysia, 17 November 2020; pp. 14–17.
11. Li, X.G.; Xie, Y.B.; Huang, M.R.; Umeyama, T.; Ohara, T.; Imahori, H. Effective Role of Eco-Friendly Acetyl Tributyl Citrate in Large-Scale Catalyst-Free Synthesis of Waterborne Polyurethanes without Volatile Organic Compounds. *J. Clean. Prod.* **2019**, *237*, 117543. [CrossRef]
12. Li, X.G.; Song, G.; Huang, M.R.; Ohara, T.; Yamada, H.; Umeyama, T.; Higashino, T.; Imahori, H. Cleaner Synthesis and Systematical Characterization of Sustainable Poly(Isosorbide-Co-Ethylene Terephthalate) by Environ-Benign and Highly Active Catalysts. *J. Clean. Prod.* **2019**, *206*, 483–497. [CrossRef]
13. Tarique, J.; Sapuan, S.M.; Abdan, K.; Ilyas, R.A.; Zainudin, E.S.; Sherwani, S.F.K.; Hazrati, K.Z. Physical and Barrier Roperties of Arrowroot (*Maranta arundinacea*) Starch Based Composites: A Review. In Proceedings of the International Conference on Sugar Palm and Allied Fibre Polymer Composites 2021, Online, 11 December 2021; pp. 180–182.
14. Asyraf, M.R.M.; Ishak, M.R.; Norrrahim, M.N.F.; Nurazzi, N.M.; Shazleen, S.S.; Ilyas, R.A.; Rafidah, M.; Razman, M.R. Recent Advances of Thermal Properties of Sugar Palm Lignocellulosic Fibre Reinforced Polymer Composites. *Int. J. Biol. Macromol.* **2021**, *193*, 1587–1599. [CrossRef]
15. Asyraf, M.R.M.; Ishak, M.R.; Syamsir, A.; Nurazzi, N.M.; Sabaruddin, F.A.; Shazleen, S.S.; Norrrahim, M.N.F.; Rafidah, M.; Ilyas, R.A.; Rashid, M.Z.A.; et al. Mechanical Properties of Oil Palm Fibre-Reinforced Polymer Composites: A Review. *J. Mater. Res. Technol.* **2022**, *17*, 33–65. [CrossRef]
16. Tarique, J.; Sapuan, S.M.; Zainudin, E.S.; Khalina, A.; Ilyas, R.A.; Hazrati, K.Z.; Aliyu, I. A Comparative Review of the Effects of Different Fibre Concentrations on Arrowroot Fibre and Other Fibre-Reinforced Composite Films. *Mater. Today Proc.* **2022**, *74*, 411–414. [CrossRef]
17. Kamaruddin, Z.H.; Jumaidin, R.; Ain, N.; Mohamd, H. Water Affinity Properties of Thermoplastic Cassava Starch/Wax Reinforced with Sugarcane Bagasse Fiber. In Proceedings of the 7th International Conference and Exhibition on Sustainable Energy and Advanced Materials (ICE-SEAM 2021), Melaka, Malaysia, 17 November 2021; pp. 1–6.
18. Khan, A.; Vijay, R.; Singaravelu, D.L.; Arpitha, G.R.; Sanjay, M.R. Extraction and Characterization of Vetiver Grass (*Chrysopogon zizanioides*) and Kenaf Fiber (*Hibiscus cannabinus*) as Reinforcement Materials for Epoxy. *Integr. Med. Res.* **2019**, *9*, 773–778. [CrossRef]
19. Engel, J.B.; Ambrosi, A.; Tessaro, I.C. Development of Biodegradable Starch-Based Foams Incorporated with Grape Stalks for Food Packaging. *Carbohydr. Polym.* **2019**, *225*, 115234. [CrossRef]
20. Jumaidin, R.; Ahmad Diah, N.; Alamjuri, R.H.; Ahmad Rushdan, I.; Yusof, F.A. Processing and Characterisation of Banana Leaf Fibre Reinforced Thermoplastic Cassava Starch Composites Ridhwan. *Polymers* **2021**, *13*, 1420. [CrossRef]
21. Asyraf, M.R.M.; Rafidah, M.; Azrina, A.; Razman, M.R. Dynamic Mechanical Behaviour of Kenaf Cellulosic Fibre Biocomposites: A Comprehensive Review on Chemical Treatments. *Cellulose* **2021**, *28*, 2675–2695. [CrossRef]

22. Asyraf, M.R.M.; Rafidah, M.; Ebadi, S.; Azrina, A.; Razman, M.R. Mechanical Properties of Sugar Palm Lignocellulosic Fibre Reinforced Polymer Composites: A Review. *Cellulose* **2022**, *29*, 6493–6516. [CrossRef]
23. Asyraf, M.R.M.; Ishak, M.R.; Norrahim, M.N.F.; Amir, A.L.; Nurazzi, N.M.; Ilyas, R.A.; Asrofi, M.; Rafidah, M.; Razman, M.R. Potential of Flax Fiber Reinforced Biopolymer Composites for Cross-Arm Application in Transmission Tower: A Review. *Fibers Polym.* **2022**, *23*, 853–877. [CrossRef]
24. Neto, J.; Queiroz, H.; Aguiar, R.; Lima, R.; Cavalcanti, D.; Banea, M.D. A Review of Recent Advances in Hybrid Natural Fiber Reinforced Polymer Composites. *J. Renew. Mater.* **2022**, *10*, 561–589. [CrossRef]
25. Syafri, E.; Sudirman; Mashadi; Yulianti, E.; Deswita; Asrofi, M.; Abral, H.; Sapuan, S.M.; Ilyas, R.A.; Fudholi, A. Effect of Sonication Time on the Thermal Stability, Moisture Absorption, and Biodegradation of Water Hyacinth (*Eichhornia crassipes*) Nanocellulose-Filled Bengkuang (*Pachyrhizus erosus*) Starch Biocomposites. *J. Mater. Res. Technol.* **2019**, *8*, 6223–6231. [CrossRef]
26. Norizan, M.N.; Shazleen, S.S.; Alias, A.H.; Sabaruddin, F.A.; Asyraf, M.R.M.; Zainudin, E.S.; Abdullah, N.; Samsudin, M.S.; Kamarudin, S.H.; Norrrahim, M.N.F. Nanocellulose-Based Nanocomposites for Sustainable Applications: A Review. *Nanomaterials* **2022**, *12*, 3483. [CrossRef] [PubMed]
27. Ramesh, P.; Durga Prasad, B.; Narayana, K.L. Characterization of Kenaf Fiber and Its Composites: A Review. *J. Reinf. Plast. Compos.* **2018**, *37*, 731–737. [CrossRef]
28. Sanyang, M.L.; Sapuan, S.M.; Jawaid, M.; Ishak, M.R.; Sahari, J. Development and Characterization of Sugar Palm Starch and Poly(Lactic Acid) Bilayer Films. *Carbohydr. Polym.* **2016**, *146*, 36–45. [CrossRef]
29. Hazrati, K.Z.; Sapuan, S.M.; Zuhri, M.Y.M.; Jumaidin, R. Preparation and Characterization of Starch-Based Biocomposite Films Reinforced by Dioscorea Hispida Fibers. *J. Mater. Res. Technol.* **2021**, *15*, 1342–1355. [CrossRef]
30. Martinez Villadiego, K.; Arias Tapia, M.J.; Useche, J.; Escobar Macías, D. Thermoplastic Starch (TPS)/Polylactic Acid (PLA) Blending Methodologies: A Review. *J. Polym. Environ.* **2022**, *30*, 75–91. [CrossRef]
31. Hafila, K.Z.; Jumaidin, R.; Ilyas, R.A.; Selamat, M.Z.; Yusof, F.A.M. Effect of Palm Wax on the Mechanical, Thermal, and Moisture Absorption Properties of Thermoplastic Cassava Starch Composites. *Int. J. Biol. Macromol.* **2022**, *194*, 851–860. [CrossRef]
32. Kamaruddin, Z.H.; Jumaidin, R.; Selamat, M.Z.; Ilyas, R.A. Characteristics and Properties of Lemongrass (Cymbopogan Citratus): A Comprehensive Review. *J. Nat. Fibers* **2021**, *19*, 8101–8118. [CrossRef]
33. Ibrahim, M.I.J.; Sapuan, S.M.; Zainudin, E.S.; Zuhri, M.Y.M. Physical, Thermal, Morphological, and Tensile Properties of Cornstarch-Based Films as Affected by Different Plasticizers. *Int. J. Food Prop.* **2019**, *22*, 925–941. [CrossRef]
34. Asyraf, M.R.M.; Nurazzi, N.M.; Norrrahim, M.N.F.; Hazrati, K.Z.; Ghani, A.; Sabaruddin, F.A.; Lee, S.H.; Shazleen, S.S.; Razman, M.R. Thermal Properties of Oil Palm Lignocellulosic Fibre Reinforced Polymer Composites: A Comprehensive Review on Thermogravimetry Analysis. *Cellulose* **2023**, *30*, 2753–2790. [CrossRef]
35. Panrong, T.; Karbowiak, T.; Harnkarnsujarit, N. Thermoplastic Starch and Green Tea Blends with LLDPE Films for Active Packaging of Meat and Oil-Based Products. *Food Packag. Shelf Life* **2019**, *21*, 100331. [CrossRef]
36. Asyraf, M.R.M.; Syamsir, A.; Supian, A.B.M.; Usman, F.; Ilyas, R.A.; Nurazzi, N.M.; Norrrahim, M.N.F.; Razman, M.R.; Zakaria, S.Z.S.; Sharma, S.; et al. Sugar Palm Fibre-Reinforced Polymer Composites: Influence of Chemical Treatments on Its Mechanical Properties. *Materials* **2022**, *15*, 3852. [CrossRef]
37. Asyraf, M.R.M.; Syamsir, A.; Zahari, N.M.; Supian, A.B.M.; Ishak, M.R.; Sapuan, S.M.; Sharma, S.; Rashedi, A.; Razman, M.R.; Zakaria, S.Z.S.; et al. Product Development of Natural Fibre-Composites for Various Applications: Design for Sustainability. *Polymers* **2022**, *14*, 920. [CrossRef] [PubMed]
38. Zadeh, K.M.; Ponnamma, D.; Al Ali Al-Maadeed, M. Date Palm Fibre Filled Recycled Ternary Polymer Blend Composites with Enhanced Flame Retardancy. *Polym. Test.* **2017**, *61*, 341–348. [CrossRef]
39. Zou, Y.; Yuan, C.; Cui, B.; Liu, P.; Wu, Z.; Zhao, H. Formation of High Amylose Corn Starch / Konjac Glucomannan Composite Film with Improved Mechanical and Barrier Properties. *Carbohydr. Polym.* **2021**, *251*, 117039. [CrossRef]
40. Hernández, V.; Ibarra, D.; Triana, J.F.; Martínez-Soto, B.; Faúndez, M.; Vasco, D.A.; Gordillo, L.; Herrera, F.; García-Herrera, C.; Garmulewicz, A. Agar Biopolymer Films for Biodegradable Packaging: A Reference Dataset for Exploring the Limits of Mechanical Performance. *Materials* **2022**, *15*, 3954. [CrossRef]
41. Andrew, J.J.; Dhakal, H.N. Sustainable Biobased Composites for Advanced Applications: Recent Trends and Future Opportunities—A Critical Review. *Compos. Part C Open Access* **2022**, *7*, 100220. [CrossRef]
42. Asyraf, M.R.M.; Syamsir, A.; Bathich, H.; Itam, Z.; Supian, A.B.M.; Norhisham, S.; Nurazzi, N.M.; Khan, T.; Rashid, M.Z.A. Effect of Fibre Layering Sequences on Flexural Creep Properties of Kenaf Fibre-Reinforced Unsaturated Polyester Composite for Structural Applications. *Fibers Polym.* **2022**, *23*, 3232–3240. [CrossRef]
43. Ilyas, R.A.; Sapuan, S.M.; Ishak, M.R. Isolation and Characterization of Nanocrystalline Cellulose from Sugar Palm Fibres (*Arenga pinnata*). *Carbohydr. Polym.* **2018**, *181*, 1038–1051. [CrossRef]
44. Wifek, M.; Saeed, A.; Rehman, R.; Nisar, S. Lemongrass: A Review on Its Botany, Properties, Applications and Active Components. *Int. J. Chem. Biochem. Sci.* **2016**, *9*, 79–84.
45. Putri, K.N.A.; Kaewpichai, S.; Keereerak, A.; Chinpa, W. Facile Green Preparation of Lignocellulosic Biosorbent from Lemongrass Leaf for Cationic Dye Adsorption. *J. Polym. Environ.* **2021**, *29*, 1681–1693. [CrossRef]
46. Ahmad, M.A.; Ahmed, N.B.; Adegoke, K.A.; Bello, O.S. Adsorptive Potentials of Lemongrass Leaf for Methylene Blue Dye Removal. *Chem. Data Collect.* **2021**, *31*, 100578. [CrossRef]

47. Kamaruddin, Z.H.; Jumaidin, R.; Rushdan, A.I.; Selamat, M.Z.; Alamjuri, R.H. Characterization of Natural Cellulosic Fiber Isolated from Malaysian Cymbopogan Citratus Leaves. *BioResources* **2021**, *16*, 7729–7750. [CrossRef]
48. Kamaruddin, Z.H.; Jumaidin, R.; Ilyas, R.A.; Selamat, M.Z.; Alamjuri, R.H.; Yusof, F.A.M. Biocomposite of Cassava Starch-Cymbopogan Citratus Fibre: Mechanical, Thermal and Biodegradation Properties. *Polymers* **2022**, *14*, 514. [CrossRef] [PubMed]
49. Technique, G. Standard Test Methods for Apparent Density, Bulk Factor, and Pourability of Plastic. *Methods* **2003**, *08*, 1–5.
50. Ilyas, R.A.; Sapuan, S.M.; Ishak, M.R.; Zainudin, E.S. Effect of Delignification on the Physical, Thermal, Chemical, and Structural Properties of Sugar Palm Fibre. *BioResources* **2017**, *12*, 8734–8754. [CrossRef]
51. Jawaid, M.; Abdul Khalil, H.P.S.; Noorunnisa Khanam, P.; Abu Bakar, A. Hybrid Composites Made from Oil Palm Empty Fruit Bunches/Jute Fibres: Water Absorption, Thickness Swelling and Density Behaviours. *J. Polym. Environ.* **2011**, *19*, 106–109. [CrossRef]
52. *ASTM D256*; Standard Test Methods for Determining the Izod Pendulum Impact Resistance of Plastics. *Annu. B. ASTM Stand.* **2019**, *10*, 1–20.
53. Zhang, Y.; Simpson, B.K.; Dumont, M.J. Effect of Beeswax and Carnauba Wax Addition on Properties of Gelatin Films: A Comparative Study. *Food Biosci.* **2018**, *26*, 88–95. [CrossRef]
54. Shakuntala, O.; Raghavendra, G.; Samir Kumar, A. Effect of Filler Loading on Mechanical and Tribological Properties of Wood Apple Shell Reinforced Epoxy Composite. *Adv. Mater. Sci. Eng.* **2014**, *2014*, 538651. [CrossRef]
55. Ibrahim, M.I.J.; Sapuan, S.M.; Zainudin, E.S.; Zuhri, M.Y.M. Potential of Using Multiscale Corn Husk Fiber as Reinforcing Filler in Cornstarch-Based Biocomposites. *Int. J. Biol. Macromol.* **2019**, *139*, 596–604. [CrossRef] [PubMed]
56. Sanhawong, W.; Banhalee, P.; Boonsang, S.; Kaewpirom, S. Effect of Concentrated Natural Rubber Latex on the Properties and Degradation Behavior of Cotton-Fiber-Reinforced Cassava Starch Biofoam. *Ind. Crop. Prod.* **2017**, *108*, 756–766. [CrossRef]
57. Kaisangsri, N.; Kerdchoechuen, O.; Laohakunjit, N. Biodegradable Foam Tray from Cassava Starch Blended with Natural Fiber and Chitosan. *Ind. Crop. Prod.* **2012**, *37*, 542–546. [CrossRef]
58. Soykeabkaew, N.; Supaphol, P.; Rujiravanit, R. Preparation and Characterization of Jute-and Flax-Reinforced Starch-Based Composite Foams. *Carbohydr. Polym.* **2004**, *58*, 53–63. [CrossRef]
59. Belhassen, R.; Boufi, S.; Vilaseca, F.; López, J.P.; Méndez, J.A.; Franco, E.; Pèlach, M.A.; Mutjé, P. Biocomposites Based on Alfa Fibers and Starch-Based Biopolymer. *Polym. Adv. Technol.* **2009**, *20*, 1068–1075. [CrossRef]
60. Syahida, N.; Fitry, I.; Zuriyati, A.; Hanani, N. Effects of Palm Wax on the Physical, Mechanical and Water Barrier Properties of Fish Gelatin Films for Food Packaging Application. *Food Packag. Shelf Life* **2020**, *23*, 100437. [CrossRef]
61. Ibrahim, M.I.J.; Sapuan, S.M.; Zainudin, E.S.; Zuhri, M.Y.M. Preparation and Characterization of Cornhusk/Sugar Palm Fiber Reinforced Cornstarch-Based Hybrid Composites. *J. Mater. Res. Technol.* **2020**, *9*, 200–211. [CrossRef]
62. Rosa, M.F.; Chiou, B.S.; Medeiros, E.S.; Wood, D.F.; Williams, T.G.; Mattoso, L.H.C.; Orts, W.J.; Imam, S.H. Effect of Fiber Treatments on Tensile and Thermal Properties of Starch/Ethylene Vinyl Alcohol Copolymers/Coir Biocomposites. *Bioresour. Technol.* **2009**, *100*, 5196–5202. [CrossRef]
63. Jumaidin, R.; Sapuan, S.M.; Jawaid, M.; Ishak, M.R.; Sahari, J. Thermal, Mechanical, and Physical Properties of Seaweed/Sugar Palm Fibre Reinforced Thermoplastic Sugar Palm Starch/Agar Hybrid Composites. *Int. J. Biol. Macromol.* **2017**, *97*, 606–615. [CrossRef]
64. Sarifuddin, N.; Ismail, H.; Ahmad, Z. Effect of Fiber Loading on Properties of Thermoplastic Sago Starch/Kenaf Core Fiber Biocomposites. *BioResources* **2012**, *7*, 4294–4306.
65. Jumaidin, R.; Saidi, Z.A.S.; Ilyas, R.A.; Ahmad, M.N.; Wahid, M.K.; Yaakob, M.Y.; Maidin, N.A.; Ab Rahman, M.H.; Osman, M.H. Characteristics of Cogon Grass Fibre Reinforced Thermoplastic Cassava Starch Biocomposite: Water Absorption and Physical Properties. *J. Adv. Res. Fluid Mech. Therm. Sci.* **2019**, *62*, 43–52.
66. Lomelí Ramírez, M.G.; Satyanarayana, K.G.; Iwakiri, S.; De Muniz, G.B.; Tanobe, V.; Flores-Sahagun, T.S. Study of the Properties of Biocomposites. Part I. Cassava Starch-Green Coir Fibers from Brazil. *Carbohydr. Polym.* **2011**, *86*, 1712–1722. [CrossRef]
67. Edhirej, A.; Sapuan, S.M.; Jawaid, M.; Zahari, N.I. Cassava/Sugar Palm Fiber Reinforced Cassava Starch Hybrid Composites: Physical, Thermal and Structural Properties. *Int. J. Biol. Macromol.* **2017**, *101*, 75–83. [CrossRef] [PubMed]
68. Jumaidin, R.; Khiruddin, M.A.A.; Asyul Sutan Saidi, Z.; Salit, M.S.; Ilyas, R.A. Effect of Cogon Grass Fibre on the Thermal, Mechanical and Biodegradation Properties of Thermoplastic Cassava Starch Biocomposite. *Int. J. Biol. Macromol.* **2020**, *146*, 746–755. [CrossRef]
69. Asim, M.; Jawaid, M.; Khan, A.; Asiri, A.M.; Malik, M.A. Effects of Date Palm Fibres Loading on Mechanical, and Thermal Properties of Date Palm Reinforced Phenolic Composites. *J. Mater. Res. Technol.* **2020**, *9*, 3614–3621. [CrossRef]
70. Saba, N.; Alothman, O.Y.; Almutairi, Z.; Jawaid, M.; Ghori, W. Date Palm Reinforced Epoxy Composites: Tensile, Impact and Morphological Properties. *J. Mater. Res. Technol.* **2019**, *8*, 3959–3969. [CrossRef]

polymers MDPI

Article

An Experimental Study on Drilling Behavior of Silane-Treated Cotton/Bamboo Woven Hybrid Fiber Reinforced Epoxy Polymer Composites

Karthik Aruchamy [1], Sathish Kumar Palaniappan [2,*], Rajeshkumar Lakshminarasimhan [3], Bhuvaneshwaran Mylsamy [4], Satish Kumar Dharmalingam [5], Nimel Sworna Ross [5] and Sampath Pavayee Subramani [6]

1 Department of Mechatronics Engineering, Akshaya College of Engineering and Technology, Coimbatore 642109, Tamil Nadu, India
2 Department of Mining Engineering, Indian Institute of Technology, Kharagpur 721302, West Bengal, India
3 Department of Mechanical Engineering, KPR Institute of Engineering and Technology, Coimbatore 641407, Tamil Nadu, India
4 Department of Mechanical Engineering, K. S. R. College of Engineering, Tiruchengode 637215, Tamil Nadu, India
5 Department of Mechanical Engineering, Saveetha School of Engineering, SIMATS University, Chennai 602105, Tamil Nadu, India
6 Department of Mechanical Engineering, K. S. Rangasamy College of Technology, Tiruchengode 637215, Tamil Nadu, India
* Correspondence: sathishiitkgp@gmail.com

Citation: Aruchamy, K.; Palaniappan, S.K.; Lakshminarasimhan, R.; Mylsamy, B.; Dharmalingam, S.K.; Ross, N.S.; Pavayee Subramani, S. An Experimental Study on Drilling Behavior of Silane-Treated Cotton/Bamboo Woven Hybrid Fiber Reinforced Epoxy Polymer Composites. *Polymers* **2023**, *15*, 3075. https://doi.org/10.3390/polym15143075

Academic Editors: R.A. Ilyas, S.M. Sapuan, Emin Bayraktar, Shukur Abu Hassan, Nabil Hayeemasae, Khubab Shaker and Chenggao Li

Received: 15 May 2023
Revised: 10 July 2023
Accepted: 16 July 2023
Published: 18 July 2023

Abstract: Machining is considered to be an important post-manufacturing process. Evaluation of machinability of natural-fiber-reinforced composites is important owing to its wide application spectrum. Current experiments focus on the drilling parameters of cotton/bamboo woven fabric reinforced epoxy composites laminates using a solid twist drill. Composites were manufactured with 45 wt.% cotton/bamboo woven fabric in epoxy resin using a compression molding method. Drilling experiments were carried out in pillar-type drilling machine and the drilling characteristics, such as thrust force, were analyzed using four process parameters like spindle speed, feed rate, drill diameter, and silane-treated fabric. Drilling experiments were carried out using the Box–Behnken Experimental Design, and the recommended drilling characteristics were analyzed using quadratic models based on response surface methodology. It was observed from the results that the thrust force is low with small drill-bit diameter, higher cutting speed, and lower feed rate, according to the response surface analysis. Surface morphology of the drilled hole suggested that a better quality of hole can be obtained at lower feed rates.

Keywords: cotton/bamboo woven fabric; silane treatment; thrust force; SEM analysis; RSM; machining operation

1. Introduction

The study of composite material from the last two decades that focuses on investigating natural-fiber-reinforced composite is an important branch of materials science [1–5]. Cotton, bamboo, bark, wood, pulp, cereal straw, bark, bagasse, corncobs, nut shells, and vegetable (e.g., coir, ramie, sisal, jute, flax, sun hemp, banana, and pineapple) are the examples of plant-based natural fibers [6–8]. The major microconstituents of these fibers are hemicelluloses, cellulose, wax, and lignin, along with a minimum percentage of extractives. The composition of fiber varies depending on their origin [9–11]. The fibers used in engineering applications mainly include synthetic fibers and natural fibers. Synthetic fibers such as carbon fiber and glass fiber have been widely used in different fields because of their lightweight and high strength, excellent mechanical properties, corrosion resistance, fatigue resistance, and so on. The above advantages can make synthetic fiber composites the main load-bearing components in engineering applications [12–14]. Also, natural fibers show several benefits compared to standard glass and carbon fibers. Natural fibers are of low cost, abundantly available, environment-friendly, easy to process, have low density, and possess reasonably good flexural and tensile modulus. By considering some key problems in synthetic fibers such as high cost, environmental problems during the preparation, and nonrenewable material, natural fibers overcome the above disadvantages and can be mixed with polymer matrix to create composites. Even if the strength of the fiber-reinforced composites is sometimes lower than the individual fiber itself, either hybridization or suitable treatments may eradicate this. However, the strength of composites, at most times, would be less than the individual fiber itself. Natural fibers with biodegradable and renewable properties make composting and incineration easier than all other synthetic and industrial fibers. In addition, the fibers contain stored atmospheric carbon dioxide and have a minimum embodied energy than industrially produced glass fibers [15–17].

Even though natural fibers have the possibility to complement synthetic fibers in polymer composite materials [15,16], natural fibers have limitations in terms of mechanical performance and absorption of humidity [17,18]. Regardless of whether thermoplastic or thermoset polymers are utilized as the matrix material, the same limitations exist. According to previous studies [4,19], the chemical incompatibility between the hydrophilic lignocellulosic molecules of the natural fiber and the hydrophobic thermoplastic molecules is the key factor in limiting the mechanical characteristics of natural-fiber-reinforced thermoplastic composites. Because of this incompatibility, achieving good fiber matrix interface bonding is challenging, resulting in inadequate load shift between the reinforced fibers and matrix. In order to obtain the requisite compatibility between the fiber and matrix, various solutions have been suggested, including chemical treatments of lignocelluloses fibers and the use of compatibilizers. Various authors stated that the different chemical processes can be used to enhance the lignocellulosic fiber compatibility with polymer matrices. The use of alkali treatment, isocyanate treatment, acetylation, permanganate treatment, peroxide treatment, benzoylation, and silane treatment were all studied. Treatments such as alkaline and silane were frequently reported among the various approaches [18,19].

The fibers were soaked in alkaline solution, most often NaOH, for a specified period of time during the alkaline treatment. It was expected to improve the mechanical bonding by enhancing the roughness of fiber surface. It increases the crystallinity of the fiber which enhances the possibility of chemical interaction with the matrix [20–22]. It also exposes more reactive functional groups of cellulose to be bonded with a polymer matrix. Alkaline treatment effects were studied by some authors on kenaf fiber [20]. Cleaning the fiber surface with an alkaline treatment using a 6% NaOH solution was found to be effective. On the other hand, a 9% NaOH solution was found to degrade the fiber surface and decreased its strength. The fibers are normally soaked in silane solution diluted in water/alcohol. As a result, silane hydrolyses itself into silanol from alcohol that contains water. The silanol combines with the OH groups of cellulose in natural fibers to generate stability in covalent linkages with cell walls that assimilate onto the fiber surface [21]. Silane treatment increases the cross-linking in the connected regions between fiber and the matrix, allowing for better interaction between the fiber and matrix [22]. Mechanical properties such as tensile strength and modulus of flax fiber reinforced with epoxy composite were found to increase, as stated by Van de Weyenburg et al. [23]. Results showed that improvement in the tensile strength and modulus was roughly 30% and 45%, respectively, when 40 vol.% of treated flax fibers were utilized as a reinforcement in epoxy matrix. This improvement was due to the combined chemical treatment of flax fibers, i.e., by using a combination of sodium hydroxide and silane coupling agent chemicals.

Natural-fiber-reinforced polymer composites are more difficult to machine because of lack of machining records and a higher level of parameter complexity [24]. Occurrence of voids, delamination, matrix, and fiber debonding are some of the examples for manufacturing and machining faults. Despite the fact that the majority of fiber-based hybrid composites are made to actual shape by using the machining techniques such as drilling, grinding, and milling, which are completely unavoidable at the time of assembly process, drilling is considered the most important machining techniques in the overall process. When compared to drilling of conventional metals and alloys, the drilling procedure for fiber-reinforced composite samples is more complex because of the anisotropic nature of samples. Due to the presence of insoluble constituents in fiber-reinforced composites, the isotropy cannot be defined as in conventional metals and alloys, which makes them more anisotropic materials. This complexity is considered to be a significant one in the drilling of fiber-reinforced composites. It finds its application in structural automobile components, kitchen cabinets, doors, furniture, electronic components, and so on. The end product is assembled with sliced composite panels to form the desired shape and dimensions, and attaching them with bolts and screws in all of the above applications. As a result, before placing it into use, drilling is performed to complete the assembly process. Hence, drilling is considered to be extremely important, because a fault in the structural integrity of the drilling process influences the quality of the final product [25].

Drilling experiments in fiber-based polymer composites have sufficiently provided an indication of the importance and research direction. On a coir–polyester composite, the influence of drill diameter, speed, and feed on torque, thrust force, and tool wear was investigated. Drilling studies on roselle–sisal hybrid composites were performed and optimized using an artificial neural network model [26,27]. Few experimenters investigated initially with peel-up delamination and later with push-down delamination for hemp/flax-fiber-based composites and also discovered that the most important elements to consider throughout the drilling process are feed rate and cutting speed [28]. In the drilling of sisal polypropylene composites, drill geometry was determined to be the most influential element [29–31]. Drilling was performed on a cotton/bamboo (CB) woven fabric reinforced composite, and thrust force was calculated for various combinations of cutting speed, feed rate, and drill shape. The cutting parameters, as well as the input parameters, were optimized using a Box–Behnken Experimental Design (BBED). When the drill diameter was increased, the results revealed that the feed rate increases with the decrease in cutting speed. Drill diameter and rate of feed influence the increment of thrust force, according to Srinivasan et al. [32]. Using BBED and analysis of variance (ANOVA), the effect of thrust force was investigated, and they observed that low feed rate, high spindle speed, and smaller drill diameter resulted in the lowest thrust force.

In spite of various studies on drilling behavior of natural fiber composites, only few studies discuss the machining behavior of treated woven fiber reinforced composites. In the current study, silane-treated CB reinforced hybrid composites were analyzed for their machinability behavior through drilling process. Cutting thrust force was evaluated with various feed rates, cutting speeds, and drill geometries. The optimum parameter was determined in order to achieve the lowest cutting thrust force by using the response surface methodology (RSM) technique. The research experiments were controlled by BBED. The data were analyzed using 3D plots and influence graphs. The drilled surface morphology was examined using a scanning electron microscopy (SEM) and the optimal parameter settings for obtaining a good quality hole and better machinability were recommended.

2. Materials and Methodology

2.1. Fabrication of CB Woven Fabric

In this paper, the CB woven fabric composites were manufactured with cotton yarn and bamboo yarn packed with corresponding warp and weft directions. In this study, cotton yarn, bamboo yarn, and CB woven fabric were collected from Pallava Spinning Mills (Pvt.) Limited, Erode District, Tamil Nadu, India, and Ganapathi Chettiar Tex, Tirupur District, Tamil Nadu, India, with good-quality yarn that has bamboo fiber length of 36 mm, fiber fineness of 1.52 dtex, linear density of 0.155 tex, moisture regain value of 11.42%, tenacity of 22.84 g/tex, and elongation of 21.2% [33]. The yarn and fabric particulars are given in Tables 1 and 2, respectively.

Table 1. Particulars of yarns.

S. No.	Properties	Cotton Yarn	Bamboo Yarn
1	Fiber length (mm)	40	36
2	Fiber fineness (dtex)	1.35	1.52
3	Linear density (tex)	0.115	0.155
4	Moisture regain value (%)	7.85	11.42
5	Tenacity (g/tex)	32.22	22.84
6	Elongation (%)	5.8	21.2

Table 2. Particulars of CB woven fabric [30,31].

S. No.	Fabric Particulars	Values
1	Fabric GSM (g)	164
2	End per inch (EPI)	69
3	Picks per inch (PPI)	54
4	Thickness (mm)	0.38

2.2. *Silane Treatment of CB Fabrics*

CB fabrics were immersed in the 1.5% concentrations of 3-aminopropyl trimethoxy silanes solution (procured from M/s. Covai seenu, Coimbatore, Tamil Nadu, India) with water and acetone (50/50 volume) for 2 h. From various screening experiments and based on the literature, the concentration of silane was taken as 1.5% [23]. The fibers were then soaked in 1.5% silane mixed with 2% acetic acid in the form of acetic anhydride for pH adjustment between 4.5 and 5, and sundried for 24 h. Then, the fibers were oven dried at 80 °C for 4 h to maintain their stiffness. Silane coupling agents were expected to effectively modify the natural fiber–polymer matrix interface and increase their interfacial strength.

2.3. *Fabrication of Composite Laminates*

Composite laminates were manufactured using a compression molding technique. Initially, wax was applied to the surface of the mold for easy removal, and the size of the mold measured 270 mm × 270 mm × 3 mm. The CB woven fabrics (2 layers each) were laid and arranged, layer after layer, with epoxy adhesive to keep woven fabrics composed. For this experiment, Araldite LY 556 epoxy resin and HY951 hardener (procured from M/s. Covai seenu, Coimbatore, Tamil Nadu, India) were used. To assist gradual and homogeneous curing of the resin, a 10:1 mix of resin and hardener was used. A composite plate was created by stacking alternate layers of resin and fabric. It was then compressed at 30 MPa for 1 h at 160 °C. Although a greater temperature will speed up the process of curing, the natural fiber's existence limited high-temperature processing. A consistent weight fraction of 45% was used to make the composite laminates. The composite included isotropic character, which is the most important requirement in various commercial applications and is the only way to ensure uniform load-carrying capability of the material.

2.4. *Drilling Process Parameters and Optimization*

Drilling operations were performed in a pillar-type drilling machine (*Make: Kirloskar, Pune, India*) using HSS drill bits with pointed ends. All the operations were carried out at ambient atmospheric conditions. In this experimental work, data acquisition systems were used along with a drill dynamometer to obtain thrust force signals. Forces developed on the drill bit axis or spindle axis during the machining process are termed as thrust force, which induces damage during the drilling of silane-treated woven CB composite laminates, as shown in Figure 1.

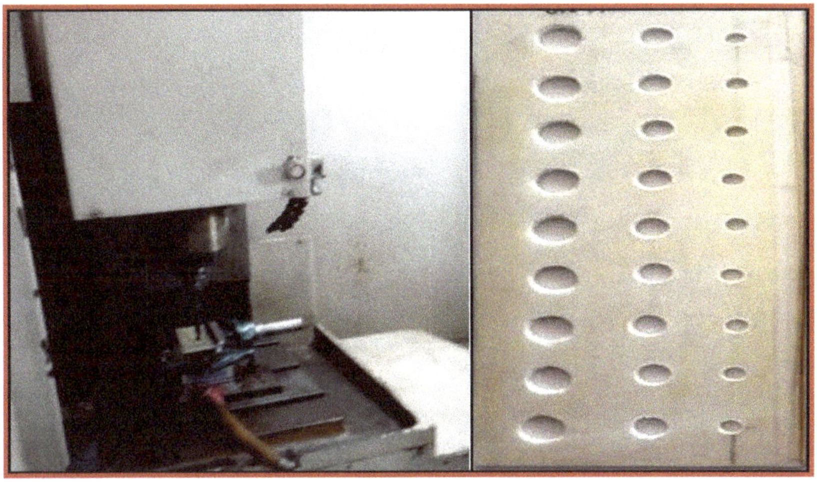

Figure 1. Drilling of CB composite laminate.

The hole-making process in composite materials is influenced by a number of variables such as spindle speed, drill diameter, drill type feed rate, drill point angle drill material, type of matrix, type of fiber reinforcement, and fiber volume fraction. These drilling parameters define and influence the thrust force, surface quality of the drilled hole, delamination near the drilled hole, and other geometrical factors such as roundness and eccentricity [34]. Many investigations have discovered that the major influencing parameters in the drilling of fiber-reinforced composite panels include feed rate [35], drill diameter [36,37], and cutting speed [31]. Drilling operations were carried out by changing the drill bit rotation speed (rpm), drill diameter (mm), and rate of feed (mm/min). Table 3 shows the various process parameters used for the drilling of treated CB fabric woven hybrid composites.

Table 3. Drilling process parameters [30,31].

S. No.	Process Parameter	Unit	Upper Limit	Lower Limit
1	Cutting speed (n)	rpm	3000	1000
2	Drill diameter (d)	mm	12	6
3	Feed rate (f)	mm/min	60	20

A scanning electron microscope (model: VEGA 3 TE SCAN) was used to study the surface topography of the drilled holes in silane-treated CB composite laminates. A thin layer of sputtered gold was coated on the surface of all the tested composite specimens to improve electrical conductivity and to obtain better images.

RSM is the collection of mathematical and statistical data to develop modeling and analyses of engineering problems [38]. Nowadays, RSM is used in the design of experiments (DOE). Also, RSM helps in identifying the correlation between input parameters to obtain the response surface. Box and Draper [39] developed a model to suit physical experiments and later for machining problems. Further, RSM also has its objective to later mine the optimization of process parameters.

Douglas C. Montgomery [40] also developed a model relating the process variables with output response, which was utilized for process prediction and control. The experiment domain's boundary must be investigated in RSM. The process parameters for this analysis were feed rate, cutting speed, and diameter of the drill, and BBED was utilized to formulate the design of experiments. BBED, an alternate way to central composite design (CCD), is the common statistical tool employed to frame the number of experiments. This approach was used to analyze experiments comprising three levels of design. The BBED can be rotated, and it is said to have no fractional factorial design, which makes it easy to understand the results [41]. In RSM, the general correlation of the control parameters and their responses is illustrated in Equation (1):

$$Y = \beta_0 + \beta_1 x_1 + \beta_2 x_2 + \beta_i x_i + \ldots\ldots + \varepsilon \quad (1)$$

where Y is the response (thrust force), β_0, β_1, β_2, ... β_i are the regression coefficients, x_1, x_2, ... x_i denote the predictor variable, namely, feed rate, diameter, and cutting speed, and ε shows the error occurring in the thin model. Generally, in RSM, quadratic response functions (Equation (2)) are used and are represented as follows:

$$Y = \beta_0 + \sum_{i=1}^{k} \beta_1 x_1 + \sum_{i=1}^{k} \beta_{ii} x_i^2 + \sum_{i<j}^{k} \beta_{ij} x_i x_j + \varepsilon \quad (2)$$

3. Results and Discussion

3.1. Thrust Force Analysis

Silane-treated CB woven fabric composites were drilled using twisted drill bits, and the thrust force signals were obtained, as shown in Figure 2. The plot was obtained during drilling at a speed of 1800 rpm and at a feed of 60 mm/min. This plot depicts different phases, as shown in the figure: initially, the thrust force was found to be low before the entry of the drill bit and during the progress of the drill bit. Next, the thrust force increased to the maximum level during the process of making the hole, i.e., full contact of the drill bit and the specimen. Then, as the drill bit reached the bottom of the specimen, it gradually reduced the thrust force and marked the exit of the drill bit where the thrust force finally attained a zero value [37].

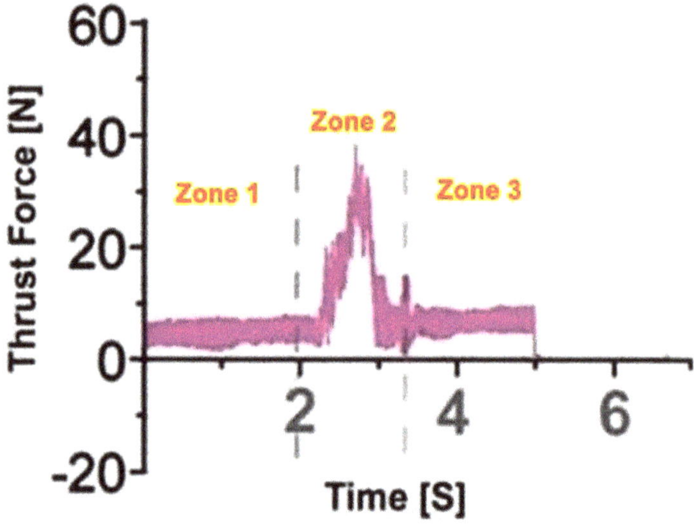

Figure 2. Average cutting forces signal that used a twist drill at 1800 rpm and 60 mm/min.

3.2. Response Surface Methodology (RSM)

By considering input variables, a quadratic equation was newly developed for determining thrust force in order to attain exact process parameters. The final model for thrust force is

$$Y(N) = 19.9525 + (4.072667 \times d) - (0.30465 \times f) + (0.004042 \times n) - (0.03583 \times d \times f) - (0.00049 \times d \times n) + (0.000183 \times f \times n) - (0.06561 \times d^2) + (0.007761 \times f^2) - (3.5E - 6 \times n^2) \quad (3)$$

Table 4 shows the experimental results for the drilling of CB woven fiber reinforced hybrid epoxy composites using BBED.

Table 4. The experimental results for the BBED.

S. No.	Run	Diameter of the Drill (mm)	Feed Rate (mm/min)	Speed of the Spindle (rpm)	Thrust Force of Cutting (N)
1	2	6	20	1800	32.43
2	16	12	20	1800	39.89
3	11	6	60	1800	49.11
4	12	12	60	1800	47.97
5	14	6	40	1200	38.99
6	10	12	40	1200	44.43
7	7	6	40	2400	33.34
8	4	12	40	2400	35.22
9	15	9	20	1200	40.11
10	1	9	60	1200	49.23
11	8	9	20	2400	29.75
12	17	9	60	2400	47.67
13	3	9	40	1800	39.12
14	13	9	40	1800	41.98
15	9	9	40	1800	39.06
16	5	9	40	1800	40.11
17	6	9	40	1800	38.91

Table 5 shows the ANOVA table for thrust force. Based on F and p-value, the RSM model can be simultaneously analyzed, considering all the design factors and the value of F, which predicts the quality of the entire design model. The very low probability factors identified by p-value have an insignificant effect on the response [42]. The higher F value and lower p-value (p-value < 0.05) significance enhanced the fit for the predicted designed model with the experimental model. The experimental result was calculated with a confidence level of 95%. Any value of p less than 0.05 indicates that the developed thrust model will be significant. However, the F value of the lack of a fitted model is 0.372, which indicates that the model is insignificant. It proves that the planned model is very suitable for the prediction of thrust force during drilling of developed hybrid composites [43]. In Table 5, it is shown that the feed rate (f) and spindle speed (N) are significant. The coefficient correlation (R-Seq) and adjusted coefficient correlation (0.98438 and 0.9643) indicate that the model is sufficient, and so the procedures were successfully utilized to find the cutting thrust force during the drilling process in composite laminates.

Figure 3a depicts the normal percentage of probability plot with respect to internally studentized residuals and it is observed from the graph that the result is grouped to form a straight line; hence, it confirms the efficiency of the developed model. Similar results were obtained by some of the researchers previously [44]. The association between the actual and predicted thrust is depicted Figure 3b. This result shows that the predicted values are in close agreement with the experimental values [35].

Figure 4 shows the interrelationship between the considered process parameters and the cutting thrust force of CB-reinforced epoxy hybrid composite laminates. Figure 4a explains the influence of the drill diameter on the thrust force. It turns out that as the diameter of the drill bit increases, the thrust force also increases [42].

As the diameter of the drill bit increases, the contact area between the drill bit and the laminate becomes larger; hence, due to higher heat generated, the thrust force increases. The influence of feed speed on thrust is depicted in Figure 4b. The results show that whenever the rate of feed increases, the thrust force also increases. As the diameter of the drill (d) and the feed (f) increases, the cross-sectional area (A) increases because of the formed chip area, $A = df/4$, and the thrust force also increases, so the axial thrust and the resistance to the developed chip increase [45]. Figure 4c shows that the increase in spindle speed decreases the thrust force, and due to higher spindle speed, the chip formation softens, thereby minimizing the heat of the material and increasing the thrust force. It could be stated from the study that the change in cutting speed and feed rate influenced the thrust force significantly, irrespective of the working environment [46].

Table 5. Analysis of variance (ANOVA) table for thrust force.

Source	d_f	Adj SS	Adj MS	F Value	*p*-Value Prob > F
Regression model	9	535.903	59.545	49.041	<0.0001
d	1	23.256	23.256	19.154	0.0032
f	1	335.405	335.405	276.236	<0.0001
n	1	89.646	89.646	73.832	<0.0001
d_f	1	18.490	18.490	15.228	0.0059
d_n	1	3.168	3.168	2.609	0.1503
n_f	1	19.360	19.360	15.945	0.0052
d^2	1	1.468	1.468	1.209	0.3079
f^2	1	40.581	40.581	33.422	0.0007
n^2	1	6.584	6.584	5.423	0.0527
Residual	7	8.499	1.214		
Lack of fit	3	1.855	0.618	0.372	0.7784
Pure error	4	6.644	1.661		
Cor. total	16	544.402			
Std. Dev.		1.102	R-Squared	0.984	
Mean		40.431	Adj R-Squared	0.964	
C.V.%		2.725	Pred R-Squared	0.926	
Press		40.065	Adeq Precision	23.245	

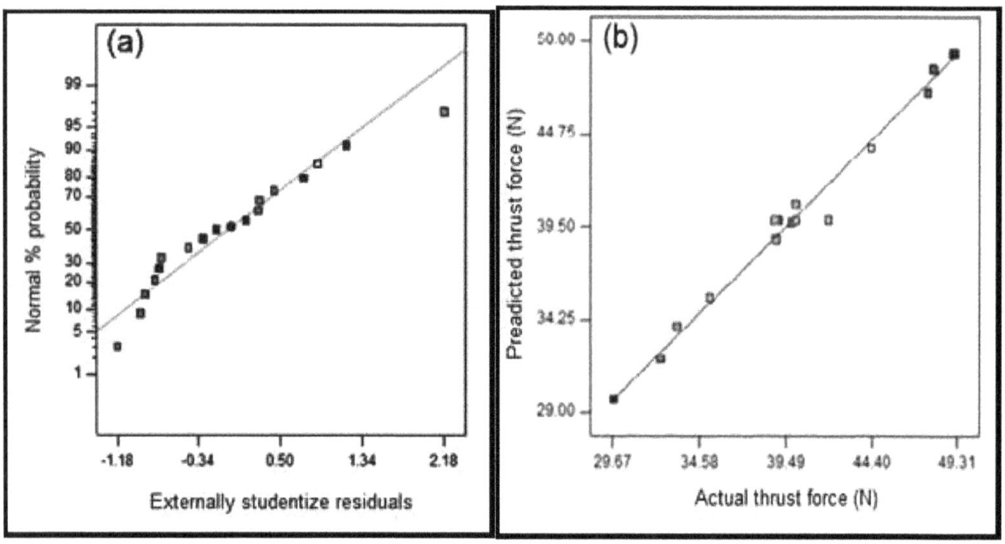

Figure 3. (**a**) Normal probability of residuals. (**b**) Predicated thrust force in the drilling of the composites.

Figure 5 shows the 3D model of the thrust force during the drilling of silane-treated CB composite laminate. The effect of the reaction among processing variables is discussed through 3D response plots. The 3D model has a graph that shows the influence of two changing variables while maintaining the constant third parameter throughout the experiments. Figure 5a indicates the variation of thrust force with the diameter of the drill and the rate of feed, keeping the cutting speed constant. The results show that the thrust force increases as the diameter of the drill bit and feed rate increase. Figure 5b shows the impact of drill diameter and cutting speed on thrust force. As a result, it was found that as the spindle speed of the CB composite laminate increased, the thrust force decreased. Figure 5c explains the plot of cutting speed and feed rate on the thrust force of the CB composite laminate. The findings show that the enhancement in feed and speed increased the thrust force of the CB composite laminate. It can be seen from the experiment that the drill diameter, feed rate, and cutting speed affected the machining operation irrespective of the working environment.

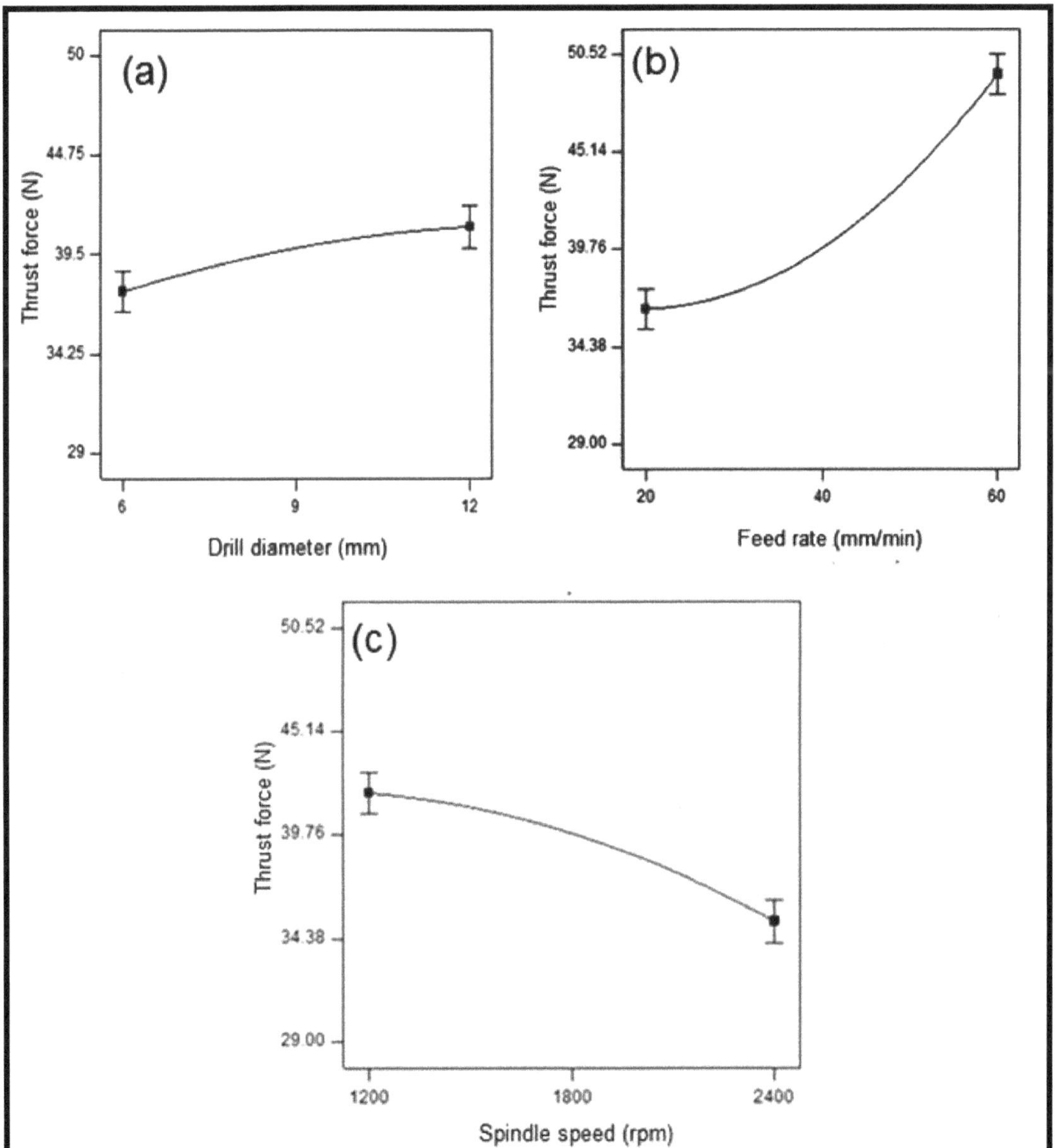

Figure 4. Influence of machining characterization on thrust force in CB fabric/epoxy composites.

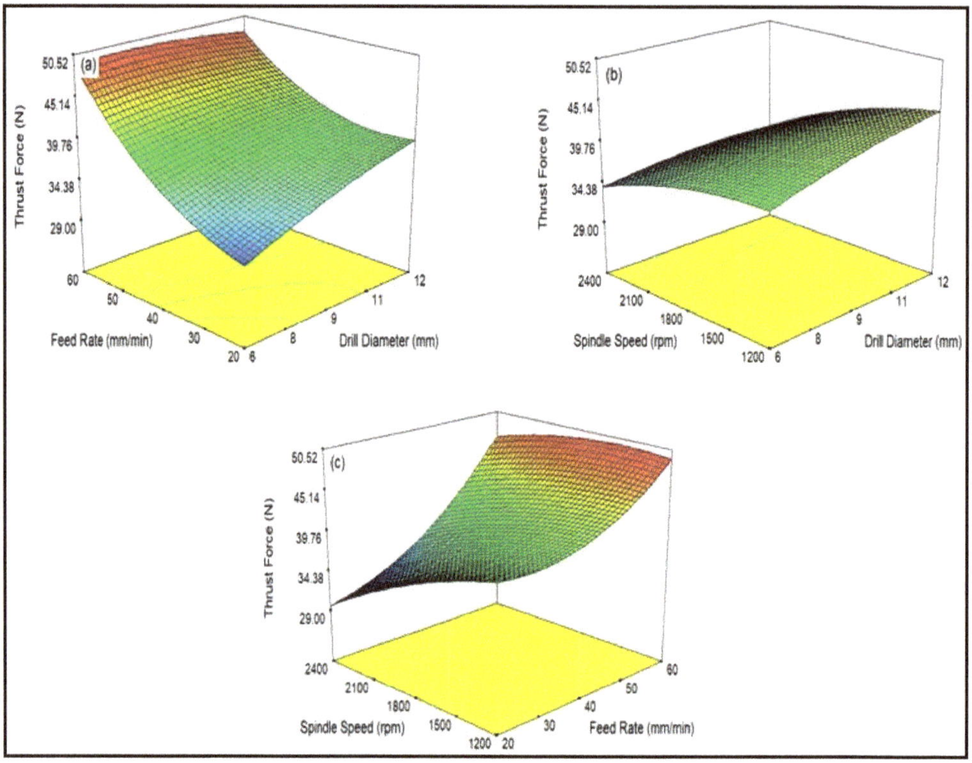

Figure 5. A 3D response plot for thrust force during drilling of silane-treated CB fabric/epoxy composites; (**a**) influence of drill diameter and feed rate on thrust force; (**b**) influence of drill diameter and spindle speed on thrust force; and (**c**) influence of feed rate and spindle speed on thrust force.

3.3. Comparison of Experimental and Thrust Force Values

In this experiment, the R-squared value of the model is greater than 90%, and so the modeling system can be effectively used to predict the different properties of composite drilling. The comparison between the RSM value and the experimental value is shown in Figure 6. It is proven that the relevant statistics between the experimental model result and the RSM predicted result can be obtained through the RSM technique.

3.4. SEM Observation

The morphology of the drilled holes in silane-treated CB composite laminates was examined using a scanning electron microscope at various locations. The SEM micrograph shows the microstructure of the drilled surface. It could be observed that debonding between fiber and matrix, surface roughness, and fiber pull-out exist in the drilled silane-treated CB fabric epoxy composite laminate using a solid twist drill. Figure 7 shows that minimum cutting speed and maximum feed rate damage the drill walls and cause large fiber pull-out and surface roughness. This is due to the effect of high thrust force applied on the drill zone of the composite laminate [47]. Figure 8 shows the morphology of the composite specimen under better cutting behavior with less drill damage during drilling when compared to high-speed cutting and minimum feed rate conditions. This is due to less contact of the drill bit with the CB composite laminates. The result indicates that the thrust force is low and has a better quality of hole at lower feed rates [33,48].

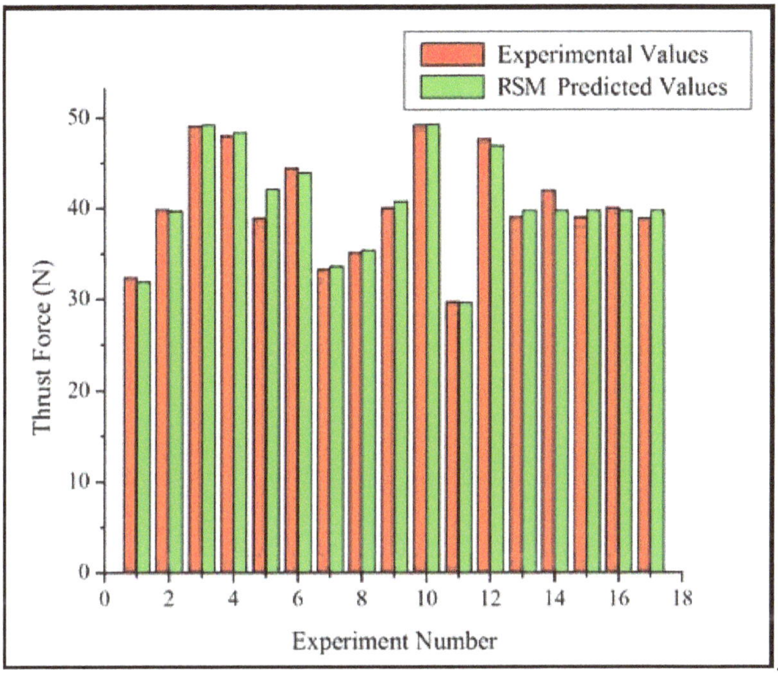

Figure 6. Comparison of experimental and RMS thrust force values.

Figure 7. SEM image of drilled hole drill (1800 rpm and 60 mm/min).

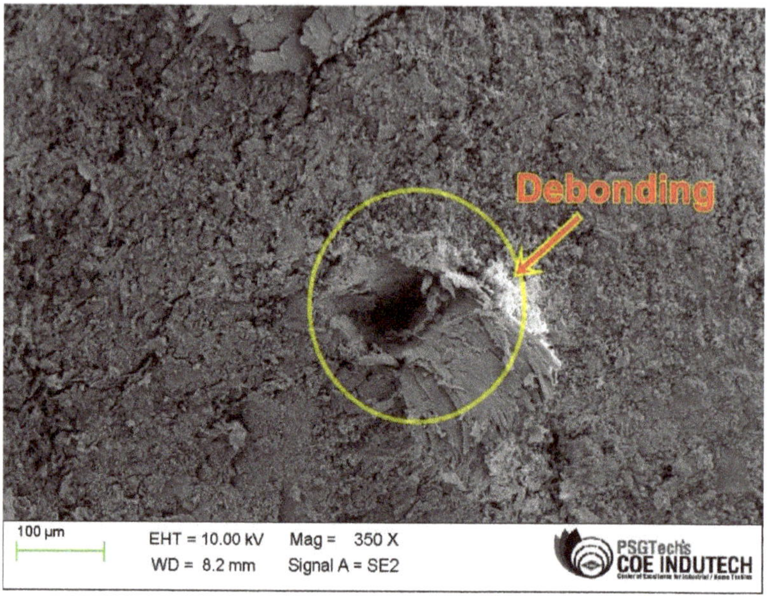

Figure 8. SEM image of drilled hole drill (2400 rpm and 20 mm/min).

4. Conclusions

Cotton/bamboo woven fabric reinforced epoxy composite laminates were manufactured using a compression molding technique and were subjected to machinability analysis. The following conclusions were obtained from the experimental study.

- From the thrust force analysis, it could be observed that during drilling, three significant phases of thrust force could be obtained. The maximum value of thrust force was during the complete contact of drill bit with the composites.
- The influence of various process parameters on the thrust force on fabric/epoxy composite can be examined by RSM. The optimum cutting parameters are identified at a minimum rate of feed (20 mm/min), maximum spindle speed (2400 rpm), and drill diameter (6 mm). The adequacy of this model is proven to be fitted for finding thrust force on the composite material.
- The size of the hole is one of the main factors and it turns out to be the reason for the increase in thrust force. When the thrust force increases, the size of the hole also increases considerably and the composite laminate thickness plays a vital role in thrust force development.
- From the response plots and 3D surface plots, it can be concluded that the thrust force increases with drill diameter and feed rate and decreases with spindle speed. Hence, lower feed rate and small drill diameter at higher spindle speed produce good-quality holes.
- From the SEM micrograph, the minimum thrust forces developed at a minimum feed rate and high cutting speed along with minimum diameter of the drill is proven. Hence, it can be concluded from the study that the woven cotton/bamboo fabrics can be machined with lower feed rates and drill diameter but at higher speeds. Such composites can be applied in low and medium load-carrying structural applications in automobile, aerospace, and marine industries, where machining of the composites is an integral part of manufacturing.

Author Contributions: Conceptualization, K.A., S.K.P. and B.M.; methodology, R.L., S.K.D. and N.S.R.; software, K.A., S.K.P. and B.M.; validation, K.A., S.K.P., R.L., B.M., S.K.D., N.S.R. and S.P.S.; formal analysis, K.A., S.K.P. and B.M.; investigation, K.A. and B.M.; resources, K.A. and B.M.; data curation, K.A.; S.K.P. and B.M.; writing—original draft preparation, K.A. and S.K.P.; writing—review and editing, S.K.P. and R.L.; visualization, S.K.D., N.S.R. and S.P.S.; supervision, S.K.P. and S.P.S.; project administration, S.K.P. and S.P.S. All authors have read and agreed to the published version of the manuscript.

Funding: This research received no external funding.

Institutional Review Board Statement: Not applicable.

Data Availability Statement: The data presented in this study are available on request from the corresponding author. The data are not publicly available due to privacy.

Acknowledgments: The authors would like to thank Akshaya College of Engineering, K. S. R. College of Engineering and K. S. Rangasamy College of Technology for providing the necessary research facilities and technical support to carry out this research work successfully.

Conflicts of Interest: The authors declare no conflict of interest.

References

1. Rajeshkumar, L.; Ramesh, M.; Bhuvaneswari, V.; Balaji, D.; Deepa, C. Synthesis and thermomechanical properties of bioplastics and biocomposites: A systematic review. *J. Mater. Chem. B* **2023**, *11*, 3307–3337. [CrossRef]
2. Sumrith, N.; Techawinyutham, L.; Sanjay, M.R.; Dangtungee, R.; Siengchin, S. Characterization of Alkaline and Silane Treated Fibers of 'Water Hyacinth Plants' and Reinforcement of 'Water Hyacinth Fibers' with Bioepoxy to Develop Fully Biobased Sustainable Ecofriendly Composites. *J. Polym. Environ.* **2020**, *28*, 2749–2760. [CrossRef]
3. Akhil, U.V.; Radhika, N.; Saleh, B.; Aravind Krishna, S.; Noble, N.; Rajeshkumar, L. A comprehensive review on plant-based natural fiber reinforced polymer composites: Fabrication, properties, and applications. *Polym. Compos.* **2023**, *44*, 2598–2633. [CrossRef]
4. Nagappan, S.; Subramani, S.P.; Palaniappan, S.K.; Mylsamy, B. Impact of alkali treatment and fiber length on mechanical properties of new agro waste *Lagenaria Siceraria* fiber reinforced epoxy composites. *J. Nat. Fibers* **2021**, *19*, 6853–6864. [CrossRef]
5. Asyraf, M.R.M.; Syamsir, A.; Zahari, N.M.; Supian, A.B.M.; Ishak, M.R.; Sapuan, S.M.; Sharma, S.; Rashedi, A.; Razman, M.R.; Zakaria, S.Z.S.; et al. Product Development of Natural Fibre-Composites for Various Applications: Design for Sustainability. *Polymers* **2022**, *14*, 920. [CrossRef]
6. Ramesh, M.; Rajeshkumar, L.; Bhuvaneswari, V. Leaf fibres as reinforcements in green composites: A review on processing, properties and applications. *Emergent Mater.* **2021**, *5*, 833–857. [CrossRef]
7. Kaliyannan, G.V.; Rathanasamy, R.; Kumar, H.K.M.; Raj, M.K.A.; Chinnasamy, M.; Pal, S.K.; Palaniappan, S.K. Natural fiber-based bio-degradable polymer composite. In *Green Sustainable Process for Chemical and Environmental Engineering and Science*; Inamuddin, A.M., Altalhi, T., Alrooqi, A., Eds.; Elsevier: San Diego, CA, USA, 2023; pp. 145–165.
8. Karthik, A.; Jeyakumar, R.; Sampath, P.S.; Soundararajan, R.; Manikandan, G.K. Study and Fabrication of Fan Blade Using Coconut Leaf Sheath Fibre/Epoxy-Reinforced Composite Materials. *J. Inst. Eng. (India) Ser. D* **2023**, 1–8. [CrossRef]
9. Ramesh, M.; Rajeshkumar, L.N.; Srinivasan, N.; Kumar, D.V.; Balaji, D. Influence of filler material on properties of fiber-reinforced polymer composites: A review. *E-Polymers* **2022**, *22*, 898–916. [CrossRef]
10. Asokan, R.; Rathanasamy, R.; Paramasivam, P.; Chinnasamy, M.; Pal, S.K.; Palaniappan, S.K. Hemp composite. In *Green Sustainable Process for Chemical and Environmental Engineering and Science*; Inamuddin, A.M., Altalhi, T., Alrooqi, A., Eds.; Elsevier: San Diego, CA, USA, 2023; pp. 89–112.
11. Ramesh, M. Flax (*Linum usitatissimum* L.) fibre reinforced polymer composite materials: A review on preparation, properties and prospects. *Prog. Mater. Sci.* **2018**, *102*, 109–166. [CrossRef]
12. Sathishkumar, T.P.; De-Prado-Gil, J.; Martínez-García, R.; Rajeshkumar, L.; Rajeshkumar, G.; Rangappa, S.M.; Siengchin, S.; Alosaimi, A.M.; Hussein, M.A. Redeemable environmental damage by recycling of industrial discarded and virgin glass fiber mats in hybrid composites—An exploratory investigation. *Polym. Compos.* **2023**, *44*, 318–329. [CrossRef]
13. Ilyas, R.; Sapuan, S.; Harussani, M.; Hakimi, M.; Haziq, M.; Atikah, M.; Asyraf, M.; Ishak, M.; Razman, M.; Nurazzi, N.; et al. Polylactic Acid (PLA) Biocomposite: Processing, Additive Manufacturing and Advanced Applications. *Polymers* **2021**, *13*, 1326. [CrossRef]
14. Asmawi, N.; Ilyas, R.A.; Roslim, M.H.M.; Rajeshkumar, L.; Abotbina, W.; Syafri, E.; Jumaidin, R.; Syafiq, R.; Rafiqah, S.A.; Ridwan, R.; et al. Cassava starch nanocomposite films reinforced with nanocellulose. *Phys. Sci. Rev.* **2023**. [CrossRef]
15. Asyraf, M.; Ishak, M.; Norrrahim, M.; Nurazzi, N.; Shazleen, S.; Ilyas, R.; Rafidah, M.; Razman, M. Recent advances of thermal properties of sugar palm lignocellulosic fibre reinforced polymer composites. *Int. J. Biol. Macromol.* **2021**, *193*, 1587–1599. [CrossRef]

16. Mylsamy, B.; Palaniappan, S.K.; Subramani, S.P.; Pal, S.K.; Aruchamy, K. Impact of nanoclay on mechanical and structural properties of treated Coccinia indica fibre reinforced epoxy composites. *J. Mater. Res. Technol.* **2019**, *8*, 6021–6028. [CrossRef]
17. Asyraf, M.; Ishak, M.; Syamsir, A.; Nurazzi, N.; Sabaruddin, F.; Shazleen, S.; Norrrahim, M.; Rafidah, M.; Ilyas, R.; Rashid, M.Z.A.; et al. Mechanical properties of oil palm fibre-reinforced polymer composites: A review. *J. Mater. Res. Technol.* **2022**, *17*, 33–65. [CrossRef]
18. Sanjeevi, S.; Shanmugam, V.; Kumar, S.; Ganesan, V.; Sas, G.; Johnson, D.J.; Shanmugam, M.; Ayyanar, A.; Naresh, K.; Neisiany, R.E.; et al. Effects of water absorption on the mechanical properties of hybrid natural fibre/phenol formaldehyde composites. *Sci. Rep.* **2021**, *11*, 13385. [CrossRef]
19. Karthik, A.; Jeyakumar, R.; Sampath, P.S.; Soundarararajan, R. *Mechanical Properties of Twill Weave of Bamboo Fabric Epoxy Composite Materials*; SAE Technical Paper; SAE International: Warrendale, PA, USA, 2022. [CrossRef]
20. Hamidon, M.H.; Sultan, M.T.; Ariffin, A.H.; Shah, A.U. Effects of fibre treatment on mechanical properties of kenaf fibre reinforced composites: A review. *J. Mater. Res. Technol.* **2019**, *8*, 3327–3337. [CrossRef]
21. Bhuvaneshwaran, M.; Subramani, S.P.; Palaniappan, S.K.; Pal, S.K.; Balu, S. Natural Cellulosic Fiber from *Coccinia Indica* Stem for Polymer Composites: Extraction and Characterization. *J. Nat. Fibers* **2021**, *18*, 644–652. [CrossRef]
22. Mohammed, M.; Rahman, R.; Mohammed, A.M.; Adam, T.; Betar, B.O.; Osman, A.F.; Dahham, O.S. Surface treatment to improve water repellence and compatibility of natural fiber with polymer matrix: Recent advancement. *Polym. Test.* **2022**, *115*, 107707. [CrossRef]
23. Van de Weyenberg, I.; Ivens, J.; De Coster, A.; Kino, B.; Baetens, E.; Verpoest, I. Influence of processing and chemical treatment of flax fibres on their composites. *Compos. Sci. Technol.* **2003**, *63*, 1241–1246. [CrossRef]
24. Gohil, P.P.; Chaudhary, V.; Patel, K. Challenges in machining of natural fibre composites. In *Manufacturing of Natural Fibre Reinforced Polymer Composites*; Springer: Cham, Switzerland, 2015; pp. 139–153. [CrossRef]
25. Mansingh, B.B.; Binoj, J.S.; Anbazhagan, V.N.; Abu Hassan, S.; Goh, K.L.; Siengchin, S.; Sanjay, M.R.; Jaafar, M.; Liu, Y. Characterization of *Cocos nucifera* L. peduncle fiber reinforced polymer composites for lightweight sustainable applications. *J. Appl. Polym. Sci.* **2022**, *139*, 52245. [CrossRef]
26. Mercy, J.L.; Sivashankari, P.; Sangeetha, M.; Kavitha, K.; Prakash, S. Genetic Optimization of Machining Parameters Affecting Thrust Force during Drilling of Pineapple Fiber Composite Plates—An Experimental Approach. *J. Nat. Fibers* **2022**, *19*, 1729–1740. [CrossRef]
27. Raj, S.S.; Dhas, J.E.R.; Jesuthanam, C. Challenges on machining characteristics of natural fiber-reinforced composites—A review. *J. Reinf. Plast. Compos.* **2021**, *40*, 41–69. [CrossRef]
28. Krishnasamy, P.; Rajamurugan, G.; Muralidharan, B.; Arbat, A.P.; Kishorkumar, B.P. Effect of S-2304 wire-mesh angle in hemp/flax composite on mechanical and twist drilling surface response analysis. *J. Ind. Text.* **2021**, *51*, 2774S–2798S. [CrossRef]
29. Bajpai, P.K.; Singh, I. Drilling behavior of sisal fiber-reinforced polypropylene composite laminates. *J. Reinf. Plast. Compos.* **2013**, *32*, 1569–1576. [CrossRef]
30. Karthik, A.; Sampath, P.S. Analysis of thrust force in drilling cotton with bamboo blended fibre-reinforced composites using Box-Behnken methodology. *Indian J. Fibre Text. Res.* **2020**, *45*, 267–273. [CrossRef]
31. Aruchamy, K.; Subramani, S.P.; Palaniappan, S.K.; Sethuraman, B.; Kaliyannan, G.V. Study on mechanical characteristics of woven cotton/bamboo hybrid reinforced composite laminates. *J. Mater. Res. Technol.* **2020**, *9*, 718–726. [CrossRef]
32. Srinivasan, T.; Palanikumar, K.; Rajagopal, K. Influence of Thrust Force in Drilling of Glass Fiber Reinforced Polycarbonate (GFR/PC) Thermoplastic Matrix Composites Using Box-behnken Design. *Procedia Mater. Sci.* **2014**, *5*, 2152–2158. [CrossRef]
33. Aruchamy, K.; Subramani, S.P.; Palaniappan, S.K.; Pal, S.K.; Mylsamy, B.; Chinnasamy, V. Effect of Blend Ratio on the Thermal Comfort Characteristics of Cotton/Bamboo Blended Fabrics. *J. Nat. Fibers* **2022**, *19*, 105–114. [CrossRef]
34. Liu, D.; Tang, Y.; Cong, W. A review of mechanical drilling for composite laminates. *Compos. Struct.* **2012**, *94*, 1265–1279. [CrossRef]
35. Palanikumar, K.; Srinivasan, T.; Rajagopal, K.; Latha, B. Thrust Force Analysis in Drilling Glass Fiber Reinforced/Polypropylene (GFR/PP) Composites. *Mater. Manuf. Process.* **2016**, *31*, 581–586. [CrossRef]
36. Saravanakumar, A.; Reddy, S.A. Optimization of process parameter in drilling of snake grass fiber reinforced composites. *Mater. Today Proc.* **2022**, *62*, 5460–5466. [CrossRef]
37. Arunachalam, S.; Narasimhan, R.L.; Palaniappan, S.K.; Ross, N.S.; Chandran, A.J.; Rangappa, S.M.; Siengchin, S. Machinability analysis of *Typha angustifolia* natural fiber reinforced composites through experimental modeling—Influence of fiber orientation. *Polym. Compos.* **2023**, *44*, 3808–3823. [CrossRef]
38. Belaadi, A.; Boumaaza, M.; Alshahrani, H.; Bourchak, M.; Jawaid, M. Drilling performance prediction of HDPE/Washingtonia fiber biocomposite using RSM, ANN, and GA optimization. *Int. J. Adv. Manuf. Technol.* **2022**, *123*, 1543–1564. [CrossRef]
39. Box, G.E.; Draper, N.R. *Empirical Model-Building and Response Surfaces*; John Wiley & Sons: Hoboken, NJ, USA, 1987.
40. Montgomery, D.C. *Design and Analysis of Experiments*; John Wiley & Sons: Hoboken, NJ, USA, 2017.
41. Dembri, I.; Belaadi, A.; Boumaaza, M.; Alshahrani, H.; Bourchak, M. Drilling performance of short Washingtonia filifera fiber–reinforced epoxy biocomposites: RSM modeling. *Int. J. Adv. Manuf. Technol.* **2022**, *121*, 7833–7850. [CrossRef]
42. Thakre, A.A.; Soni, S. Modeling of burr size in drilling of aluminum silicon carbide composites using response surface methodology. *Eng. Sci. Technol. Int. J.* **2016**, *19*, 1199–1205. [CrossRef]

43. Barreno-Avila, E.; Moya-Moya, E.; Pérez-Salinas, C. Rice-husk fiber reinforced composite (RFRC) drilling parameters optimization using RSM based desirability function approach. *Mater. Today Proc.* **2022**, *49*, 167–174. [CrossRef]
44. Anjinappa, C.; Y. J, M.; Ahmed, O.S.; Abbas, M.; Alahmadi, A.A.; Alwetaishi, M.; Alzaed, A.N. Experimental Study and an RSM Modelling on Drilling Characteristics of the Sheep Horn Particle Reinforced Epoxy Composites for Structural Applications. *Processes* **2022**, *10*, 2735. [CrossRef]
45. Aruchamy, K.; Mylsamy, B.; Palaniappan, S.K.; Subramani, S.P.; Velayutham, T.; Rangappa, S.M.; Siengchin, S. Influence of weave arrangements on mechanical characteristics of cotton and bamboo woven fabric reinforced composite laminates. *J. Reinf. Plast. Compos.* **2023**, 07316844221140350. [CrossRef]
46. Mahdi, A.; Turki, Y.; Habak, M.; Salem, M.; Bouaziz, Z. Experimental study of thrust force and surface quality when drilling hybrid stacks. *Int. J. Adv. Manuf. Technol.* **2020**, *107*, 3981–3994. [CrossRef]
47. Yallew, T.B.; Kumar, P.; Singh, I. A study about hole making in woven jute fabric-reinforced polymer composites. *Proc. Inst. Mech. Eng. Part L J. Mater. Des. Appl.* **2016**, *230*, 888–898. [CrossRef]
48. Nassar, M.M.A.; Arunachalam, R.; Alzebdeh, K.I. Machinability of natural fiber reinforced composites: A review. *Int. J. Adv. Manuf. Technol.* **2017**, *88*, 2985–3004. [CrossRef]

polymers

MDPI

Article

Evaluation of Physico-Mechanical Properties on Oil Extracted Ground Coffee Waste Reinforced Polyethylene Composite

Hoo Tien Nicholas Kuan [1,*], Ming Yee Tan [1], Mohamad Zaki Hassan [2] and Mohamed Yusoff Mohd Zuhri [3,4]

[1] Department of Mechanical and Manufacturing Engineering, Faculty of Engineering, Universiti Malaysia Sarawak, Kota Samarahan 94300, Malaysia
[2] Razak Faculty of Technology and Informatics, Universiti Teknologi Malaysia, Jalan Sultan Yahya Petra, Kuala Lumpur 54100, Malaysia
[3] Advanced Engineering Materials and Composites Research Centre, Department of Mechanical and Manufacturing Engineering, Universiti Putra Malaysia, Serdang 43400, Malaysia
[4] Laboratory of Biocomposite Technology, Institute of Tropical Forestry and Forest Products, Universiti Putra Malaysia, Serdang 43400, Malaysia
* Correspondence: khtnicholas@unimas.my

Abstract: The current work discusses ground coffee waste (GCW) reinforced high-density polyethylene (HDPE) composite. GCW underwent two types of treatment (oil extraction, and oil extraction followed by mercerization). The composites were prepared using stacking HDPE film and GCW, followed by hot compression molding with different GCW particle loadings (5%, 10%, 15% and 20%). Particle loadings of 5% and 10% of the treated GCW composites exhibited the optimum level for this particular type of composite, whereby their mechanical and thermal properties were improved compared to untreated GCW composite (UGC). SEM fracture analysis showed better adhesion between HDPE and treated GCW. The FTIR conducted proved the removal of unwanted impurities and reduction in water absorption after the treatment. Specific tensile modulus improved for OGC at 5 vol% particle loading. The highest impact energy absorbed was obtained by OGC with a 16% increment. This lightweight and environmentally friendly composite has potential in high-end packaging, internal automotive parts, lightweight furniture, and other composite engineering applications.

Keywords: polyethylene; composite materials; ground coffee waste; impact; TGA

Citation: Kuan, H.T.N.; Tan, M.Y.; Hassan, M.Z.; Zuhri, M.Y.M. Evaluation of Physico-Mechanical Properties on Oil Extracted Ground Coffee Waste Reinforced Polyethylene Composite. *Polymers* **2022**, *14*, 4678. https://doi.org/10.3390/polym14214678

Academic Editor: Swarup Roy

Received: 28 September 2022
Accepted: 12 October 2022
Published: 2 November 2022

1. Introduction

The escalating demand for synthetic fiber has led to over exploration of natural resources causing environmental pollution as they are resistant to degradation. The production and disposal treatment processes of synthetic fiber contribute to environmental pollution as well. In order to protect the earth, natural fiber-reinforced biodegradable polymer matrix and bio-composite have gained attention from researchers worldwide as they are green and eco-friendly materials. Natural fiber is an eco-friendly resource obtained from wood [1], bamboo [2], and agricultural waste [3–6] such as straw [7] that comprises cellulose, hemicellulose, lignin and aromatics, waxes and other lipids, ash, and water-soluble compounds [8]. It is biodegradable, sustainable, economically viable, low density and possesses good specific strengths and moduli [9–14]. Both the natural filler and polymer matrix used in this study are fully biodegradable at the end of their life cycles.

Coffee beans are seeds of the genus flowering plant, Coffea, a member of the botanical family of *Rubiaceace* which comprises almost 500 genera and more than 6500 species [8]. They are normally distributed in tropical and sub-tropical regions. Coffee is the world's second largest traded commodity, and the top five world leading countries in coffee production are Brazil, Vietnam, Indonesia, Colombia, and Ethiopia [15]. In accordance with the International Coffee Organization, world coffee consumption was about 9 million tons or 149.82 million bags (60 kg per bag) in 2014, an increment of 1.8% since 2013 [16].

Meanwhile, the solid waste generated by coffee production is about one ton annually [17]. Concerted efforts are underway to cultivate sustainable practices for the disposal of residues from processing, such as GCW.

Oxo-biodegradable plastic is a commodity plastic containing highly degradable plastic additive (TDPA) and 2–3% transition metals. The metal salts catalyze the degradation process while speeding it up to degrade abiotically at the end of its useful life in the presence of oxygen faster than ordinary plastic [18–20]. The European Standards Organization, CEN, defines oxo-biodegradation as degradation resulting from oxidative and cell-mediated phenomena, either simultaneously or successively [21]. In addition, the useful life of product made using biodegradable plastic could be programmed at manufacture, and the existing technology, machinery and workforce in factories can manufacture it at no extra cost. Biodegradable plastic retains all the advantages of conventional plastic but displays the required additional characteristic of faster degradation after use and disposal by natural mechanisms and with no unwanted environmental consequences [22–24].

Fermoso et al. [25] used thermogravimetric analysis (TGA) and mass spectrometry (MS) to carry out pyrolysis tests on spent coffee grounds in a nitrogen atmosphere with heating rates ranging from 5 to 100 °C/min. From the test, they found that the main gases produced during ground coffee pyrolysis are those containing oxygen: H_2O, CO and CO_2 within a temperature range between 250 and 425 °C. When a zeolite is used as a catalyst, they found a significant increase in the amount of low molecular weight hydrocarbons such as CH_4, C_2H_4 and C_2H_6 [25]. Cai et al. [26] used the thermogravimetric technique together with Fourier transform infrared spectroscopy to analyze the thermal decomposition of some plant residues used in the preparation of tea. They used two mixed N_2/O_2 and CO_2/O_2 working atmospheres and found that the main gases resulting from the combustion process are: O_2, H_2O, CH_4, CO, CO_2, NH_3 and HCN [26]. GCW has a structure that is very similar to components often utilized as filler particles in wood polymer composites. This by-product stands out due to its high protein content (17–18%), which may improve the characteristics of polymer composites. Moreover, because of the presence of these functional groups, it may provide extra opportunities for interfacial bonding within the polar polymeric matrix. Table 1 summarizes the outcomes of previous studies on the mechanical and thermal properties of coffee waste element reinforcement in polymer composites. Even though the hydrophilic nature of GCW may reduce the composite performance due to their poor matrix compatibility and tendency to absorb water, surface treatments may increase the compatibility of the matrix and GCW reinforcement. For example, the treated waste coffee fiber using steam explosion treatment could enhance the thermostability of the composite, increase the fiber crystallinity index and finally be suitable for use as a green composite product [27]. Also, according to Zhao et al. [28], treating the natural fiber using sodium hydroxide (NaOH) increased the interfacial interaction between the matrix and the fiber by up to 300%. Goncalves et al. [29] stated that a NaOH concentration of less than 5 wt% could improve the interfacial bonding between coffee waste and the polyurethane matrix. However, increasing NaOH concentrations up to 10 wt% worsens coffee husk fiber by eroding and deteriorating the structure. Generally, the NaOH solutions ionize the hydroxyl groups in the substrate and chemically alter the fiber structure by dissolving hydrogen bonds. The effectiveness of this process is determined by elements such as response time, heat applied, and solution concentration.

Furthermore, the proportion of GCW in the structure is an essential component that influences the mechanical performance of composites. Mechanical properties increase if high-strength GCW content increases. However, if the quantity of this GCW exceeds its ideal level, the mechanical capabilities may deteriorate. This is owing to the belief that as fiber volume increases, matrix composition decreases, resulting in poorer interfacial interaction. Table 1 reveals that almost all the investigations used a filler ratio of 1–20 wt% and, in some cases, up to 30 wt% of fiber loading. As can be seen from the table, an improvement of the elastic modulus of coffee silverskin-reinforced PP was obtained at the higher loading of fiber, as reported by Dominici et al. [30]. Similar findings were also reported by Sarasini et al. [31] and Hejna et al. [32]. When coffee silverskin was added to a green polymer, such as poly (butylene succinate) or poly (butylene adipate-co-terephthalate)/poly (3-hydroxybutyrate-co-3-hydroxyvalerate), the tensile strength of the composite structure increased. However, this effect was not significantly improved for stronger poly (lactic acid) matrices [33]. Nevertheless, a great deal of research solely examined the effects of coffee silverskin on morphological, mechanical, and thermal efficiency in composite technology. It was noted that the stiffness and crystallinity of polymer matrices were improved by adding coffee silverskin. However, the effects of GCW on the physico-mechanical properties and thermal stability of composites have not been reported.

Table 1. Previous reported work on the mechanical and thermal characteristics of coffee waste-reinforced composites.

Fiber	Matrix	Parametric Evaluation	Particle/ Fiber Loading (wt%.)	Finding	Reference
Coffee chaff and spent ground coffee	PP	Different type of coffee	26	Coffee chaff provided greater thermal stability and was suitable for composite reinforcement.	Zarrinbakhsh et al., 2016 [34]
Coffee silverskin	PBAT-PHBV	Different filler content	10, 20 and 30	Increasing the filler content enhanced Young's modulus but decreased tensile strength.	Sarasini et al., 2018 [31]
Coffee silverskin	PLA and PBS	Different filler content	10, 20 and 30	Adding fillers to both matrices improved tensile characteristics.	Totaro et al., 2019 [33]
Coffee silverskin	PE	Different filler content	10, 20 and 30	Improvement in the elastic modulus and a reduction in strain at maximum stress were observed with the increase in fiber fraction.	Dominici et al., 2019 [30]
Green coffee cake	PP	Steam explosion treatment	Nil	The addition of treated fiber increased the thermostability of the composite.	*de* Brito et al., 2020 [27]
Coffee silverskin	HDPE	Different filler content	1, 2, 5, 10 and 20	Maximum tensile strength was obtained at 20 wt%. particle loading.	Hejna et al., 2021 [32]
Coffee husk fiber waste	PU	Effect of NaOH concentration	5, 10 and 20	Increased tensile strength by lowering the NaOH content.	Gonçalves et al., 2021 [29]

PP: polypropylene, PBAT: poly(butylene adipate-co-terephthalate), PHBV: poly(3-hydroxybutyrate-co-3-hydroxyvalerate), PLA: poly(lactic acid), PBS: poly(butylene succinate), PE: polyethylene, HDPE: high-density polyethylene, PU: polyurethane.

The aim of this study is to reuse the waste from the coffee beverage industry while reinforcing it with HDPE to produce green and eco-friendly lightweight composites. Oil extracted from GCW using the Soxhlet method can be used to produce biodiesel and for the synthesis of bio-plastic or other potential applications, while the extracted GCW can be used as reinforcement in composite fabrication. The GCW underwent oil extraction and a combination of oil extraction followed by alkaline treatment. The composites were prepared using different particle loadings and were characterized by tensile, impact, FTIR, SEM, TGA, DSC and water absorption.

2. Materials and Methods

2.1. Materials

Ground coffee waste (GCW) (Arabica) was obtained from a local cafe. The GCW was washed with distilled water, dried in an oven at 105 °C for 24 h to 1–2% moisture content, and then sieved to a particle size of 850 μm. Table 2 shows the chemical composition of GCW from previous work conducted by other researchers. Commercial bio-degradable high-density polyethylene (HDPE) thin film (density of 0.96 g/mL at 23 °C, melt flow index of 0.35 g/10 min, 190 °C/5 kg) was used as the polymer matrix. GCW was extracted using the Soxhlet method in a solvent, n-hexane. Approximately 200 g of dried GCW was extracted in a Soxhlet extractor with 800 ml of n-hexane at a reaction temperature between 80 °C and 95 °C for 3 h. The oil-extracted GCW (OG) was dried at 105 °C for 24 h to evaporate n-hexane. After that, the OG was soaked in 1% concentration of NaOH aqueous solution (0.25 molarity) at room temperature for 24 h. Then, the NaOH-treated OG (ONG) was rinsed with distilled water until it reached pH 7. Both the OG and ONG were dried in an oven at 105 °C for 24 h before use.

Table 2. Chemical composition of GCW.

Composition of GCW (%)			Reference
Cellulose	**Hemicellulose**	**Lignin**	
13.02	-	26.52	[35]
13.85	-	19.84	[35]
13.8	-	33.6	[36]
12.4 ± 0.79	39.1 ± 1.94	23.9 ± 1.7	[37]
8.6	36.7	-	[38]

2.2. Methods

2.2.1. Compounding and Compression Molding

The dimensions of HDPE thin film used were 230 mm × 230 mm. Each laminate was prepared by sieving GCW in between plies of HDPE films. The compounding of GCW and HDPE were placed in a mold and covered. The composite was prepared using a hot compression molding method at 3 MPa for approximately 15 min until it reached 150 °C. The hot compression mold was then left to cool to room temperature for approximately 2 h before the laminate was taken out. Table 3 shows a summary of all the composites investigated in this research. The results of HDPE and UGC from the previous study were compared, whereby the earlier study performed on modified ground coffee waste showed an enhancement of the properties as compared to untreated ground coffee waste composite [39,40]. OGC is oil-extracted GCW-reinforced HDPE composite (Figure 1) and ONGC is oil-extracted + NaOH GCW-reinforced HDPE composite with different particle loadings.

Table 3. Summary of type of composites investigated in this study.

Composite	Sample	Stacking Sequence	Volume of GCW (%)	Volume of HDPE (%)	Reference
HDPE	HDPE	150PE	0	100	[39]
UGC5	Untreated GCW5/HDPE	150PE + 5UG	5	95	[39]
UGC10	Untreated GCW10/HDPE	150PE + 10UG	10	90	[39]
UGC15	Untreated GCW15/HDPE	150PE + 15UG	15	85	[39]
UGC20	Untreated GCW20/HDPE	150PE + 20UG	20	80	[39]
OGC5	Oil Extraction GCW5/HDPE	150PE + 5OG	5	95	
OGC10	Oil Extraction GCW10/HDPE	150PE + 10OG	10	90	
OGC15	Oil Extraction GCW15/HDPE	150PE + 15OG	15	85	
OGC20	Oil Extraction GCW20/HDPE	150PE + 20OG	20	80	
ONGC5	Oil Extraction + NaOH5/HDPE	150PE + 5ONG	5	95	
ONGC10	Oil Extraction + NaOH10/HDPE	150PE + 10ONG	10	90	
ONGC15	Oil Extraction + NaOH15/HDPE	150PE + 15ONG	15	85	
ONGC20	Oil Extraction + NaOH20/HDPE	150PE + 20ONG	20	80	

UG = Untreated ground coffee waste, OG = Oil-extracted ground coffee waste, ONG = Oil-extracted + NaOH ground coffee waste.

Figure 1. OGC, oil-extracted GCW-reinforced HDPE composite laminate.

2.2.2. Composite Property Testing

A. Fourier Transform Infrared Spectroscopy (FT-IR)

FTIR was analyzed using an IRAffinity-1 from Shimadzu, Kyoto, Japan. Approximately 20 mg of GCW each were put on the ATR (Attenuated Total Reflectance). Each sample was subjected to 20 scans in the range of 600 cm^{-1}–4000 cm^{-1}.

B. Scanning Electron Microscopy (SEM)

SEM (Hitachi TM3030) with 15 kV electrons was used to investigate all the samples. Before observation, the samples were coated with a thin layer of gold using a JEOL JFC-600 auto fine coater.

C. Thermal Analysis

I. Thermogravimetric Analysis (TGA) TGA was performed via Netzsch TG 200 F3 Tarsus (Selb, Germany). Approximately 20 mg of the composite was prepared and put into a crucible. A temperature range from 30 °C to 600 °C was used with a heating rate of 10 °C min^{-1} in a nitrogen atmosphere. The mass loss, initial degradation temperature (T_{onset}), maximum degradation temperature (T_{max}), and final degradation temperature (T_{final}) were determined.

II. Differential Scanning Calorimetry (DSC) DSC measurements were performed using Perkin Elmer equipment, DSC 8000 model (MA, USA). Approximately 20 mg of the composite was prepared. A temperature range from 50 °C to 180 °C was used with a heating rate of 20 °C min^{-1} in a nitrogen atmosphere. The sample was heated from 50 °C to 180 °C, then maintained at 180 °C for 2 min; the second cycle was performed with a cooling rate of 20 °C min^{-1} until it reached 50 °C; while for the third cycle, the sample was heated from 50 °C to 180 °C at a heating rate of 20 °C min^{-1}. The crystalline melting temperature (T_m) and degree of crystallinity (X_c) of the HDPE were obtained by considering the second heating curves.

D. Mechanical Properties of the Prepared Composites

i. Tensile Testing Tensile properties were examined using a Shimadzu universal testing machine, model AG-300K IS MS (Kyoto, Japan). Tensile tests were carried out in accordance with ASTM using a load cell of 50 kN at a crosshead rate displacement of 1 mm/min [41]. Tensile strength and tensile modulus were based on initial sample dimensions and the results were averaged over five measurements.

ii. Impact Testing Dart drop impact properties were characterized using Instron 9250 HV (U.S.) in accordance with ASTM D3763-15. A 12.88 mm hemisphere head of 3.4727 kg was employed and a free-falling initial drop height was set at 0.50 m. A total of five different samples were subjected to impact testing. The composites can be tested to assess their resistance to falling weight [42].

E. Water Absorption

A water absorption test was performed according to ASTM D570-98. Test specimens were prepared at a size of 76.2 mm × 25.4 mm by the thickness of the composite. The specimens were immersed entirely in a container of distilled water maintained at a temperature of 23 ± 1 °C for the test. The specimens were weighed after 2 h and then repeatedly every 24 h after that. The specimens were removed from distilled water one at a time, wiped with a dry cloth and weighed to the nearest 0.001 g. The percentage of water absorption was measured by [43]:

$$percentage\ of\ water\ absorption,\ \% = \frac{W_{final} - W_{initial}}{W_{initial}} \times 100\% \quad (1)$$

3. Results and Discussion

3.1. FT-IR Spectroscopy

The FT-IR spectroscopy in Figure 2 shows exemplary peaks of untreated GCW (UG), oil-extracted GCW (OG) and oil-extracted + NaOH GCW (ONG), respectively, at 3351 cm^{-1}, 2923 cm^{-1}, 2859 cm^{-1}, 1733 cm^{-1}, 1444 cm^{-1}, 1367 cm^{-1}, 1238 cm^{-1}, 1120 cm^{-1} and 1031 cm^{-1}. The absorption band of 3351 cm^{-1} belongs to the absorption of –OH stretching vibrations on the GCW. The –OH compound consisted of the presence of water, aliphatic primary and secondary alcohol found in cellulose, hemicellulose, lignin, extractives and carboxylic acids in extractives [44,45]. The decreasing area of this peak represented the removal of lipids. Peaks at 2859 cm^{-1} and 2923 cm^{-1} are recognized as asymmetric and symmetric stretching of C–H bonds in aliphatic chains that are ascribed to lipids [46]. After both treatments, the amount of lipids was reduced. The characteristic GCW peaks at 1733 cm^{-1} and 1238 cm^{-1} were credited to the acetylated xylan (hemicellulose), aldehyde lignin or carbonyl lipid, pectin and wax [47,48]. A peak of 1444 cm^{-1} assigned to benzene ring stretching in lignin and $=CH_2$ vibration in polyoses (Hemicellulose) [49], while 1120 cm^{-1} attributed to the aromatic C–H in the plane deformation mode of the guaiacyl/syringyl units of lignin [44] disappeared after both treatments. A peak of 1367 cm^{-1} shown as the in-plane C–H bending in hemicellulose reduced after the treatments [50]. During the 600 cm^{-1}–1100 cm^{-1} peak, the –OH concentration increased from UG to OG followed by ONG, revealing this was a more active site available to intermingle between filler/matrix interfaces. Table 4 demonstrates the details of band wavelength and the associated chemical group.

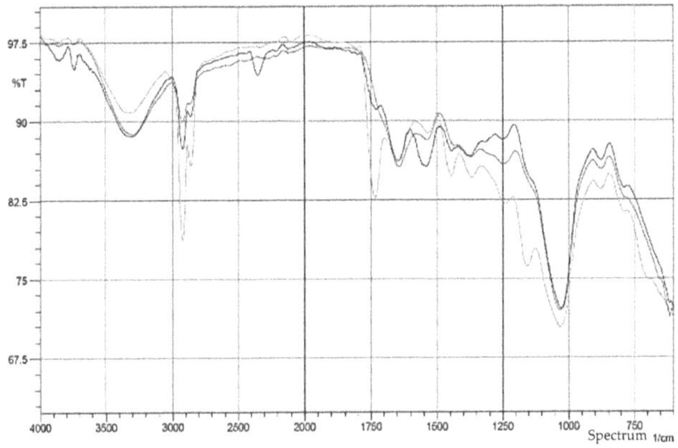

Figure 2. FTIR Spectra of a) UG (green), b) OG (red) and c) ONG (black).

Table 4. FTIR results of UG, OG and ONG.

Band Wavelength (cm^{-1})	Associated Chemical Group	UG	OG	ONG
3351	–OH intensity	High	Low	Low
2923	Lignin peak	Present	Reduced	Reduced
2859	Lignin peak	Present	Reduced	Reduced
1733	Hemicellulose, Lignin, Lipid, pectin-wax peak	Present	-	-
1444	Lignin, hemicellulose peak	Present	-	-
1367	Hemicellulose peak	Present	Reduced	Reduced
1238	Hemicellulose, Lignin peak	Present	Reduced	Reduced
1120	Lignin peak	Present	-	-
600–1100	–OH intensity	Low	High	High

3.2. TGA Analysis

Figure 3 shows the emblematic TGA and DTG curve of HDPE, untreated GCW/HDPE (UGC), oil-extracted GCW/HDPE (OGC) and oil-extracted + NaOH/HDPE (ONGC) composites based on particle loadings of 10%. As shown in Tables 5 and 6, the decomposition began at a lower temperature in the composites than in the neat HDPE due to the presence of GCW. However, the degradation temperature of OGC and ONGC were higher than HDPE at 50% weight loss. The process of oil extractions and alkaline treatment had improved their thermal stabilities. The thermal stability increased in UGC, OGC and ONGC, respectively. The percentage of charred residue decreased from UGC to OGC and followed by ONGC. This showed the treatment effectiveness as more unwanted impurities were removed leading to less charred residue. Weight losses before 100 °C were due to the moisture absorption in the sample. The first step of degradation can be observed in the range of 250 °C–400 °C, marking the degradation of hemicellulose, cellulose and small amount of lignin [51]. The second step of degradation, which is also the main decomposition of GCW, can be observed at approximately 440 °C that contributed to the dramatic degradation of lignin, as shown in DTG curve [52]. The degradation temperature of the materials was increased in the sequence of hemicellulose < cellulose < lignin.

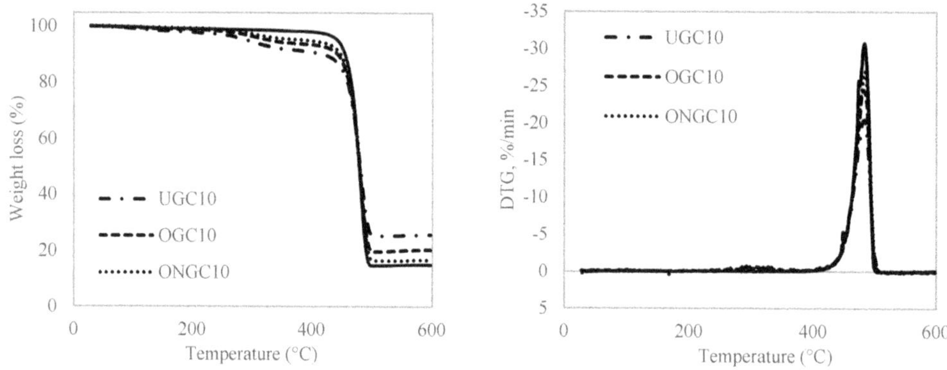

(a) (b)

Figure 3. (**a**) Thermogravimetric analysis (TGA) thermogram; and (**b**) Differential thermogravimetric graphics (DTG) thermogram of HDPE (smooth line), UGC, OGC and ONGC (10% particle loading).

Table 5. Summary of TGA data for HDPE, UGC, OGC and ONGC (10% particle loading).

Composites	T_{onset} (°C)	T_{max} (°C)	T_{final} (°C)	Residue (%)	Reference
HDPE	461.8	474.4	493.2	15.13	[39]
UGC10	460.4	481.9	492.4	25.80	[39]
OGC10	460.4	481.9	492.4	20.47	
ONGC10	459.1	482.0	492.8	17.71	

Table 6. Degradation temperature of HDPE, UGC, OGC and ONGC (10% particle loading).

Composites	Temperature at 10% Weight Loss (°C)	Temperature at 15% Weight Loss (°C)	Temperature at 25% Weight Loss (°C)	Temperature at 50% Weight Loss (°C)
HDPE	456.52	462.48	469.74	479.60
UGC10	422.11	450.39	464.64	478.01
OGC10	444.50	456.08	466.70	479.88
ONGC10	448.79	458.33	467.87	480.29

3.3. DSC Analysis

The sample was heated at a heating rate of 20 °C·min^{-1} initially, from 50 °C to 180 °C. When the sample reached 180 °C, it was then left to cool off to 50 °C in approximately 2 min. In the second cycle, the sample was reheated again to 180 °C. The curves exhibited endothermic peaks, as shown in Figure 4. These peaks are attributed to the melting of crystalline domains of the HDPE matrix. Table 7 shows the important data of DSC measurement. Thermal characteristics such as crystalline melting temperature, T_m, melting enthalpy of composite according to the content of HDPE in the composite, ΔH_m, and degree of crystallinity, X_c, were identified. The content of HDPE in the composite, ΔH_m, was calculated by using the area under the curve; while X_c was determined based on the ratio of ΔH_m to $\Delta H^0{}_m$, melting enthalpy of 100% crystalline HDPE (293 Jg^{-1}) [53]. With regard to melt temperature, the peaks were located in the same temperature range of 137 °C–138 °C. In all composite formulations, X_c increased with the incorporation of GCW content. In this case, GCW acted as nucleating agent that produced crystals to increase the percentage of X_c in the composites. The nucleating effect was due to the low filler content. Furthermore, the removal of lignin from the GCW after treatment also contributed to an increase in crystallinity [54].

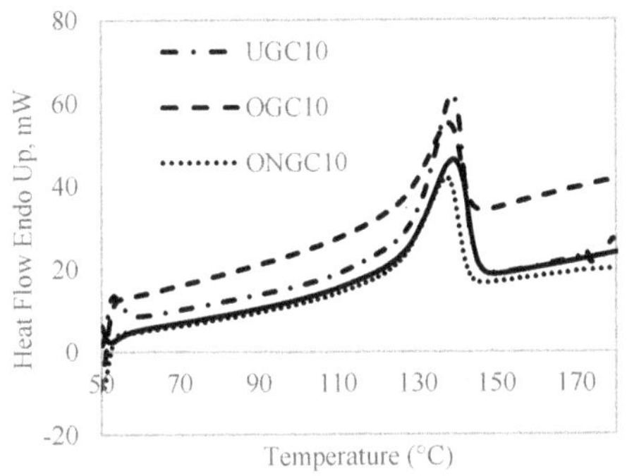

Figure 4. DSC curves of HDPE (smooth line), UGC10, OGC10 and ONGC10.

Table 7. Summary of DSC data of HDPE, UGC10, OGC10 and ONGC10.

Composites	T_m (°C)	ΔH_m (J/g)	X_c (%)
HDPE	138.92	142.44	48.61
UGC10	138.85	133.27	50.54
OGC10	137.72	140.67	53.34
ONG10	137.35	140.76	53.38

3.4. Tensile

The tensile properties were not as good after the inclusion of GCW. Reinforcement with a uniform circular cross-section area and with an aspect ratio more than its critical value would normally improve the tensile strength. However, the irregular shape of GCW fillers with an aspect ratio of less than two cannot support the stress transferred from the polymer matrix [55]. Therefore, the tensile strength reduced significantly. Compared to UGC, both treated GCW/HDPE composites showed an improvement in tensile strength. The tensile strength of the composites was maximized when reinforced with OGC > ONGC > UGC, respectively (Figure 5a). The tensile strength of OGC increased and reached an optimum of 16.47 MPa at a particle loading of 10%, an increment of 27% as compared to UGC, then declined. A large amount of lipids in OG was removed thus exposing a coarser surface with cellulose micro fibrils revealed, as verified by FTIR (Figure 2) and SEM (Figure 10). The exposed cellulose micro fibrils improved the effectiveness of the contact surface area thus enabling better impregnation of GCW by the HDPE matrix. On the other hand, increasing the particle loading led to the decreasing tensile strength of ONGC (Figure 5a). During a particle loading of 5%, the tensile strength was slightly better than UGC at 15.49 MPa, with 8% improvement. This may be due to less exposure of cellulose micro fibrils on its surface after the treatment. In addition, ONG became porous and weaker after alkaline treatment and this led to its having poor properties.

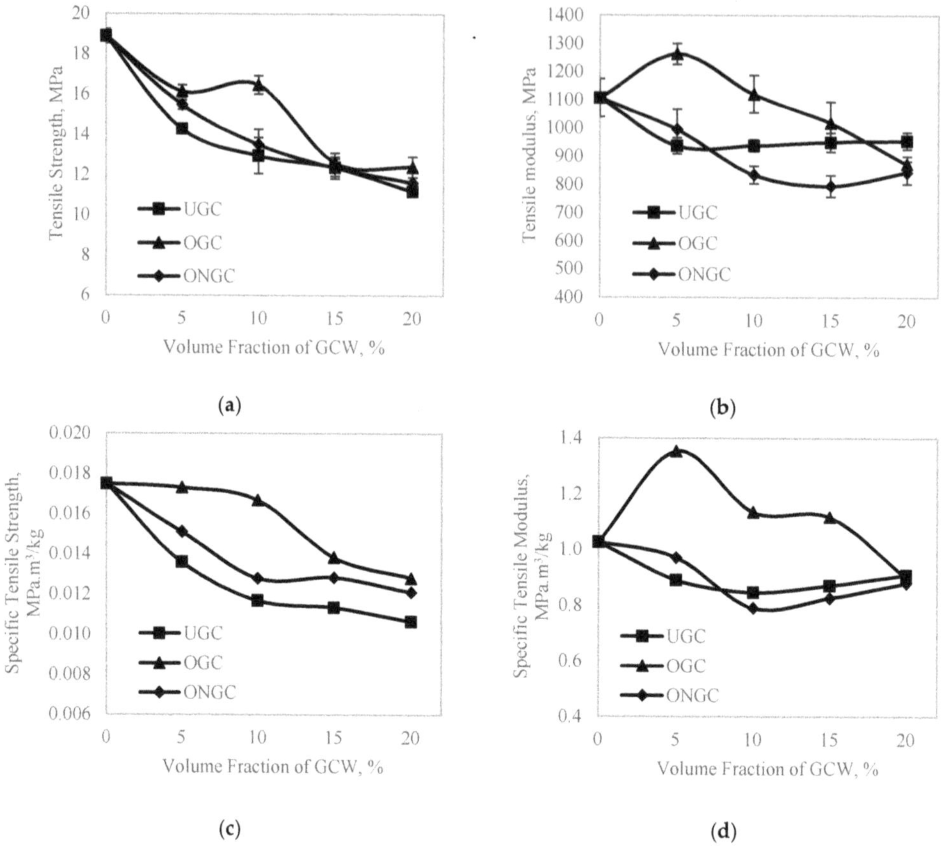

Figure 5. (**a**) Tensile Strength (**b**) Tensile Modulus (**c**) Specific Tensile Strength (**d**) Specific Tensile Modulus of HDPE, UGC, OGC and ONGC of different particle loadings.

As observed in Figure 5b, OGC showed a slight improvement as compared to neat HDPE during 5% and 10% of particle volume fraction reaching 1263 MPa (+14%) and 1122 MPa (+1.4%), respectively. The chemical modification proved to provide a better homogeneous dispersion of GCW. The tensile properties showed a proportional increment with particle loadings until 10%. This was due to an enhancement of load transfer between the matrix and GCW interface. Nevertheless, the composites declined in tensile properties beyond optimum values (>10%) due to the agglomeration of GCW. Agglomerations create flaws and voids between GCW and the matrix thus diminishing the tensile properties [56]. Other than that, aspect ratio played a very important role. GCW had a very low aspect ratio of less than two due to insufficient stress transferred; therefore, it displayed a low tensile modulus. In contrast, the tensile modulus obtained for ONGC was lower than HDPE as ONG became weak due to the treatment.

Figure 5c displays the specific tensile strength, known as strength to weight ratio. All of the treated GCW showed lower specific tensile strength. Particle loadings of 5% and 10% for OGC showed a slight decrease to 0.0173 MPa.m^3/kg (−1.1%) and 0.0167 MPa·m^3/kg (−4.6%), respectively. Figure 5d illustrates the specific tensile modulus, known as stiffness to weight ratio. The specific tensile modulus of neat HDPE was 1.0335 MPa·m^3/kg. The specific tensile modulus improved for every particle loading except for particle loadings of more than 15% for OGC. The specific tensile modulus had improved by 32%. The inclusion of GCW contributed to a lighter weight composite. The specific tensile properties improved slightly as the composites exhibited a lower density. Conversely, the specific tensile properties for ONGC did not improve due to the poor condition of ONG. Another researcher reported the same trend of results using oil extraction oil palm empty fruit brunch composite [57]. Tensile, flexural and toughness properties were improved significantly more than those without extraction.

3.5. Impact Test

A dart drop weight impact test was performed to study the impact response of UGC, OGC and ONGC with a 10% particle loading. Figures 6 and 7 show the results. The ascending curve in Figure 6 signifies the stiffness of the composites. The stiffness of the composites increased from UGC, to OGC to ONGC. The maximum impact force corresponds to the onset of material damage. The maximum load was sustained by the composite before fracture increasing in the sequence of UGC < ONGC (+14%) < OGC (+28%), respectively. Meanwhile, the energy to maximum load increased from UGC < ONGC (+11%) < OGC (+51%) as shown by the dotted line in Figure 6. The decreasing curve shows the properties of the materials. The ductility of the composites experienced a similar trend due to the energy of the maximum load. Total energy absorbed increased in the sequence of UGC < ONGC (+7%) < OGC (+16%), respectively. Hydrophilic UG and hydrophobic HDPE had poor interaction thus yielding a lower impact resistance. Both applied treatments had improved impact resistance. After oil extraction, the filler and matrix had a better adhesion and this gave a better uniform distribution of stress transfer. Therefore, more load and energy were required to perforate the samples. The treatments between matrix and reinforcement had significantly improved the toughness of the composite while showing the ability to absorb more energy and deform plastically before fractured. ONGC yielded lower results than OGC. The reason was supported by the results discussed earlier where ONG itself had become weaker. Overall, the impact resistance of the composites enhanced in the sequence of UGC < ONGC < OGC. Figure 8 displays the post-impact fracture behavior of the perforated composites. The failure implicated plastic deformation of the HDPE and debonding of the GCW filler within the composites. With similar particle loading percentages, it seems that the treated GCW composites had better plastic deformation than UGC. The bottom surface of the impacted samples had a bulging effect. Delamination only occurred at the perforated area. A similar impact result trend was obtained by another researcher [58].

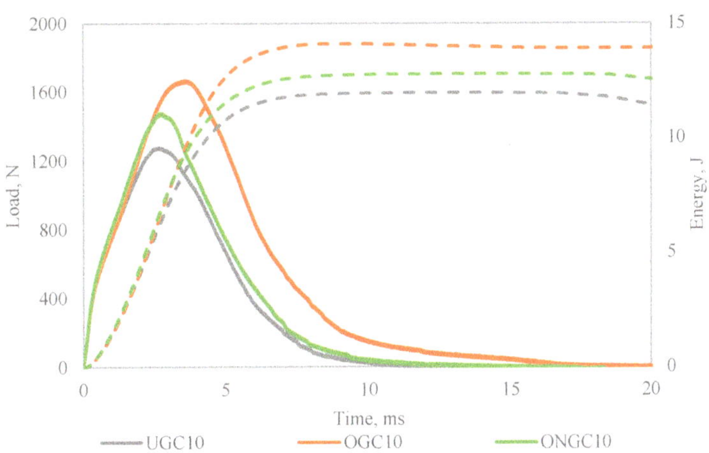

Figure 6. Load-Energy-Time graph of impact test for UGC, OGC and ONGC.

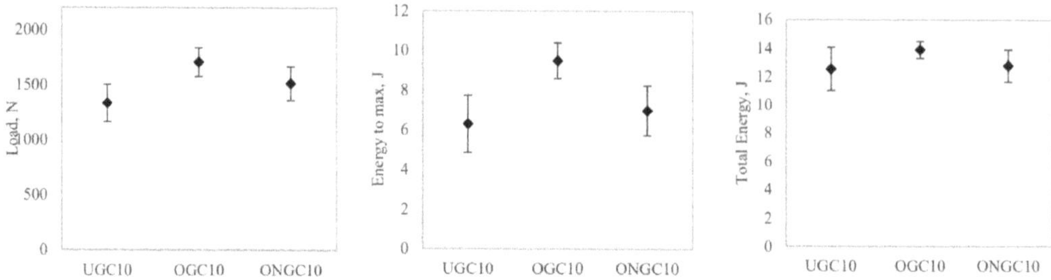

(a) (b) (c)

Figure 7. Effect of filler on (**a**) load, (**b**) energy to maximum, (**c**) total energy of GCW/HDPE composite at 10% particle loading.

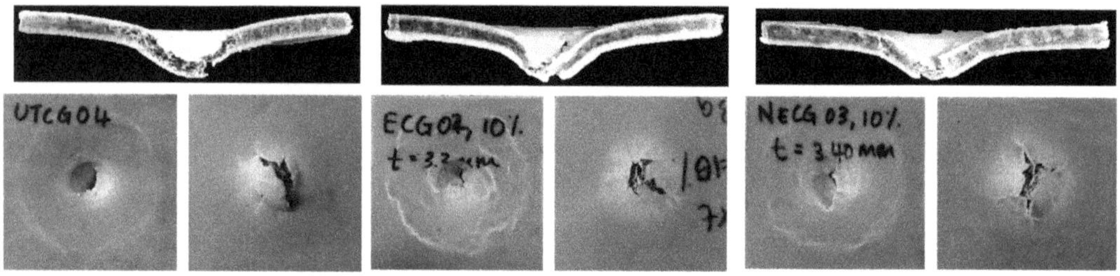

(a) (b) (c)

Figure 8. Photograph of fracture behavior of perforated samples (**a**) UGC, (**b**) OGC, (**c**) ONGC (10% particle loading).

3.6. Water Absorption

Figure 9 depicts the water absorption proficiency of HDPE, UGC, OGC and ONGC. The water absorption results showed that the non-polar HDPE demonstrated the lowest water absorption. In comparison to similar particle loadings of 10%, the percentage of water absorption increased in HDPE, OGC, ONGC and UGC at 0.05%, 3.22%, 3.80%, 5.22%, respectively, with an improvement of 38% and 27% in UGC. The treated GCW reinforced composites exhibited a lower water absorption rate as it became more hydrophobic after the treatments. Furthermore, a better interfacial adhesion of GCW/HDPE reduced the width of the interface area and diminished water penetration to the material inner parts [59].

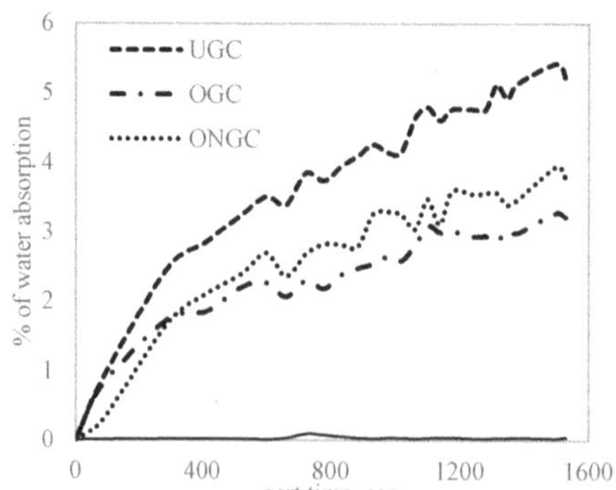

Figure 9. The water absorption proficiency percent mass gain for all the composites with 10% particle loading in comparison with HDPE (smooth line).

3.7. Morphological Studies

Figure 10 shows the surface morphology of UG, OG and ONG. All of the GCW particles had a dendritic appearance. The SEM pictograph in Figure 10 illustrates that the modifications conducted on GCW had increased its surface roughness. This had been proven with the FTIR and TGA with the removal of hemicellulose, lignin, pectin, wax, lipid and impurities on the GCW surface. As seen in Figure 10b, a coarser surface was revealed, and cellulose micro fibrils were exposed on the OG surface after extracting the oil. In Figure 10c, ONG showed uneven concave and rougher surfaces with less cellulose micro fibrils exposure. Additionally, ONG was more fragile, as proven by its mechanical properties.

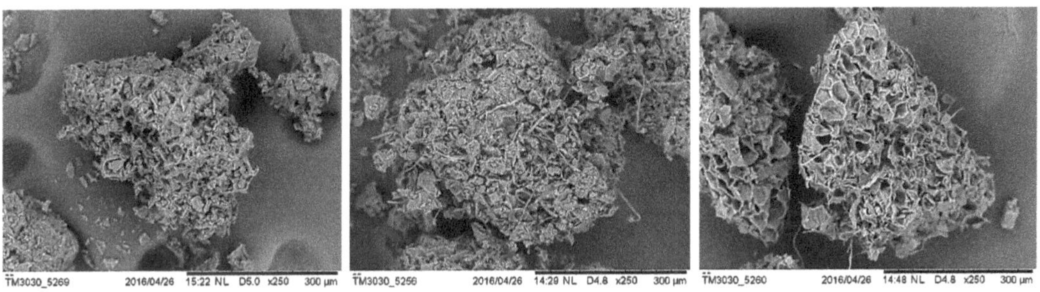

(a) (b) (c)

Figure 10. SEM micrograph of GCW particles of (**a**) UG, (**b**) OG and (**c**) ONG.

Figure 11 shows the SEM micrographs conforming to fracture surfaces from the tensile test at 10% particle loading for UGC, OGC and ONGC. For UGC, the fractured surface showed a gap in between the UG and HDPE, which resulted in poor fiber/matrix adhesion. For oil extraction application, a clearer and better interface can be seen between OG and HDPE with a layer of lipid being removed that led to a coarser surface of OG. The adhesion of ONGC was improved to a better impregnation of matrix in ONG. However, ONG became fragile and porous and thus yielded lower mechanical properties.

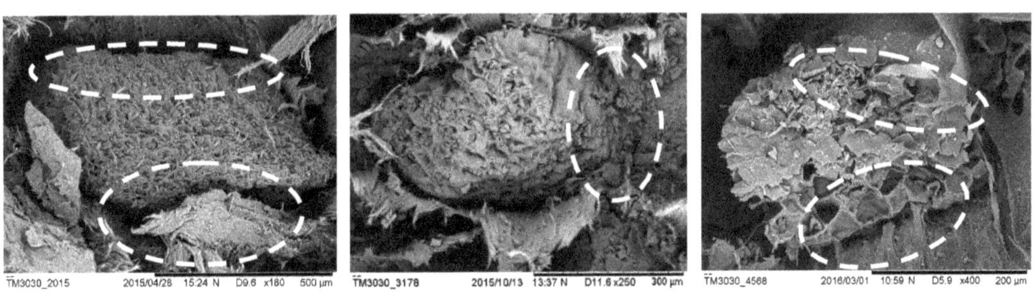

(a) (b) (c)

Figure 11. SEM micrograph of breaking surface of (**a**) UGC, (**b**) OGC, (**c**) ONGC (10% particle loading).

4. Conclusions

The tensile and impact properties, thermal analysis and analytical analysis of untreated and treated GCW/HDPE were evaluated. The characterization results show OGC to be the most impressive among all the fabricated composites. FTIR, SEM and TGA showed the effectiveness of the chemical modification performed on GCW by removing impurities, pectin, wax, hemicellulose, and lignin. Both treatments also produced a more hydrophobic GCW as water absorption rate had decreased. Overall, particle loadings of 5% and 10% exhibited the optimum level for this particular type of composite since mechanical properties had improved as compared to UGC. The incorporation of GCW into biodegradable HDPE aimed to improve the recyclability of waste in order to turn it into a value-added product, minimize ecological damage and reduce the production cost. The composites fabricated are able to degrade themselves at the end of their life cycle without leaving any negative effect on the environment. This particular type of composite has potential for application in consumables, packaging, gardening, furniture parts and automotive components.

Author Contributions: Methodology, H.T.N.K. and M.Y.T.; validation, H.T.N.K., M.Z.H. and M.Y.M.Z.; formal analysis, H.T.N.K. and M.Y.T.; data curation, M.Y.T.; writing—review and editing, H.T.N.K., M.Y.T., M.Z.H. and M.Y.M.Z.; supervision, H.T.N.K. All authors have read and agreed to the published version of the manuscript.

Funding: This research was funded by Universiti Malaysia Sarawak (UNIMAS) under the Cross Disciplinary Research Grant (F02/CDRG/1823/2019).

Data Availability Statement: Not applicable.

Acknowledgments: The authors gratefully acknowledge Universiti Malaysia Sarawak (UNIMAS) for providing the opportunity and support to conduct this research through Cross Disciplinary Research Grant (F02/CDRG/1823/2019).

Conflicts of Interest: The authors declare no conflict of interest.

References

1. Stark, N.M.; Rowlands, R.E. Effects of wood fiber characteristics on mechanical properties of wood/polypropylene composites. *Wood Fibre Sci.* **2003**, *35*, 167–174.
2. Liu, W.; Xie, T.; Qiu, R.; Fan, M. N-methylol acrylamide grafting bamboo fibers and their composites. *Compos. Sci. Technol.* **2015**, *117*, 100–106. [CrossRef]
3. Kuan, H.T.N.; Lee, M.C. Tensile Properties of Pandanus Atrocarpus based Composites. *J. Appl. Sci. Process Eng.* **2014**, *1*, 39–44. [CrossRef]
4. Santulli, C.; Kuan, H.T.; Sarasini, F.; De Rosa, I.; Cantwell, W. Damage characterisation on PP-hemp/aluminium fibre–metal laminates using acoustic emission. *J. Compos. Mater.* **2012**, *47*, 2265–2274. [CrossRef]
5. Lim, S.F.; Pah PY, L.; Chua, S.N.; Kuan, H.T.N. Optimisation of Water Soluble Essential Oil from Lemongrass Leaves using Steam Distillation. *Pertanika J. Sci. Technol.* **2017**, *25*, 137–143.
6. Kuan, H.T.N.; Lee, M.C.; Khan, A.A.; Sawawi, M. The low velocity impact properties of pandanus fiber composites. In *Materials Science Forum*; Trans Tech Publications Ltd.: Zurich, Switzerland, 2017; Volume 895, pp. 56–60.
7. Fávaro, S.L.; Lopes, M.S.; Neto, A.G.V.D.C.; de Santana, R.R.; Radovanovic, E. Chemical, morphological, and mechanical analysis of rice husk/post-consumer polyethylene composites. *Compos. Part A Appl. Sci. Manuf.* **2010**, *41*, 154–160. [CrossRef]
8. Mohanty, A.K.; Misra, M.; Drzal, L.T. Surface modifications of natural fibers and performance of the resulting biocomposites: An overview. *Compos. Interfaces* **2001**, *8*, 313–343. [CrossRef]
9. García, D.; López, J.; Balart, R.; Ruseckaite, R.; Stefani, P. Composites based on sintering rice husk–waste tire rubber mixtures. *Mater. Des.* **2007**, *28*, 2234–2238. [CrossRef]
10. Panthapulakkal, S.; Sain, M.; Law, S. Effect of coupling agents on rice-husk-filled HDPE extruded profiles. *Polym. Int.* **2004**, *54*, 137–142. [CrossRef]
11. Arbelaiz, A.; Fernández, B.; Ramos, J.A.; Retegi, A.; Llano-Ponte, R.; Mondragon, I. Mechanical properties of short flax fibre bundle/polypropylene composites: Influence of matrix/fibre modification, fibre content, water uptake and recycling. *Compos. Sci. Technol.* **2005**, *65*, 1582–1592. [CrossRef]
12. Sawawi, M.; Sahari, S.K.; Kuan, N.; Yusof, M.; Andew, M.; Salleh, S.F.; Sapawi, R. Optimisation of electrospinning parameter for Poly (L-lactic) acid (PLLA) electrospun nanofiber. In *MATEC Web of Conferences*; EDP Sciences: Ulis, France, 2017; Volume 87, p. 02012.
13. Kuan, H.T.N.; Sarawak, U.M.; Kee, S.L.P.; Yusof, M.; Ting, S.N.; Ng, C.K. Mechanical properties study of durio zibethinus skin fiber reinforced polyethylene composite. *J. Sustain. Sci. Manag.* **2022**, *17*, 152–158. [CrossRef]
14. Kuan, H.T.N.; Tan, M.Y.; Shen, Y.; Yahya, M.Y. Mechanical properties of particulate organic natural filler-reinforced polymer composite: A review. *Compos. Adv. Mater.* **2021**, *30*, 26349833211007502. [CrossRef]
15. Edwin, R.; Sanker, P.; Sekar, T.; Munusamy, T.S. *Botany*; Government of Tamil Nadu: Chennai, India, 2006.
16. International Coffee Organization. Coffee Consumption in East and Southeast Asia: 1990–2012. 2014. Available online: icc-112-4e-consumption-asia.pdf (accessed on 30 October 2019).
17. International Coffee Organization. World Coffee Consumption. 2015. Available online: http://www.ico.org/prices/new-consumption-table.pdf (accessed on 27 October 2015).

18. Tokimoto, T.; Kawasaki, N.; Nakamura, T.; Akutagawa, J.; Tanada, S. Removal of lead ions in drinking water by coffee grounds as vegetable biomass. *J. Colloid Interface Sci.* **2005**, *281*, 56–61. [CrossRef] [PubMed]
19. Joseph, S.; Sreekala, M.S.; Oommen, Z.; Koshy, P.; Thomas, S. A comparison of the mechanical properties of phenol formaldehyde composites reinforced with banana fibres and glass fibres. *Compos. Sci. Technol.* **2002**, *62*, 1857–1868. [CrossRef]
20. Noremylia, M.B.; Hassan, M.Z.; Ismail, Z. Recent advancement in isolation, processing, characterization and applications of emerging nanocellulose: A review. *Int. J. Biol. Macromol.* **2022**, *206*, 954–976. [CrossRef]
21. Haris, N.; Hassan, M.; Ilyas, R. Crystallinity, Chemical, Thermal, and Dynamic Mechanical Properties of Rice Husk/Coco Peat Fiber Reinforced ABS Biocomposites. *J. Nat. Fibers* **2022**, *20*, 1–12. [CrossRef]
22. Siracusa, V.; Rocculi, P.; Romani, S.; Rosa, M.D. Biodegradable polymers for food packaging: A review. *Trends Food Sci. Technol.* **2008**, *19*, 634–643. [CrossRef]
23. Suhot, M.A.; Hassan, M.Z.; Aziz SA, A.; Md Daud, M.Y. Recent progress of rice husk reinforced polymer composites: A review. *Polymers* **2021**, *13*, 2391. [CrossRef]
24. Gho, J.G.; Ballingham, E.P.N.; Chiellini, E.; Swift, G.; Wiles, D.; Katsaros, N. EPI's Oxo-biodegradable Plastics Guide-book. *Vancouver, Canada,* **2008**.
25. Fermoso, J.; Mašek, O. Thermochemical decomposition of coffee ground residues by TG-MS: A kinetic study. *J. Anal. Appl. Pyrolysis* **2018**, *130*, 358–367. [CrossRef]
26. Cai, H.; Liu, J.; Xie, W.; Kuo, J.; Buyukada, M.; Evrendilek, F. Pyrolytic kinetics, reaction mechanisms and products of waste tea via TG-FTIR and Py-GC/MS. *Energy Convers. Manag.* **2019**, *184*, 436–447. [CrossRef]
27. De Brito, E.B.; Tienne LG, P.; Cordeiro, S.B.; Marques MD, F.V.; Monteiro, S.N. The influence of steam explosion treatment of green coffee cake on the thermal and mechanical properties of reinforced polypropylene matrix compo-sites. *J. Mater. Res. Technol.* **2020**, *9*, 4051–4060. [CrossRef]
28. Zhao, X.; Lawal, T.; Rodrigues, M.; Geib, T.; Vodovotz, Y. Value-Added Use of Invasive Plant-Derived Fibers as PHBV Fillers for Biocomposite Development. *Polymers* **2021**, *13*, 1975. [CrossRef] [PubMed]
29. Gonçalves, B.M.M.; Camillo, M.D.O.; Oliveira, M.P.; Carreira, L.G.; Moulin, J.C.; Fantuzzi Neto, H.; de Oliveira, B.F.; Pereira, A.C.; Monteiro, S.N. Surface Treatments of Coffee Husk Fiber Waste for Effective Incorporation into Polymer Biocomposites. *Polymers* **2021**, *13*, 3428. [CrossRef]
30. Dominici, F.; García, D.G.; Fombuena, V.; Luzi, F.; Puglia, D.; Torre, L.; Balart, R. Bio-Polyethylene-Based Composites Reinforced with Alkali and Palmitoyl Chloride-Treated Coffee Silverskin. *Molecules* **2019**, *24*, 3113. [CrossRef] [PubMed]
31. Sarasini, F.; Tirillò, J.; Zuorro, A.; Maffei, G.; Lavecchia, R.; Puglia, D.; Dominici, F.; Luzi, F.; Valente, T.; Torre, L. Recycling coffee silverskin in sustainable composites based on a poly (butylene adipate-co-terephthalate)/poly (3-hydroxybutyrate-co-3-hydroxyvalerate) matrix. *Ind. Crops Prod.* **2018**, *118*, 311–320. [CrossRef]
32. Hejna, A.; Barczewski, M.; Kosmela, P.; Mysiukiewicz, O.; Kuzmin, A. Coffee Silverskin as a Multifunctional Waste Filler for High-Density Polyethylene Green Composites. *J. Compos. Sci.* **2021**, *5*, 44. [CrossRef]
33. Totaro, G.; Sisti, L.; Fiorini, M.; Lancellotti, I.; Andreola, F.N.; Saccani, A. Formulation of Green Particulate Composites from PLA and PBS Matrix and Wastes Deriving from the Coffee Production. *J. Polym. Environ.* **2019**, *27*, 1488–1496. [CrossRef]
34. Zarrinbakhsh, N.; Wang, T.; Rodriguez-Uribe, A.; Misra, M.; Mohanty, A.K. Characterization of wastes and coproducts from the coffee industry for composite material production. *BioResources* **2016**, *11*, 7637–7653. [CrossRef]
35. Pujol, D.; Liu, C.; Gominho, J.; Olivella, M.; Fiol, N.; Villaescusa, I.; Pereira, H. The chemical composition of exhausted coffee waste. *Ind. Crop. Prod.* **2013**, *50*, 423–429. [CrossRef]
36. Caetano, N.S.; Silvaa, V.F.M.; Mata, T.M. Valorization of coffee grounds for biodiesel production. *Chem. Eng. Trans.* **2012**, *26*, 267–272.
37. Ballesteros, L.F.; Teixeira, J.A.; Mussatto, S.I. Chemical, Functional, and Structural Properties of Spent Coffee Grounds and Coffee Silverskin. *Food Bioprocess Technol.* **2014**, *7*, 3493–3503. [CrossRef]
38. Mussatto, S.I.; Machado, E.M.S.; Martins, S.; Teixeira, J.A. Production, Composition, and Application of Coffee and Its Industrial Residues. *Food Bioprocess Technol.* **2011**, *4*, 661–672. [CrossRef]
39. Tan, M.Y.; Kuan, H.T.N.; Lee, M.C. Characterization of Alkaline Treatment and Fiber Content on the Physical, Thermal, and Mechanical Properties of Ground Coffee Waste/Oxobiodegradable HDPE Biocomposites. *Int. J. Polym. Sci.* **2017**, *2017*, 6258151. [CrossRef]
40. Tan, M.Y.; Kuan, H.T.N.; Khan, A.A. Tensile Properties of Ground Coffee Waste Reinforced Polyethylene Composite. *Mater. Sci. Forum* **2016**, *880*, 73–76. [CrossRef]
41. *ASTM D3039/D3039M*; Standard Test Method for Tensile Properties of Polymer Matrix Composite Materials. ASTM International: West Conshohocken, PA, USA, 2000.
42. *ASTM D3763*; ASTM D3763: Standard Test Method for High Speed Puncture Properties of Plastics Using Load and Displacement Sensors. ASTM International: West Conshohocken, PA, USA, 2012.
43. *ASTM D570-98*; ASTM D570-10: Standard Test Method for Water absorption of Plastics. ASTM International: West Conshohocken, PA, USA, 2010.
44. Khan, M.A.; Ali, K.M.I.; Basu, S.C. IR studies of wood plastic composites. *J. Appl. Polym. Sci.* **1993**, *49*, 1547–1551. [CrossRef]
45. Kolboe, S.; Ellefsen, O. Infrared investigations of lignin. A discussion of some recent results. *Tappi* **1962**, *45*, 163–166.

46. Cremer, D.R.; Kaletunç, G. Fourier transform infrared microspectroscopic study of the chemical microstructure of corn and oat flour-based extrudates. *Carbohydr. Polym.* **2003**, *52*, 53–65. [CrossRef]
47. Faix, O. Classification of Lignins from Different Botanical Origins by FT-IR Spectroscopy. *Holzforschung* **1991**, *45*, 21–28. [CrossRef]
48. Matuana, L.M.; Balatinecz, J.J.; Sodhi, R.N.S.; Park, C.B. Surface characterization of esterified cellulosic fibers by XPS and FTIR Spectroscopy. *Wood Sci. Technol.* **2001**, *35*, 191–201. [CrossRef]
49. Pervan, S.; Prekrat, S.; Gorisek, Z.; Straze, A.; Humar, M. Effect of Steaming on Colour and Chemistry of Cherrywood (*Prunus avium* L.). In *Wood Structure and Properties '06*; Kurjatko, S., Kudela, J., Lagana, R., Eds.; Arbora Publisher: Zvolen, Slovakia, 2006; pp. 331–336.
50. Fan, M.; Dai, D.; Huang, B. Fourier transform infrared spectroscopy for natural fibres. In *Fourier Transform—Materials Analysis*; IntechOpen: London, UK, 2012; pp. 45–68.
51. Castro, D.; Ruvolo-Filho, A.; Frollini, E. Materials prepared from biopolyethylene and curaua fibers: Composites from biomass. *Polym. Test.* **2012**, *31*, 880–888. [CrossRef]
52. Brebu, M.; Vasile, C. Thermal degradation of Lignin—A review. *Cellul. Chem. Technol.* **2010**, *44*, 353–363.
53. Perez, I.S.B.; Manrich, S.; Manrich, S. The effect of different block copolymers on post consumer HDPE/HIPS Blends: Phase morphology and thermal properties. *Polímeros* **2008**, *18*, 207–214. [CrossRef]
54. Davachi, S.M.; Kaffashi, B.; Davoodi, S.; Pouresmaeel-Selakjani, P.; Bakhtiari, S.; Mohammadi-Rovshandeh, J.; Yousefi, A. Investigating the Effect of Treated Rice Straw in PLLA/Starch Composite: Mechanical, Thermal, Rheological, and Morpho-logical Study. *Adv. Polym. Technol.* **2018**, *37*, 5–16. [CrossRef]
55. Fernandes, E.M.; Mano, J.F.; Reis, R.L. Hybrid cork–polymer composites containing sisal fibre: Morphology, effect of the fibre treatment on the mechanical properties and tensile failure prediction. *Compos. Struct.* **2013**, *105*, 153–162. [CrossRef]
56. Yang, H.S.; Kim, H.J.; Son, J.; Park, H.J.; Lee, B.J.; Hwang, T.S. Rice-husk flour filled polypropylene composites; mechanical and morphological study. *Compos. Struct.* **2004**, *63*, 305–312. [CrossRef]
57. Rozman, H.D.; Tay, G.S.; Kumar, R.N.; Abusamah, A.; Ismail, H.; Ishak, Z.A.M. The effect of oil extraction of the oil palm empty fruit bunch on the mechanical properties of polypropylene—Oil palm empty fruit bunch—Glass fibre hybrid composites on the mechanical properties fibre hybrid. *Polym. Plast. Technol. Eng.* **2001**, *40*, 103–115. [CrossRef]
58. Anbukarasi, K.; Kalaiselvam, S. Study of effect of fibre volume and dimension on mechanical, thermal, and water ab-sorption behaviour of luffa reinforced epoxy composites. *Mater. Des.* **2015**, *66*, 321–330. [CrossRef]
59. Youssef, A.M.; El-Gendy, A.; Kamel, S. Evaluation of corn husk fibers reinforced recycled low density polyethylene composites. *Mater. Chem. Phys.* **2015**, *152*, 26–33. [CrossRef]

polymers

MDPI

Article

Investigation of the Contact Interface between Natural Fibre Metal Laminates under Tension Using Finite Element Analysis (FEA)

Chun Han Song, Khaled Giasin *, Abu Saifullah and Antigoni Barouni *

Advanced Polymers and Composites (APC) Research Group, School of Mechanical and Design Engineering, University of Portsmouth, Portsmouth PO1 3DJ, UK
* Correspondence: khaled.giasin@port.ac.uk (K.G.); antigoni.barouni@port.ac.uk (A.B.)

Abstract: Fibre Metal Laminates (FMLs) consist of layers of metals combined with layers of fibre-reinforced composites bonded together to create a laminate. The behaviour of a Fibre Metal Laminate (FML) with natural fibre composites has been investigated in this study with a specific focus on the performance of the laminate under uniaxial tension. The integration of aluminium layers with natural fibre flax/pp layers at different fibre orientations has been numerically modelled and analysed, by investigating the contact interface between natural fibre metal laminates (NFML) using finite elements (FE) implemented in ABAQUS/Explicit. The finite element model was developed by the isotropic-hardening behaviour of metal layers, the built-in Hashin damage model and cohesive surface-based behaviour for the interface. The results of the simulation included stress–strain response, failure sequences, delamination effect and ultimate tensile strength. It was found that those results are significantly affected by the layup sequence, giving a significant advantage to the unidirectional laminate, when the uniaxial loading is taken into consideration. This advantage is measured as a 41.9% reduction of the ultimate tensile strength when the flax fibres are oriented at [0/90] configuration between the aluminium layers and a 30% reduction when the fibres are oriented at [±45] angles.

Keywords: fibre metal laminates; natural fibre composites; progressive damage analysis; failure mechanism; finite element analysis; numerical analysis; tension; delamination

Citation: Song, C.H.; Giasin, K.; Saifullah, A.; Barouni, A. Investigation of the Contact Interface between Natural Fibre Metal Laminates under Tension Using Finite Element Analysis (FEA). *Polymers* **2022**, *14*, 4650. https://doi.org/10.3390/polym14214650

Academic Editors: R. A. Ilyas, S. M. Sapuan, Emin Bayraktar, Shukur Abu Hassan, Nabil Hayeemasae and Khubab Shaker

Received: 30 September 2022
Accepted: 26 October 2022
Published: 1 November 2022

1. Introduction

FMLs have become very common in the aerospace industry during the past decades. It is a hybrid material that is built up from metal layers and fibre-reinforced composite layers, taking advantage of the benefits of the metallic layers, such as their high bearing strength and impact resistance, as well as the good fatigue resistance and high stiffness-to-weight ratio of the composites [1,2]. The main benefit of the FMLs is sourced from the bridging effect between the composite layer and metal layer, which can significantly slow down the fatigue crack growth rate. The bridging effect is taking place between the composite layer and the delamination plane. When the metal layer cracks, the composite layer will continue to hold and reduce the stress intensity [3–6]. Hence, the stress concentration will reduce, which can lead to a slow crack growth rate. This improved fatigue behaviour of FMLs makes them ideal for aircraft structures. Asundi et al. [7] found that using glass fibre-reinforced aluminium laminates as a material of the fuselage skin can reduce the weight by 15–25% without compromising the performance. The most successful application is used as the upper fuselage skin material of the Airbus A380. FMLs in aerospace applications might experience constant tensile and compressive loading [8], making the mechanical and failure behaviour of FMLs under tension very critical, which can include matrix cracking, fibre breakage and debonding of layers [9].

Over the years, the fibre-reinforced composite materials have been used in a variety of engineering applications and domains, where the tensile fatigue behaviour is investigated for glass fibre laminates [10], and the compressive strength is improved on glass fibre composites by winding additional layers around the longitudinal glass fibres [11], and the development of data-driven models and neural network models has been attempted for the prediction of the flexural performance of fibre-reinforced concrete columns [12,13]. Various attempts have taken place in order to investigate the performance of fibre metal laminates under different loading scenarios such as bending, tensile, compression, impact and shear [14]. Khalili et al. [15] investigated the mechanical properties of fibre metal laminates with different lay-ups of laminates and metal layers by three different testing scenarios: bending, impact and tensile. The results indicated that the damage tolerance limit and energy absorption of FMLs are superior to fibre-reinforced composites. Detailed work has been done by Sharma et al. [16] on quasi-static testing of titanium-based glass fibre-reinforced laminates. The tensile behaviour of four different layups was evaluated. The overall thickness of fibre metal laminates was fixed while changing the position and thickness of the metal layers in the stack. From the results, the layup sequence rarely affected the initial modulus of the fibre metal laminates. Moussavi et al. [17] combined classical laminate theory with the elastic-plastic behaviour of the aluminium layer to predict the stress–strain behaviour of fibre metal laminates. The fibre metal laminates model was loaded by uniform tensile force along the fibre direction while no cohesive behaviour was considered between any two layers. They concluded that FMLs with zero orientation fibre layer show improvement in maximum tensile strength. In another experiment, Hashemi et al. [18] investigated the performance of 3D glass fibre-based fibre metal laminates under tensile loading, when the laminates were manufactured under a different set of parameters (i.e., temperature, pressure and time), with the pressure contributing more to the mechanical properties of the laminate.

Numerical simulations are increasingly being used to study the performance of fibre metal laminates. Soltani et al. [19] simulated the tensile behaviour of FMLs under in-plane loading using Finite Element (FE) modelling, proposing a model able to analyse GLARE laminates in structural applications. Sharma et al. [20] investigated the tensile response of FMLs using three different specimens with constant volume but different configurations of metal and composite layers. Stress–strain curves of the FMLs were presented and indicated the failure stage of each specimen. It was stated that the configuration of metal and laminate layers would affect the tensile behaviour of the FML, where the existence of a metal layer between the composite layers demonstrates weaker performance than without a metal layer between the composite layers. A work that discusses the progressive damage and failure mechanisms of open-hole fibre metal laminates was done by Du et al. [21]. The performance of three different carbon fibre-reinforced PEEK prepregs layups, namely unidirectional, cross-ply and quasi-isotropic layups, was investigated. It was found that the failure of the FMLs began from fibre breakage. After the laminates met the yield point, the fibres would break first, followed by the failure of the matrix. The facture of metal layers took place once the delamination entirely occurred between the metal and laminate layer. Furthermore, higher-order shear deformation theory has been implemented to predict the shear strains through the thickness of composite laminated plates, with the use of FE modelling [22] as well as 3D deformations and progressive failures have been predicted for conventional fibre metal laminates using a user subroutine [23].

In the last few years, the production of sustainable products has gained more attention from manufacturers and researchers, leading them to integrate natural fibre composites into different applications [24,25]. Natural fibre composites are lightweight, easier to produce from natural resources and exhibit reduced environmental impact compared to synthetic fibres, making them a very promising replacement to synthetic fibres in demanding applications, such as automotive [26]. Fidelis et al. [27] investigated the tensile strength of several natural fibre composites, including jute, sisal, curaua, coir and piassava, concluding that the curaua fibres have the highest young modulus and tensile strength among other natural fibres. The use of natural fibre-reinforced composites has been increasingly used in semi-structural applications, where impact toughness is important [28], as well as in more demanding applications with the cycle loading performance of the material is critical [29]. However, the use of natural fibres is still limited to applications where the demand for structural integrity is high and critical, such as aerospace. The combination of natural fibre composites with metallic layers, to provide a sustainable alternative to FMLs, could replace the conventional FMLs in aerospace and improve the carbon footprint of the structures significantly, bringing a novelty to aerospace materials. Several implications can appear during this attempt, with the highest would be the compatibility of the two materials, with a greater focus on the delamination between the metallic and composite layers. Therefore, a preliminary numerical study is essential to understand the behaviour of these novel materials.

This paper aims to perform a finite element analysis for simulating a natural fibre metal laminate (NFML) composite structure and its failure behaviour under tension. Hence, the main focus is to study the damage initiation and damage evolution of the composite structure under tensile loading. In the simulation, cohesive-surface-based behaviour was applied to observe the delamination and the Hashin 2D criterion was involved to capture the damage initiation and evolution in the composite. Three types of layup configurations were considered, such as unidirectional, cross-ply and quasi-isotropic, to study the effect of fibre orientation on the failure modes, which is an important factor affecting the strength of the composite. The results obtained from the simulation such as the stress distribution and damage evolution of each layer will be discussed in the results section.

2. Materials and Methods

2.1. Metallic Layers

The metallic layer used in this study is Aluminium 2024-T3. The elastic and plastic behaviour of the metallic layer was modelled by using isotropic hardening. The isotropic hardening data for Al 2024-T3 are listed in Table 1 [20].

Table 1. Isotropic hardening data for Aluminium 2024-T3.

Flow Stress (Mpa)	294	335	362	381	399	415	430	443	459	469	480
Plastic Strain (%)	0	1.1	2.2	3.2	4.4	5.5	6.7	7.8	9.2	10.1	11.2

2.2. Composite Layers

Two different composite materials were involved in this study; E-glass/Epoxy composite laminates and flax/polypropylene composite. E-glass/Epoxy composite laminates were modelled in the first part of the simulation to validate the developed FE model using the work of Sharma [20] and the properties are shown in Table 2. The second part of this study investigated the performance of natural fibre flax/polypropylene composite. The material properties used for natural fibre flax/polypropylene composite are listed in Table 3 [30].

Table 2. Material properties of the E-glass/Epoxy composite.

Mechanical Properties	Value
Longitudinal modulus E_1 (Mpa)	30,500
Transverse modulus E_2 (Mpa)	4030
Major Poisson′s ratio V_{12}	0.29
In − plane shear modulus G_{12} (Mpa)	2080
Shear modulus G_{13} (Mpa)	2080
Shear modulus G_{23} (Mpa)	1440
Longitudinal tensile strength X_T (Mpa)	700
Longitudinal compressive strength X_C (Mpa)	300
Transverse tensile strength Y_T (Mpa)	39
Transverse compressive strength Y_C (Mpa)	128
Longitudinal shear strength S_{12} (Mpa)	50
Transverse shear strength S_{23} (Mpa)	50
Longitudinal tensile fracture energy G_{1t} (mJ)	92
Longitudinal compressive fracture energy G_{1C} (mJ)	79
Transverse tensile fracture energy G_{2t} (mJ)	0.22
Transverse compressive fracture energy G_{2C} (mJ)	0.61

Table 3. Material properties of the flax/polypropylene.

Mechanical Properties	Value
Longitudinal modulus E_1 (Mpa)	22,980
Transverse modulus E_2 (Mpa)	3030
Transverse modulus E_3 (Mpa)	3030
Major Poisson′s ratio V_{12}	0.38
Poisson′s ratio V_{13}	0.38
Poisson′s ratio V_{23}	0.7
In − plane shear modulus G_{12} (Mpa)	1040
Shear modulus G_{13} (Mpa)	1040
Shear modulus G_{23} (Mpa)	1060
Longitudinal tensile strength X_T (Mpa)	334.85
Longitudinal compressive strength X_C (Mpa)	246.34
Transverse tensile strength Y_T (Mpa)	31.6
Transverse compressive strength Y_C (Mpa)	72.1
Longitudinal shear strength S_{12} (Mpa)	18.48
Transverse shear strength S_{23} (Mpa)	18.33
Longitudinal tensile fracture energy G_{1t} (mJ)	80
Longitudinal compressive fracture energy G_{1C} (mJ)	80
Transverse tensile fracture energy G_{2t} (mJ)	0.2
Transverse compressive fracture energy G_{2C} (mJ)	1

For the prediction of the damage initiation and evolution, a cohesive-surface-based behaviour method was implemented, using the material properties summarised in Table 4 [30].

Table 4. Material properties of the cohesive interface.

Mechanical Properties	Value
$K_n = K_s = K_t$	10^6 N/mm^3
t_n^0	39 Mpa
t_s^0	50 Mpa
t_t^0	50 Mpa
G_{IC}	0.22 N/mm
G_{IIC}	0.36 N/mm
G_{IIIC}	0.36 N/mm
β	1.45

2.3. *Configuration of NFMLs*

To validate the developed FE model in this work, a validation study was initially conducted with already existing literature. The fibre metal laminate with glass fibre composite layers, named GLARE 3/2-0.4, was used for the initial validation study [19].

2.3.1. Glare 3/2-0.4

The Glare3/2-0.4 laminate was made of three sheets of aluminium 2024-T3 alloy alternating with two layers of E-glass/Epoxy composite laminates. The schematic diagram of the specimen is shown in Figure 1. Due to the symmetry of the geometry and loading of the specimen, only a quarter of the model was simulated. The total thickness of the specimen is 3.7 mm, while the aluminium thickness is 0.4 mm each and the composite layer is 0.625 mm each.

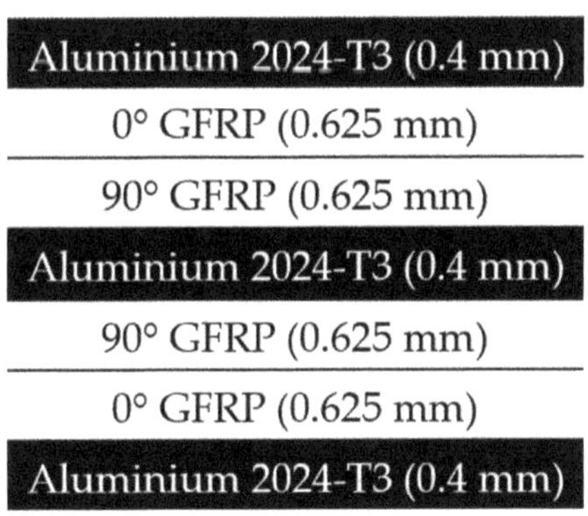

Figure 1. Schematic diagram of GLARE 3/2-0.4.

2.3.2. Aluminium/Flax Fibre Metal Laminates

The main work for this study was conducted on a natural fibre metal laminate (NFML), where flax-fibre-reinforced laminates were used combined with Aluminium layers. The material is subjected to tensile loadings at one end of the specimen. The effect of lay-up configuration on fibre metal laminates with flax fibre reinforcement is subsequently investigated in this study. Three different laminate configurations were used for this investigation, where three aluminium layers and two embedded flax-fibre/polypropylene composite layers were used for each configuration. The laminate configurations used include unidirectional, cross-ply and biaxial configurations. The thickness of the aluminium and flax/pp layers is listed in Table 5. The schematics of the fibre metal laminates with different stacking sequences are presented in Figure 2. The size and geometry of the model remain the same as described in ASTM D3039.

Table 5. Details of fibre metal laminates considered.

Laminate Code	Lay-Up	Fibre Type	Al Thickness (mm)	Flax/PP Thickness (mm)
A	[AL/0/0/AL/0/0/AL]	FLAX	0.4	0.625
B	[AL/0/90/AL/90/0/AL]	FLAX	0.4	0.625
C	[AL/+45/−45/AL/+45/−45/AL]	FLAX	0.4	0.625

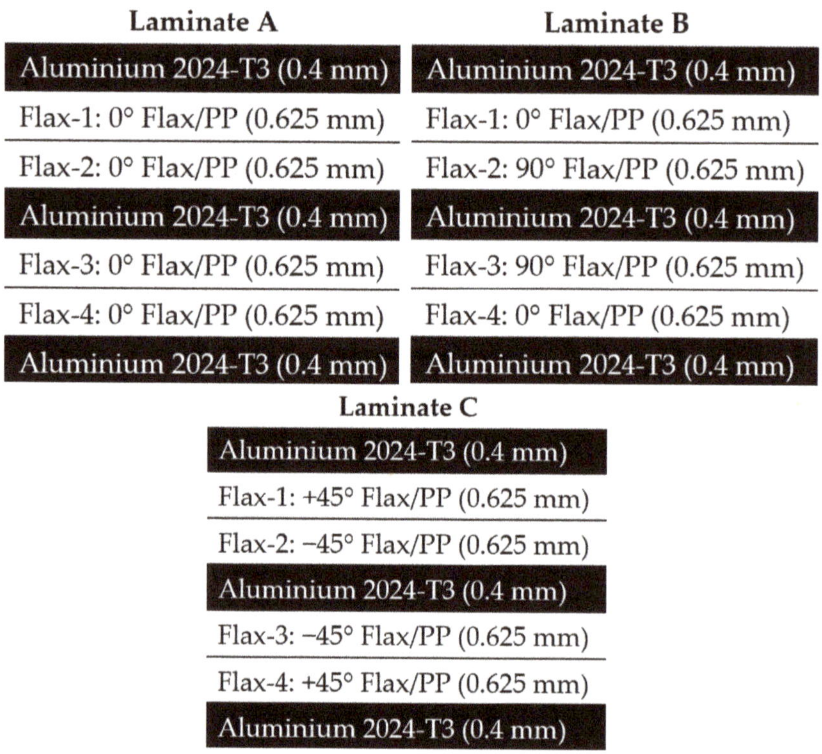

Figure 2. The various layup of Aluminium/Flax fibre metal laminates.

3. FE Modelling Approach

For the investigation of the tensile behaviour of NFMLs, an FE model was developed using ABAQUS/Explicit, using the geometry as described in the experimental testing standard ASTM D3039. The dog-bone-shaped geometry was used where one-quarter of the geometry was being modelled, due to symmetry reasons, as shown in Figure 3.

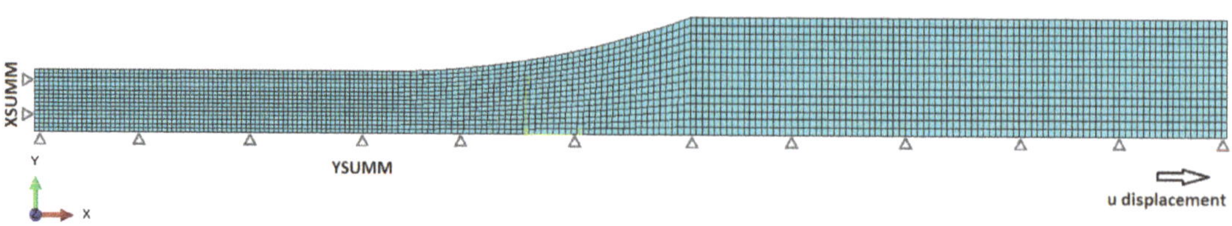

Figure 3. The FE model of the quarter dog-bone specimen.

FMLs will undergo different types of failure modes during the deformation process, such as delamination between two layers, fibre breakage, fracture of the metal layer, and the debonding process in the laminates layer and matrix cracking [31]. The cohesive surface-based behaviour was used for this numerical modelling to define the interaction between two surfaces. The cohesive-surface-based behaviour method uses traction-separation theory to define cohesive interaction between two surfaces without having cohesive elements in the geometry.

Traction-separation behaviour involves three stages, namely, the linear elastic behaviour, the damage initiation and the damage evolution. Linear elastic behaviour with uncoupled traction separation was used, meaning shear direction cohesive forces are not affected by pure normal separation and normal direction cohesive forces are not affected by pure shear slip. Hence, the uncoupled behaviour relation is written as follows.

$$t = \begin{Bmatrix} t_n \\ t_s \\ t_t \end{Bmatrix} = \begin{bmatrix} K_{nn} & 0 & 0 \\ 0 & K_{ss} & 0 \\ 0 & 0 & K_{tt} \end{bmatrix} \begin{Bmatrix} \delta_n \\ \delta_s \\ \delta_t \end{Bmatrix} \tag{1}$$

where t_n, t_s, t_t indicate the nominal traction vectors in the orthogonal direction. The value of penalty stiffness for the traction separation behaviour is denoted as K, and relative displacement is denoted as δ. Both are involved in the orthogonal direction.

Damage initiation represents the beginning of the degradation of the contact surface under loading conditions. The degradation of the material starts when the contact stresses meet the damage initiation criterion defined by the user. A value of 1 in the damage initiation criterion represents the failure of the contact surface. The quadratic stress criterion was chosen for this simulation. It involves a peak value of the contact stress to determine the degradation.

$$\left(\frac{t_n}{t_n^0}\right)^2 + \left(\frac{t_s}{t_s^0}\right)^2 + \left(\frac{t_t}{t_t^0}\right)^2 = 1 \tag{2}$$

where t_n^0, t_s^0 and t_t^0 represent the interface strength parameter in the orthogonal direction.

Damage evolution indicates the degradation of the material, which is cohesive stiffness after fulfilling the initiation criteria. The damage evolution is according to softening laws. Either displacement or fracture energy can determine the softening laws for simulating the delamination. The Benzeggagh–Kenane fracture criterion is applied to the simulation by assuming the same value of fracture energy in the first and second shear direction ($G_s^C = G_t^C$). All parameters of the cohesive-surface-based behaviour method used in this study were summarized in Table 2.

To predict the behaviour of composite layers, a constitutive model is used to describe the response of linear elastic behaviour of fibre-reinforced material. The relationship between stress and strain before damage initiates is defined as $\sigma = C_d\varepsilon$. The damaged elasticity matrix replaces the elasticity matrix once damage initiation happens. The damaged elasticity matrix C_d is given below.

$$[C_d] = \begin{bmatrix} \left(1-d_f\right)E_1 & \left(1-d_f\right)(1-d_m)v_{21}E_1 & 0 \\ \left(1-d_f\right)(1-d_m)v_{12}E_2 & (1-d_m)E_2 & 0 \\ 0 & 0 & (1-d_s)G_{12}D \end{bmatrix} \tag{3}$$

$$D = 1 - \left(1-d_f\right)(1-d_m)v_{12}v_{21} \tag{4}$$

$$d_f = d_f^t \ if\ \sigma_{11} \geq 0,\ or\ d_f^c\ if\ \sigma_{11} < 0 \tag{5}$$

$$d_m = d_m^t \ if\ \sigma_{22} \geq 0,\ or\ d_m^c\ if\ \sigma_{22} < 0 \tag{6}$$

$$d_s = 1 - \left(1-d_f^t\right)\left(1-d_f^c\right)\left(1-d_m^t\right)\left(1-d_m^c\right) \tag{7}$$

where the subscripts f, m, s represent fibre, matrix and shear, respectively, and the subscript d is damage. E_1 and E_2 are Young's modulus in the fibre and matrix direction, respectively, G_{12} and v_{12} are the shear modulus and Poisson's ratio in the X-Y direction.

Equation (3) above defines how the damage of fibre and matrix changes the material's stiffness according to four different failure modes, as described in the 2D Hashin failure criterion. ABAQUS offers the Hashin 2D criterion to predict the four different failure modes of the shell element. Failure of the composite is defined as the element reaching the yield stress and leading to damage occurrences. Maximum stress or strain theory will be needed to determine the initial damage to the matrix or fibre [32]. The purpose of using the Hashin-based damage criterion is to evaluate different failure modes in a different direction [33]. The general form of four different damage initiation modes is defined below.

Fibre tension $\sigma_{11} \geq 0$

$$F_f^t = \left(\frac{\sigma_{11}}{X^t}\right)^2 + \alpha\left(\frac{\sigma_{12}}{S^L}\right)^2 \tag{8}$$

Fibre compression $\sigma_{11} < 0$

$$F_f^C = \left(\frac{\sigma_{11}}{X^C}\right)^2 \tag{9}$$

Matrix tension $\sigma_{22} \geq 0$

$$F_m^t = \left(\frac{\sigma_{22}}{Y^t}\right)^2 + \left(\frac{\sigma_{12}}{S^L}\right)^2 \tag{10}$$

Matrix compression $\sigma_{22} < 0$

$$F_m^C = \left(\frac{\sigma_{22}}{2S^t}\right)^2 + \left[\left(\frac{Y^C}{2S^T}\right)^2 - 1\right]\left(\frac{\sigma_{22}}{Y^C}\right) + \left(\frac{\sigma_{12}}{S^L}\right)^2 \tag{11}$$

where F_f^t, F_f^C indicate failure mode of fibre in tension and compression; F_m^t, F_m^C indicate the failure mode of the matrix in tension and compression. σ_{11}, σ_{22} and σ_{12} represent the effective stress tensor. X^t and X^C are the tensile strength and compressive strength of laminate in the longitudinal direction while Y^t and Y^C are the tensile strength and compressive strength of laminate in the transverse direction. S^L and S^T, respectively, represent the shear strength of the laminate in the longitudinal and transverse directions.

Due to the symmetry of the geometry and loading of the specimen, only a quarter of the specimen was modelled. The FE model is subjected to tensile loadings at one end of the specimen, whereas encastre boundary conditions were applied on the other end. The built-in fibre-reinforced material damage model will be used in this simulation. A mesh convergence study was performed to determine the minimum element size. ABAQUS offers general contact in ABAQUS/Explicit solver, where each interface of two layers was defined. For the interface between metal and composite layers, the friction of 0.3 was used, while a friction value of 0.5 was used as input for the interface between the composite layers [34]. The element type used for the metal layer was eight-node solid brick elements (C3D8R) and for the composite layer was the eight-node continuum shell element (SC8R). ABAQUS explicit option was selected for this simulation as it is more suitable for highly nonlinear analysis and can reduce the required time. In the GLARE 3/2-0.4 FE model, no damage evolution was applied. However, the model for aluminium/flax fibre metal laminates does involve a ductile damage model. Once the damage initiation has been satisfied, Hillerborg's fracture energy damage evolution criterion will be introduced. This ductile damage model includes the fracture strain and fracture energy to modify the failure of metal.

4. Results and Discussion

To initially validate the developed FE model, the strain–stress curves were compared to the results presented in [20] for three different material types: (a) a 0.4 mm thick aluminium plate, as shown in Figure 4a, (b) a cross-ply GFRP laminate, as shown in Figure 4b, (c) a GLARE 3/2-0.4 laminate, as shown in Figure 4c. All three curves exhibit very good agreement with the current FE model. Also, by comparing the ultimate tensile strength, the percentage error is 0.958%. As can be seen, all curves agree very well, especially around the initiation of delamination, a fact which enhances the use of the current FE model for further simulations.

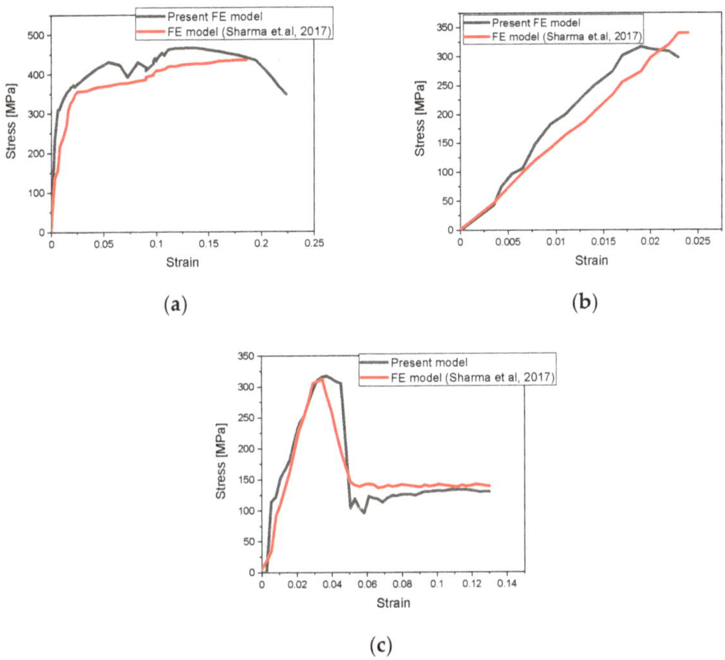

Figure 4. Stress–strain curves of Present FE model and Sharma et al. FE model [20] for (**a**) AL-2024 T3 alloy; (**b**) of [0°/90°/90°/0°] GFRP; (**c**) GLARE 3/2-0.4.

The stress–strain curves, failure sequences, delamination effect, ultimate tensile strength and effect of composite layup configurations will be discussed in this section.

The stress–strain response for each configuration A, B and C, under tensile loading is described in Figures 5 and 6. According to Figure 5, laminate A with unidirectional fibre orientation has the highest tensile strength, followed by laminate C, and laminate B exhibits the lowest UTS among the chosen laminates. Specimen A has the highest tensile strength mainly due to the fibre orientation being aligned to the tensile load in the longitudinal direction. Figure 5 shows a significant difference in ultimate tensile strength when changing the layup of the lamina. For example, laminate B with a cross-ply configuration demonstrates a reduction of 41% of ultimate tensile strength compared to laminate A.

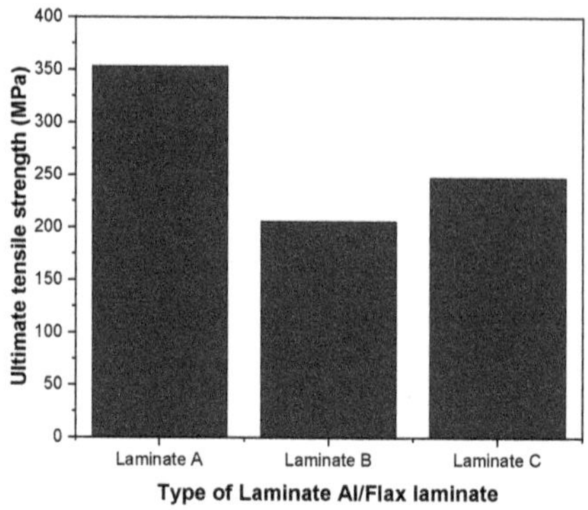

Figure 5. Comparison of ultimate tensile strength of the three laminate configurations.

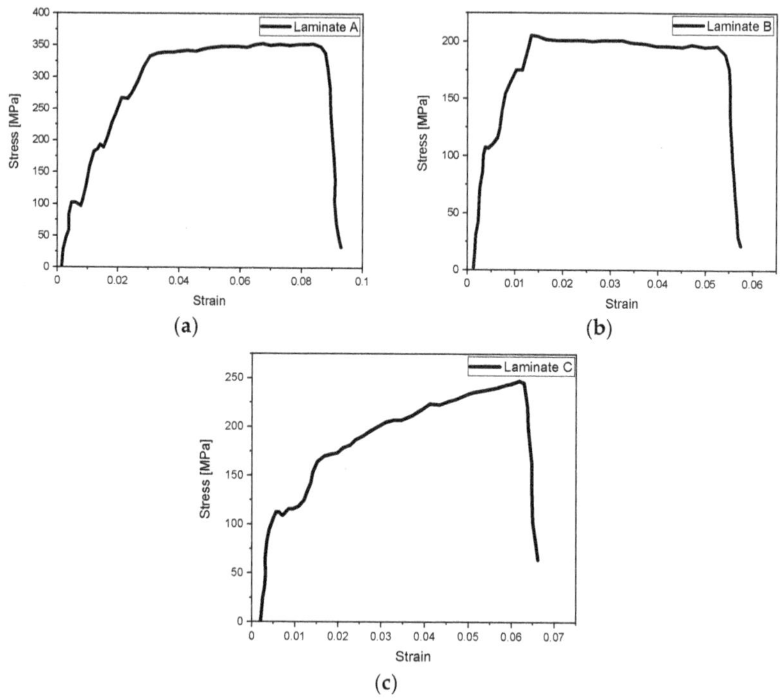

Figure 6. Stress–strain curves of (**a**) laminate A; (**b**) laminate B; (**c**) laminate C.

Due to the same metal distribution and metal volume, the curves of the three specimens in Figure 6 look similar but fail at different strain values. For laminate A, at the beginning of loading, the maximum stresses are mainly concentrated on the aluminium layers and only appear on the edges of the composite layers. The stress slowly grows into the whole structure and is distributed evenly. This is because of the characteristics of 0°-degree fibre orientation which increases the tensile loading capability of the laminate. At this point, the longitudinal stresses on the laminate will continue to rise and reach the ultimate tensile strength. Once the ultimate tensile strength is reached, cracks will appear in the metal layer and the whole laminate will fail.

Figure 7 presents the results for the damage evolution in both matrix and fibres in tension for each of the composite layers of laminate A at 2.6 mm displacement, with the naming of the flax layers following the convention of Figure 2. The damage evolution started with the matrix cracking that was first initiated at both 0° flax composite layers, as shown in Figure 7. At the same time, the fibres of both 0° flax composite layers are still capable of holding stress due to their characteristics. After the matrix failure, the fibre starts to break. This led to the failure of the specimen. As shown from the stress distribution in Figure 8, at 2.8 mm displacement, the middle aluminium layer has already failed, and cracking starts to appear at the 0° Flax-2 layer too. For the interfaces between the layers, which are numbered in Figure 9, the delamination at the interface between each layer after the failure of the specimen is shown in Figure 10. The red-coloured areas in these contours represent the highly delaminated areas of each interface (with a value of 1.0 of the damage indicator) whereas the blue-coloured areas indicate no delamination (value 0.0 of the damage indicator). The edges of the layer between the top aluminium and bottom aluminium layers have less delamination than those in between the two composite layers and mid-aluminium layers.

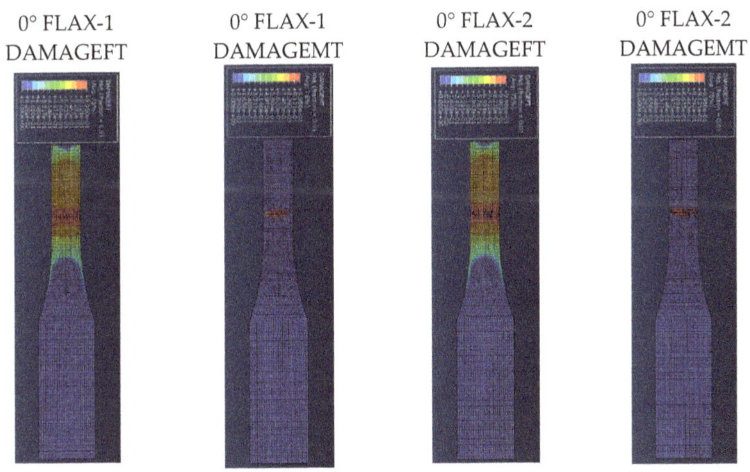

Figure 7. Damage evolution for fibres and matrix of top composite layers of Laminate A at 2.6 mm displacement.

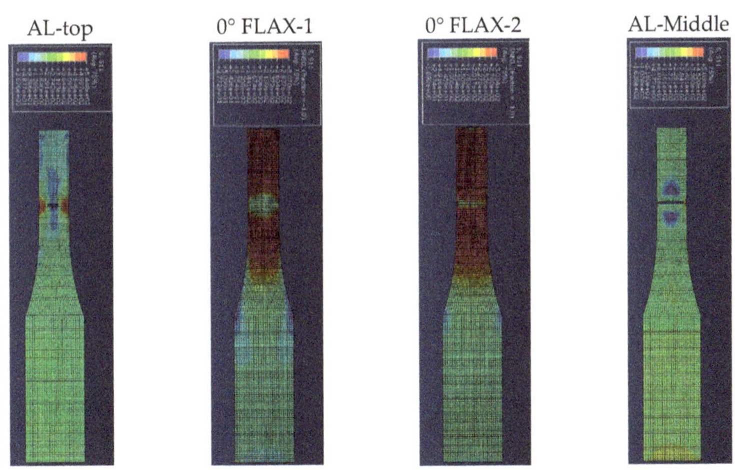

Figure 8. Stress distribution of the four top layers of laminate A at 2.8 mm displacement.

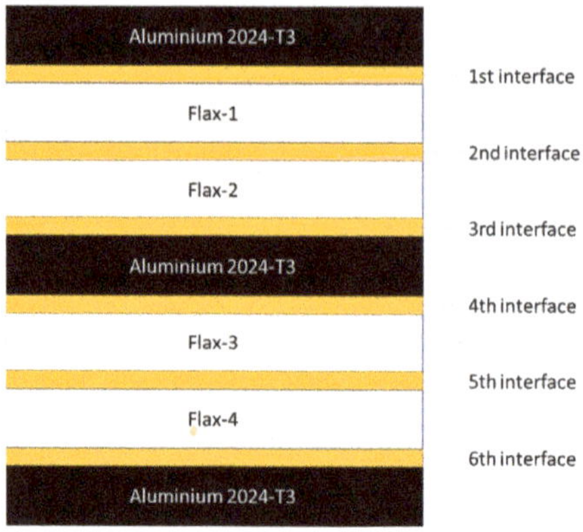

Figure 9. Schematic of interfaces between layers of the NFML.

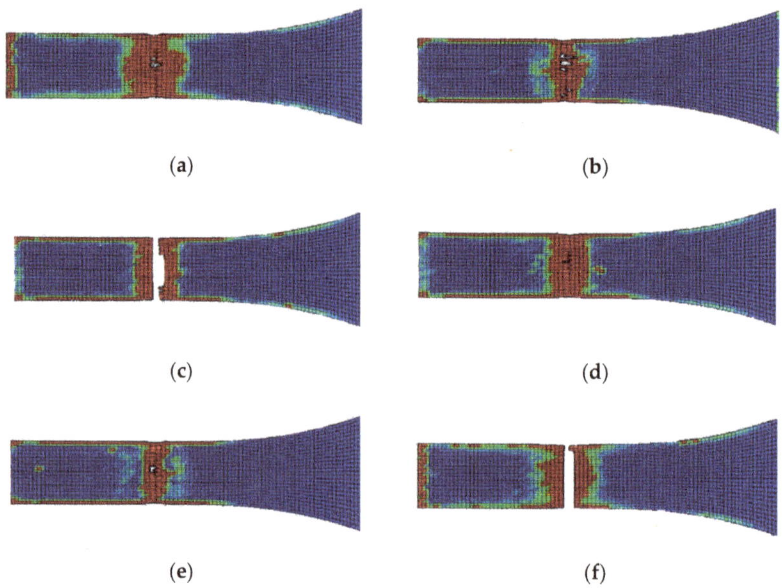

Figure 10. Delamination of each layer of laminate A after failure; (**a**) 1st interface; (**b**) 2nd interface; (**c**) 3rd interface; (**d**) 4th interface; (**e**) 5th interface; (**f**) 6th interface.

The stress distribution of laminate B is mainly focused on the 0° ply and aluminium layers as shown in Figure 11 at a displacement of 0.5 mm. The longitudinal stress carried by the aluminium layer and 0° ply is around 371 MPa and 216 MPa, respectively, while the 90° ply only carried 1.6 MPa of stress, due to the matrix cracking occurring at the 90° ply during the early stage of tension loading. With further loading, the propagation of matrix cracking will be distributed across the 90° ply. The fibres belonging to the 0° ply will start to break, and the direction of propagation is perpendicular to the fibre direction, as seen in Figure 12.

The edge delamination formation occurred first between the 90° flax composite layer and the mid-aluminium layer of laminate B (Figure 13c). Unlike laminate A, the delamination happens randomly of the edges during the early stage. After the delamination occurred on the edges of the 90° flax composite and mid-aluminium layers, it started to distribute into adjacent layers. The 1st and 6th interfaces have delamination on the end edge of the laminate while other interfaces do not, as shown in Figure 13. Again here, the red-coloured areas in these contours represent the highly delaminated areas of each interface (with a value of 1.0 of the damage indicator) whereas the blue-coloured areas indicate no delamination (value 0.0 of the damage indicator). The area of delamination across interfaces 1 and 6, that is, between Al and 0° composite layer layers, is larger than the delamination observed across interfaces 3 and 4, that is, between Al and 90° composite layers. It can be concluded that the primary failure starts from matrix cracking of 90° ply, and fibre tension damage in the 0° layer occurred during the early stage of loading. Once the matrix cracking of 90° ply started, the 90° ply lost its ability to withstand the stress. Then the fibre of the 0° ply and aluminium layer will be responsible for holding the stress. Once it reached the ultimate strain, the specimen failed.

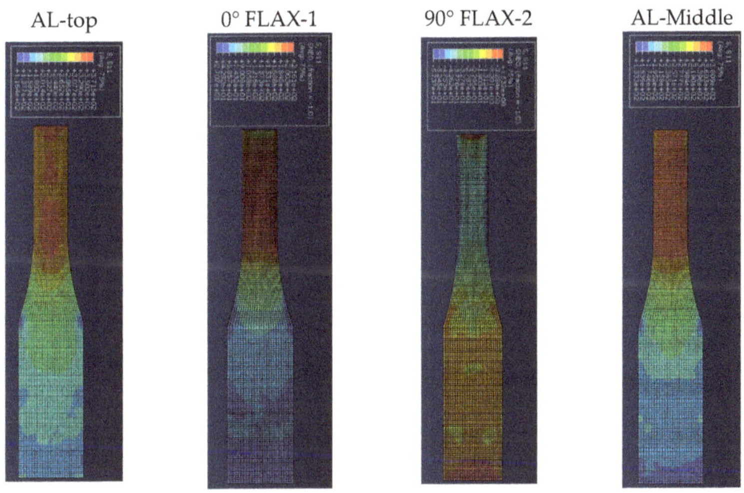

Figure 11. Stress distribution of each top layer of laminate B at 0.5 mm displacement and.

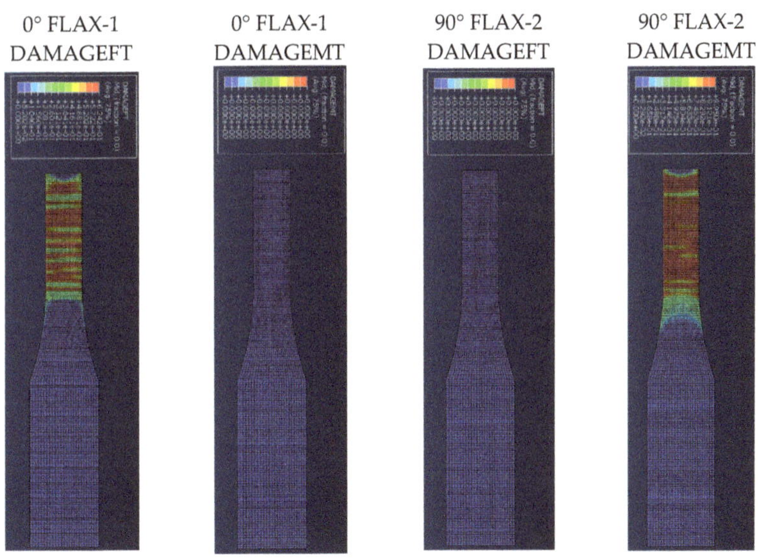

Figure 12. Damage evolution of each composite layer of the laminate B at 2 mm displacement.

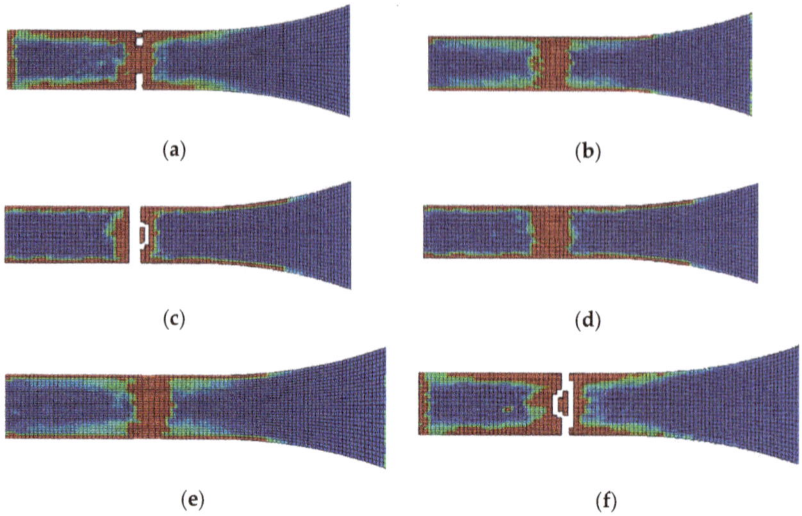

Figure 13. Delamination of each layer of laminate B after failure; (**a**) 1st interface; (**b**) 2nd interface; (**c**) 3rd interface; (**d**) 4th interface; (**e**) 5th interface; (**f**) 6th interface.

The stress distribution of laminate C follows the fibre orientation of the composite layer. Figure 14 shows the displacement at 3 mm, where the stress concentration of the 45° flax layer and the −45° flax layer is aligned to the fibre direction. Figure 14 shows that the stresses at the aluminium layers are concentrated around the edges. It is worth observing that, when the angle of the principal loading direction is more than 45 degrees compared to the fibre orientation, the metal layers are dominating the load carrying for the whole FML, which can be also verified from Figure 14, with the stresses developed on the Al layers being around 460 MPa, as opposed to circa 205 MPa for the composite layers. The results presented here show that the aluminium layers carry almost double the magnitude of stress under tensile loading compared with the composite layers. Figure 15 shows the different damage modes developed on the flax layers of the NFML. It is evident that the matrix cracking is the first model to be observed and is extended along the ±45° direction, as expected.

Unlike laminate B, the delamination of laminate C does not have sequences. It rapidly grows on the edge of the specimen. Figure 16 shows that the delamination occurs at the failure area and follows the 45° fibre orientation pattern. The red-coloured areas in these contours represent the highly delaminated areas of each interface (with a value of 1.0 of the damage indicator) whereas the blue-coloured areas indicate no delamination (value 0.0 of the damage indicator). The damage evolution of laminate C, starting with matrix damage, occurred at both 45° and −45° flax layers. Fibre breakages happened when the specimen failed. The failure of the specimen also follows the fibre orientation.

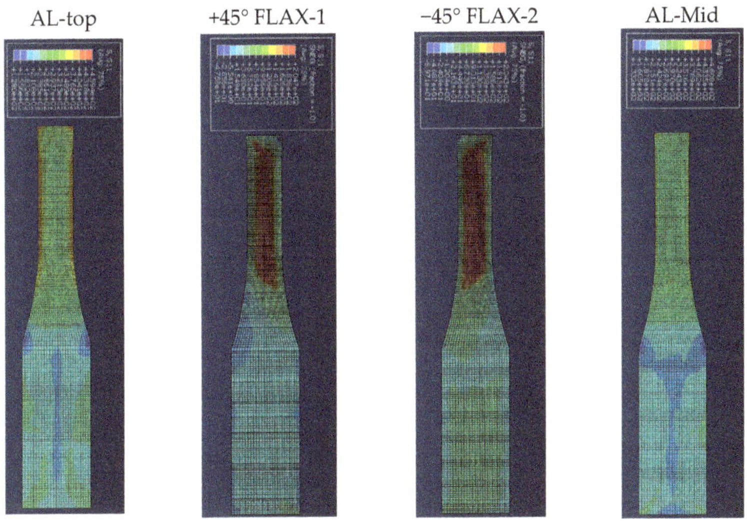

Figure 14. Stress distribution of each top layer of laminate C at 3 mm displacement.

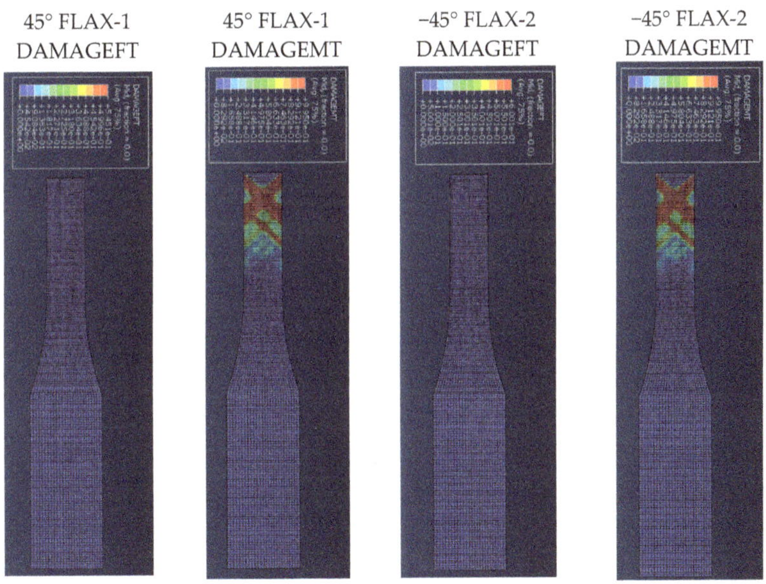

Figure 15. Damage evolution of each top composite layer of specimen C at 1 mm displacement.

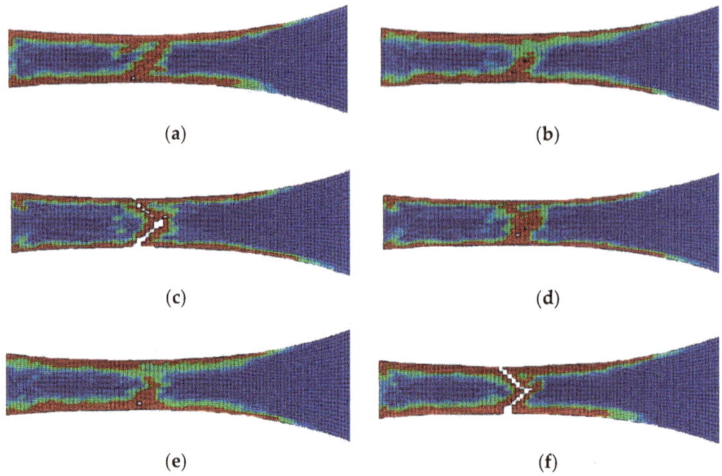

Figure 16. Delamination of each layer of laminate C after failure; (**a**) 1st interface; (**b**) 2nd interface; (**c**) 3rd interface; (**d**) 4th interface; (**e**) 5th interface; (**f**) 6th interface.

In summary, the three laminate configurations with different layup orientations are being investigated in this section when loaded under tension, where it was found that the stress distributions of each laminate are greatly affected by fibre orientation. For example, all layers of laminate A are capable of carrying load due to the longitudinal direction, while the 90° flax layer of laminate B does not contribute to the longitudinal loading. In addition, the stress distribution of laminate C is along either 45° or −45° fibre orientation. The maximum stresses developed in each laminate are shown in Figure 17, where the first 4 layers are presented, that is, the top and middle aluminium layers as well as the flax layers in between (Figure 9), with different orientations for each laminate. It is evident from this bar chart that both the aluminium layers are carrying most of the load in all three laminates, with the 0° flax layers being mostly loaded, as compared to the other orientations. It is also interesting to observe that the 90° fibre orientation carries the lowest load, due to being perpendicular to the loading direction. Delamination often occurs on the edges of the structure and the breakage area but only the interface between the 0° fibre and aluminium layers shows delamination on the fixed end side.

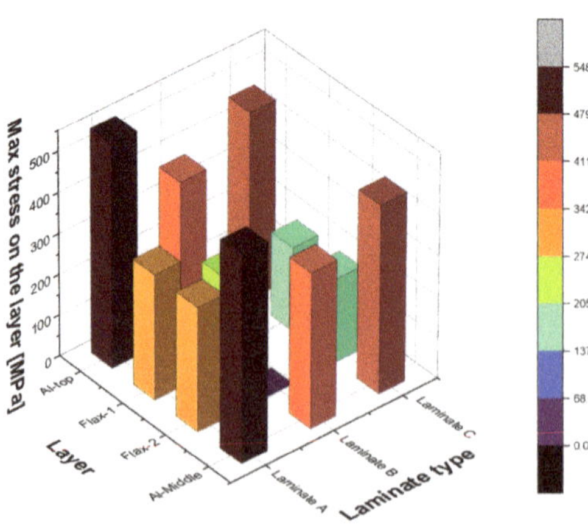

Figure 17. Maximum stress comparison between all three laminate types for the first four layers of each laminate.

5. Conclusions

In this paper, the performance of fibre metal laminates reinforced with natural fibre composite layers (NFMLs) under tension is investigated and studied under different laminate configurations using an FE model developed in Abaqus. The importance of understanding fully the behaviour of these novel materials is essential, leading to their wider adoption into secondary and primary engineering applications. The combined benefit of the proposed NFMLs in terms of their lower carbon footprint and their overall reduced CO_2 emissions has motivated this study to provide further insight into their performance. The approach used in this paper comprises a selected set of laminate configurations combining Aluminium layers with flax fibre-reinforced composite layers at different fibre orientations, in order to investigate the effect of the layup on the tensile behaviour of the material. The study is particularly focused on the contact interface between the natural fibre composite and the metal layers under tension loading. The finite element modelling approach is meso-level, which means each layer is modelled individually. The meso-level modelling approach is sufficient to analyse the behaviour of layers and the interaction between them. The developed FE model used the Hashin-2D built-in damage of composite layer and cohesive surface-based criterion in Abaqus. The damage in the interface and failure mechanisms of the fibre metal laminates was discussed in the results part of aluminium/flax fibre metal laminates. The damage to the interface, such as delamination, was also discussed and related to the fibre orientation. Three different types of layup sequences were investigated in this study and showed a different behaviour under tension loading, with the Laminate A exhibiting a higher toughness in tension, due to the fibre orientation. As expected, the fibre orientation plays a critical role in the strength and performance of the composite laminate, with the unidirectional laminate performing better under tension. However, the delamination in the interfaces of the unidirectional laminate (Figure 10) exhibits a greater extend compared to the cross-ply laminate (Figure 13). The debonding between the metallic and the composite layers is of great importance and it's evident from the stress distribution analysis of all three laminates examined in the present study that the aluminium layers are mainly responsible for carrying a greater load compared to the composite layers.

The limitations observed in this study are related to the prediction of the full stress state of the laminate, due to the use of the 2D Hashin model. However, the results produced are of great importance, as they offer an initial understanding of the performance of the novel materials. The feasibility of this study could be further improved in a future paper, to accommodate a 3D constitutive model for the full prediction of the behaviour of such NFMLs. Only one type of natural fibre composite was tested in this study. A comparative study is required in configuration with two or more natural fibre composites within the fibre metal laminates. Additionally, multiple alternative metal materials can also be applied, which can bring in more room for exploring a new combination of fibre-metal laminates. Lastly, the feasibility of applying natural fibre composite to the aerospace structural component is also worth investigating.

Author Contributions: Conceptualization, A.B.; methodology, A.B. and C.H.S.; software, C.H.S.; validation, C.H.S.; formal analysis, C.H.S. and A.B.; investigation, C.H.S.; data curation, C.H.S.; writing—original draft preparation, C.H.S.; writing—review and editing, A.B., K.G. and A.S.; visualization, C.H.S.; supervision, A.B. All authors have read and agreed to the published version of the manuscript.

Funding: This research received no external funding.

Institutional Review Board Statement: Not applicable.

Data Availability Statement: The data used in the manuscript can be found in the work of Aransáez Ortega, M. Numerical Simulation of the Impact Response of Natural Fibre Composites. 2019. Available online: https://trepo.tuni.fi/handle/123456789/27698. Any other data regarding the FE model can be requested by contacting the corresponding author Antigoni Barouni.

Conflicts of Interest: The authors declare no conflict of interest.

References

1. Sinmazçelik, T.; Avcu, E.; Bora, M.Ö.; Çoban, O. A review: Fibre metal laminates, background, bonding types and applied test methods. *Mater. Des.* **2011**, *32*, 3671–3685. [CrossRef]
2. Chang, P.-Y.; Yang, J.-M. Modeling of fatigue crack growth in notched fiber metal laminates. *Int. J. Fatigue* **2008**, *30*, 2165–2174. [CrossRef]
3. Chang, P.-Y.; Yeh, P.-C.; Yang, J.-M. Fatigue crack initiation in hybrid boron/glass/aluminum fiber metal laminates. *Mater. Sci. Eng. A* **2008**, *496*, 273–280. [CrossRef]
4. Vlot, A. Impact loading on fibre metal laminates. *Int. J. Impact Eng.* **1996**, *18*, 291–307. [CrossRef]
5. Marissen, R. Fatigue Crack Growth in ARALL. A Hybrid Aluminium-Aramid Composite Material: Crack Growth Mechanisms and Quantitative Predictions of the Crack Growth Rates. Ph.D. Thesis, Mechanical Maritime and Materials Engineering, Mekelweg, Delft, 1988. Available online: https://repository.tudelft.nl/islandora/object/uuid%3A9e27adfd-7ed8-4edf-b0ed-71d5b711f6ac (accessed on 23 May 2021).
6. Vogelesang, L.B.; Vlot, A. Development of fibre metal laminates for advanced aerospace structures. *J. Mater. Process. Technol.* **2000**, *103*, 1–5. [CrossRef]
7. Asundi, A.; Choi, A.Y.N. Fiber metal laminates: An advanced material for future aircraft. *J. Mater. Process. Technol.* **1997**, *63*, 384–394. [CrossRef]
8. Chen, Y.; Wang, Y.; Wang, H. Research Progress on Interlaminar Failure Behavior of Fiber Metal Lami-nates. *Adv. Polym. Technol.* **2020**, *2020*, 3097839. [CrossRef]
9. Karpov, E.V.; Demeshkin, A.G. Strain and Fracture of Glass-Fiber Laminate Containing Metal Layers. *J. Appl. Mech. Tech. Phys.* **2018**, *59*, 699–705. [CrossRef]
10. Ferdous, W.; Manalo, A.; Yu, P.; Salih, C.; Abousnina, R.; Heyer, T.; Schubel, P. Tensile Fatigue Behavior of Polyester and Vinyl Ester Based GFRP Laminates—A Comparative Evaluation. *Polymers* **2021**, *13*, 386. [CrossRef]
11. Liu, Y.; Zhang, H.-T.; Tafsirojjaman, T.; Dogar, A.U.R.; AlAjarmeh, O.; Yue, Q.-R.; Manalo, A. A novel technique to improve the compressive strength and ductility of glass fiber reinforced polymer (GFRP) composite bars. *Constr. Build. Mater.* **2022**, *326*, 126782. [CrossRef]
12. Almasabha, G.; Tarawneh, A.; Saleh, E.; Alajarmeh, O. Data-Driven Flexural Stiffness Model of FRP-Reinforced Concrete Slender Columns. *J. Compos. Constr.* **2022**, *26*, 04022024. [CrossRef]
13. Tarawneh, A.; Almasabha, G.; Murad, Y. ColumnsNet: Neural Network model for construction interaction dia-grams and slenderness limit for FRP-RC columns. *J. Struct. Eng.* **2022**, *148*, 04022089. [CrossRef]
14. Jawaid, M.; Thariq, M.; Saba, N. *Mechanical and Physical Testing of Biocomposites, Fibre-Reinforced Composites and Hybrid Composites*; Woodhead Publishing: Cambridge, UK, 2019. [CrossRef]
15. Khalili, S.M.R.; Mittal, R.; Kalibar, S.G. A study of the mechanical properties of steel/aluminium/GRP laminates. *Mater. Sci. Eng. A* **2005**, *412*, 137–140. [CrossRef]
16. Sharma, A.P.; Velmurugan, R. Uni-axial tensile response and failure of glass fiber reinforced titanium laminates. *Thin-Walled Struct.* **2020**, *154*, 106859. [CrossRef]
17. Moussavi-Torshizi, S.E.; Dariushi, S.; Sadighi, M.; Safarpour, P. A study on tensile properties of a novel fiber/metal laminates. *Mater. Sci. Eng. A* **2010**, *527*, 4920–4925. [CrossRef]
18. Hashemi, S.J.; Sadooghi, A.; Rahmani, K.; Davarzani, F.; Akbari, S. Investigation on the mechanical behavior of fiber-metal laminates based on polyvinyl chloride reinforced by 3D glass fibers. *Mater. Today Commun.* **2020**, *25*, 101273. [CrossRef]
19. Soltani, P.; Keikhosravy, M.; Oskouei, R.H.; Soutis, C. Studying the Tensile Behaviour of GLARE Laminates: A Finite Element Modelling Approach. *Appl. Compos. Mater.* **2010**, *18*, 271–282. [CrossRef]
20. Sharma, A.P.; Khan, S.H.; Parameswaran, V. Experimental and numerical investigation on the uni-axial tensile response and failure of fiber metal laminates. *Compos. Part B Eng.* **2017**, *125*, 259–274. [CrossRef]
21. Du, D.; Hu, Y.; Li, H.; Liu, C.; Tao, J. Open-hole tensile progressive damage and failure prediction of carbon fiber-reinforced PEEK–titanium laminates. *Compos. Part B Eng.* **2016**, *91*, 65–74. [CrossRef]
22. Madenci, E.; Özütok. Variational approximate for high order bending analysis of laminated composite plates. *Struct. Eng. Mech.* **2020**, *73*, 97–108. [CrossRef]
23. Jin, K.; Wang, H.; Tao, J.; Du, D. Mechanical analysis and progressive failure prediction for fibre metal laminates using a 3D constitutive model. *Compos. Part A Appl. Sci. Manuf.* **2019**, *124*, 105490. [CrossRef]
24. Shahzad, A. Hemp fibre and its composites—A review. *J. Compos. Mater.* **2012**, *46*, 973–986. [CrossRef]
25. Dhakal, H.N.; Zhang, Z.Y.; Richardson, M.O.W.; Errajhi, O.A.Z. The low velocity impact response of non-woven hemp fibre reinforced unsaturated polyester composites. *Compos. Struct.* **2007**, *81*, 559–567. [CrossRef]
26. Karus, M.; Kaup, M. Natural fibres in the European automotive industry. *J. Ind. Hemp* **2002**, *7*, 117–129. [CrossRef]
27. Alves Fidelis, M.E.; Pereira, T.V.C.; Gomes, O.d.F.M.; de Andrade Silva, F.; Toledo Filho, R.D. The effect of fiber morphology on the tensile strength of natural fibers. *J. Mater. Res. Technol.* **2013**, *2*, 149–157. [CrossRef]
28. Ahmed, M.M.; Dhakal, H.N.; Zhang, Z.Y.; Barouni, A.; Zahari, R. Enhancement of impact toughness and damage behaviour of natural fibre reinforced composites and their hybrids through novel improvement techniques: A critical review. *Compos. Struct.* **2021**, *259*, 113496. [CrossRef]

29. Barouni, A.; Lupton, C.; Jiang, C.; Saifullah, A.; Giasin, K.; Zhang, Z.Y.; Dhakal, H. Investigation into the fatigue properties of flax fibre epoxy composites and hybrid composites based on flax and glass fibres. *Compos. Struct.* **2022**, *281*, 115046. [CrossRef]
30. Aransáez Ortega, M. Numerical Simulation of the Impact Response of Natural Fibre Composites. 2019. Available online: https://trepo.tuni.fi/handle/123456789/27698 (accessed on 23 May 2021).
31. Camanho, P.P.; Davila, C.G. Mixed-Mode Decohesion Finite Elements for the Simulation of Delamination in Composite Materials. 2022. Available online: https://ntrs.nasa.gov/citations/20020053651 (accessed on 1 June 2022).
32. Silva Botelho, T.; Bayraktar, E.; Botelho, T. Experimental and numerical study of damage initiation mechanism in elastomeric composites. *J. Achiev. Mater. Manuf. Eng.* **2009**, *36*, 65–70.
33. Lapczyk, I.; Hurtado, J.A. Progressive damage modeling in fiber-reinforced materials. *Compos. Part A Appl. Sci. Manuf.* **2007**, *38*, 2333–2341. [CrossRef]
34. Shi, Y.; Pinna, C.; Soutis, C. Impact Damage Characteristics of Carbon Fibre Metal Laminates: Experiments and Simulation. *Appl. Compos. Mater.* **2020**, *27*, 511–531. [CrossRef]

polymers

Article

High-Efficiency Carbon Fiber Recovery Method and Characterization of Carbon FIBER-Reinforced Epoxy/4,4′-Diaminodiphenyl Sulfone Composites

Yong-Min Lee [1,2], Kwan-Woo Kim [1,*] and Byung-Joo Kim [3,*]

1 Convergence Research Division, Korea Carbon Industry Promotion Agency, Jeonju 54852, Republic of Korea
2 Department of Carbon Convergence and Composite Materials Engineering, Jeonbuk National University, Jeonju 54896, Republic of Korea
3 Department of Carbon-Nanomaterials Engineering, Jeonju University, Jeonju 55069, Republic of Korea
* Correspondence: kkw1988@kcarbon.or.kr (K.-W.K.); kimbyungjoo@jj.ac.kr (B.-J.K.); Tel.: +82-63-219-3713 (K.-W.K.); +82-63-220-3293 (B.-J.K.)

Abstract: Globally, the demand for carbon fiber-reinforced thermosetting plastics for various applications is increasing. As a result, the amount of waste from CFRPs is increasing every year, and the EU Council recommends recycling and reuse of CFRPs. Epoxy resin (EP) is used as a matrix for CFRPs, and amine hardeners are mainly used. However, no research has been conducted on recycling EP/4,4′-diaminodiphenyl sulfone (DDS)-based CFRP. In this study, the effect of steam and air pyrolysis conditions on the mechanical properties of re-cycled carbon fiber (r-CF) recovered from carbon fiber-reinforced thermosetting (epoxy/4,4′-diaminodiphenyl sulfone) plastics (CFRPs) was investigated. Steam pyrolysis enhanced resin degradation relative to N_2. The tensile strength of the recovered r-CF was reduced by up to 35.12% due to oxidation by steam or air. However, the interfacial shear strength (IFSS) tended to increase by 9.18%, which is considered to be due to the increase in functional groups containing oxygen atoms and the roughness of the surface due to oxidation. The recycling of CFRP in both a steam and an air atmosphere caused a decrease in the tensile strength of r-CF. However, they were effective methods to recover r-CF that had a clean surface and increased IFSS.

Keywords: carbon fiber-reinforced plastic (CFRP); recycling; epoxy; DDS; mechanical properties

Citation: Lee, Y.-M.; Kim, K.-W.; Kim, B.-J. High-Efficiency Carbon Fiber Recovery Method and Characterization of Carbon FIBER-Reinforced Epoxy/4,4′-Diaminodiphenyl Sulfone Composites. *Polymers* **2022**, *14*, 5304. https://doi.org/10.3390/polym14235304

Academic Editors: R.A. Ilyas, S.M. Sapuan, Emin Bayraktar, Shukur Abu Hassan, Nabil Hayeemasae and Khubab Shaker

Received: 14 November 2022
Accepted: 2 December 2022
Published: 4 December 2022

Publisher's Note: MDPI stays neutral with regard to jurisdictional claims in published maps and institutional affiliations.

1. Introduction

Carbon fiber-reinforced plastics (CFRPs) are high-performance composite materials composed of carbon fibers and thermosetting or thermoplastic polymer matrices [1–3]. A global goal is to achieve net-zero carbon dioxide emissions by 2050 [4]. Therefore, studies on improving fuel efficiency have been conducted in the fields of aerospace [5,6], ships [7,8], and automobiles [9,10] by building vehicle bodies and parts with CFRP, which is light, high intensity, and high elasticity for light weight [11–13]. As CFRP can replace metals due to the properties mentioned above, as well as the advantages of excellent corrosion resistance [14,15] and durability [16], demand has also increased in structural applications such as renewable energy [17,18], sports [19], and civil engineering and architecture [20–22]. As the production of and demand for CFRP increases, the amount of waste also increases yearly [23–25]. In the European Union, automobiles generate 8-9 million tons of waste each year at the end of their lifespan. Accordingly, the European Parliament and the Council adopted a directive on end-of-life vehicles in 1997 to address this issue. This directive requires that waste be reused and recovered. Thus, giving priority to reuse and recycling is the basic principle of the directive [26]. Meanwhile, delivery of approximately 40,000 aircraft is expected by 2038, with an annual air traffic growth rate of at least 4.3%. Of these aircraft, 56% are expected to accommodate market growth, and the remaining deliveries are expected to replace approximately 75% of current aircraft over the next 20 years. This will result in the retirement of approximately 19,000 commercial aircraft over the next 18 years. Due to regulations, OEMs consider the end of life of an aircraft at the design stage, and research on aircraft recycling is underway in industry and academia [27]. CFRP is used in 53% of the Airbus a350 XWB, and in 50% of the Boeing 787 [28]. The recent launch of clean aviation to achieve climate neutrality has driven the goal of replacing 75% of existing civil aircraft with hydrogen- and electric-powered aircraft [29]. By 2050, the aviation industry is expected to collect approximately 500,000 tons of end-of-life (EOL) waste from CFRP [30]. CFRP has been disposed of by landfill in the past. However, disposal by landfill is restricted or prohibited because it causes environmental problems due to the lifespan of polymers [31–33]. Therefore, various recycling methods for CFRP have recently been developed. There are three typical recycling methods for CFRP: mechanical, chemical, and thermal. Mechanical recycling is the simple process of physically crushing CFRP. However, CF and resin are not separated during the crushing, which is a drawback, and long fibers are broken during the grinding process, which results in poor physical properties [34]. The chemical method is a method that uses a solvent to decompose important chemical bonds of the used thermosetting resin. Acids [35–37], hydrogen peroxide [34,38], and supercritical fluids [39–41] are used as solvents. However, acids take a long time to decompose CFRP, and hydrogen peroxide is difficult to manage due to its high reactivity. In addition, although supercritical fluids have a high CF recovery rate and can recover CF with a clean surface and excellent physical properties, high temperatures and high-pressure conditions are required, and the cost of r-CF recovery is high. In addition, large-capacity and continuous processes are difficult, and environmental problems may occur due to processing with solvents. Conventional thermal recycling is a simple and continuous process that decomposes resin and recovers CF at a high temperature of approximately 500 °C or higher in N_2 or air atmosphere. However, a large amount of energy is consumed, and undecomposed impurities generated during the process are adsorbed on the CF surface, leaving a residue. In addition, pyrolysis produces CO and CO_2, which can lead to environmental problems [42]. To compensate for these shortcomings, Kim et al. suggested a decomposition method using superheated steam in 2017 [23]. Superheated steam can uniformly apply heat to CFRP and penetrate the resin to decompose rapidly. In addition, unlike other pyrolysis methods, it is possible to recover in the form of tar and thus generate less CO_2. Epoxy is the most commonly used thermosetting resin in CFRP production, and amine curing agents are used primarily for epoxy curing. Among the commonly used 4,4′-diaminodiphenyl methane (DDM) and 4,4′-diaminodiphenyl sulfone (DDS), many studies have been conducted on recycling methods of EP/DDM-based CFRP, but EP/DDS-based CFRP recycling methods have not been studied. In this study, EP/DDS-based CFRPs were pyrolysis in steam and air atmosphere. The temperature conditions in each atmosphere were essential to remove the residual epoxy on the CF surface. Optimal process conditions such as atmosphere gas type, temperature, and holding time were derived, and the mechanical properties of the recovered r-CF were investigated.

2. Materials and Methods

For the CFRPs used in this study, polyacrylonitrile-based unidirectional carbon fiber (CF, T700, Toray, Tokyo, Japan) was used as a reinforcing material, and an amine curing agent (4,4′-diaminodiphenyl sulfone, DDS, Sigma-Aldrich, Saint Louis, MO, USA, AHEW of 62.08 g/eq, T_m of 172–175 °C) and a bifunctional epoxy resin (EP, diglycidyl ether of bisphenol A, Kukdo Chem., Seoul, Republic of Korea, YD-128, EEW of 184–190 g/eq, a viscosity of 11,500–13,500 cps at 25 °C) were mixed in an equivalent ratio and then impregnated using the hand lay-up method. At this time, the impregnated EP was 35 ± 2 wt.%. Curing was performed at 190 °C (10 °C/min) for 1 h based on the differential scanning calorimeter (DSC) results shown in Figure 1. The structures of EP and DDS are shown in Figure 2.

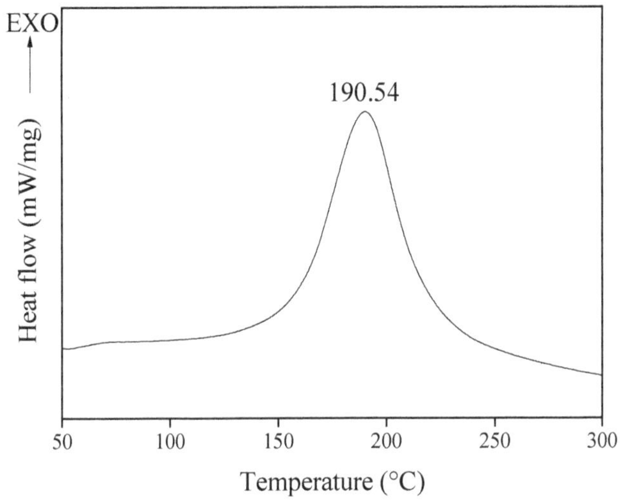

Figure 1. Curing temperature of epoxy/4,4′-diaminodiphenyl sulfone mixture using differential scanning calorimetry.

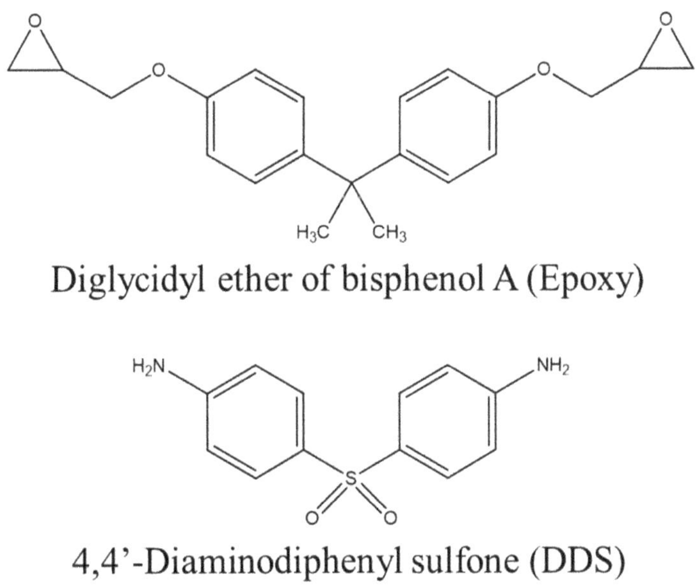

Figure 2. Chemical structures of epoxy and 4,4′-diaminodiphenyl sulfone.

2.1. CFRP Pyrolysis

A quartz tube furnace (length of 950 mm, inner diameter of 8 mm) was used for the CFRP pyrolysis. The fabricated CFRP was placed on an alumina crucible and loaded into a quartz tube. The heating steam was generated at 170 °C using triple-distilled water (5 cc/min) in a pre-heater. CFRP was heated to 500–600 °C at a rate of 10 °C/min and pyrolyzed for 180 min, followed by post-treatment at 500–600 °C for 10 min with air (O_2-21% under N_2; 200 cc/min). Finally, recycled carbon fiber (r-CF) was recovered after cooling to room temperature. The nomenclature for r-CF according to the pyrolysis conditions is shown in Table 1.

Table 1. Nomenclature for recycled carbon fibers and carbon fiber-reinforced plastics pyrolysis conditions.

Sample	Step 1				Step 2			
	Temperature (°C)	Heating Rate (°C/min)	Hold Time (min)	Steam (cc/min)	Temperature (°C)	Heating Rate (°C/min)	Hold Time (min)	Air (cc/min)
Ar-CF	-	-	-	-	-	-	-	-
50 55 60	500 550 600	10	180	5	-	-	-	-
50-50-10 50-55-10 50-60-10	500	10	180	5	500 550 600	10	10	200
55-60-10 60-60-10	550 600	10	180	5	600	10	10	200

2.2. Fixed Pyrolysis Conditions

The curing temperature of the EP/DDS mixture used as a matrix was measured using a differential scanning calorimeter (DSC, DSC 3, Mettler Toledo, Columbus, OH, USA). Approximately 7 mg of the mixture was placed in a 100 μL aluminum pan and heated from 30 °C to 300 °C at a rate of 10 °C/min under an N_2 atmosphere (50 cc/min).

The thermal decomposition temperatures of the CF and EP were measured using a simultaneous thermal analyzer (STA, TGA/DSC 3+, Mettler Toledo). Approximately 4 mg of the samples were placed in a 70 μL aluminum pan and heated from room temperature to 500, 550, and 600 °C at a rate of 10 °C/min under an N_2 and air atmosphere (50 cc/min).

2.3. Physical Properties

The tensile properties of a single fiber of Ar-CF and r-CF were measured using a universal testing machine (UTM, Seron, Seoul, Republic of Korea) according to ASTM D3379. At this time, the gauge length of the CF was set to 10 mm, and the draw-off clamp speed was set to 1 mm/min. After placing the CF between the grips of the UTM, both sides of the sample window were cut at the midpoint of the gauge. For each sample, at least 50 CFs were measured; a schematic diagram of the CF tensile strength sampling is shown in Figure 3a. In the tensile test, an error may occur if the carbon fiber is placed diagonally rather than in the center of the window.

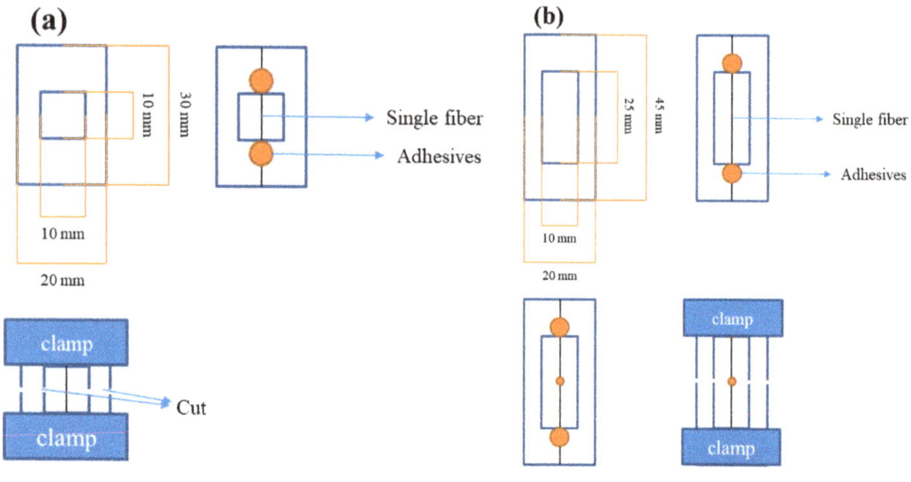

Figure 3. Schematics of physical properties test; (a) tensile strength and (b) interfacial shear strength.

The interfacial shear strength (IFSS) between the Ar-CF and r-CF and the EP resin was also measured using UTM. A single fiber was placed at the center of a paper window and adhered with an EP bond, and EP droplets with a diameter of approximately 100–200 μm were placed using a sharp needle. The EP droplet is a difunctional epoxy (EP, diglycidyl ether of bisphenol A, Kukdo Chem., YD-128, EEW of 184–190 g/eq, a viscosity of 11,500–13,500 cps at 25 °C) and an amine curing agent (4,4′-diaminodiphenyl methane, Sigma-Aldrich, Saint Louis, MO, USA, AHEW of 49.565 g/eq, Tm of 89 °C) was used. It was mixed for 20 min on a hot plate at 80 °C, and the temperature was maintained during the droplet operation. EP droplets were cured at 150 °C (10 °C/min) for 1 h. At least 25 CFs were measured for each sample, and the schematic diagram for the IFSS sampling of CFs is shown in Figure 3b. The IFSS test can be considered a failure if the carbon fiber is located diagonally to the window or if the epoxy bond is not fully cured and the carbon fiber rather than the EP droplet is pulled out.

2.4. Component Analysis

Functional groups on the Ar-CF and r-CF surfaces were detected in the 400–4000 cm^{-1} range using Fourier-transform infrared spectroscopy (FT-IR, Nicolet iS10, Thermo Scientific, Waltham, MA, USA). FT-IR samples were prepared as discs by grinding CF and potassium bromide (KBr, Sigma-Aldrich, Saint Louis, MO, USA) together and applying a clamp force of 7 tons for 2 min using a hydraulic press (CrushIR, PIKE Technologies, Madison, Wisconsin, WI, USA).

The surface properties of Ar-CF and r-CF were observed using field emission scanning electron microscopy (FE-SEM, S-4800, Hitachi, Tokyo, Japan). Ar-CF and r-CF were fixed in the SEM sample holder using carbon tape, and all images were obtained at 10^4 magnifications at 1.0×10^{-5} torr and an acceleration voltage of 15 kV.

3. Results and Discussion

3.1. Pyrolysis Conditions of EP/DDS and Carbon Fiber

CF and cured EP/DDS thermal decomposition behaviors were measured by STA under a gas (N_2 or air) atmosphere, and the thermal decomposition behavior under a steam atmosphere was measured by a self-made TGA furnace. The results are shown in Figure 4, and a schematic diagram is shown in Figure 5. The mass of CF was decreased by 2.81% (N_2), 4.58% (steam), and 7.62% (air) at 600 °C. In addition, EP showed a mass reduction of 71.09% (N_2), 85.40% (steam), and 96.64% (air) at 600 °C. The mass loss was the least under N_2 atmosphere and the degree of damage to the CF was therefore also small. On the other hand, the degree of damage to CF was large under the air atmosphere because the mass loss was the highest. Under the steam atmosphere, the decrease in mass was between that of N_2 and air. Even though the amount of EP char remaining after treatment showed a slight difference (approximately 14.31%) between N_2 and steam, it exerted a significant influence during the post-treatment (air) time. The post-treatment can have a significant impact on CF damage. In conclusion, it is considered that the CF damage can be minimized by shortening the duration of air post-treatment because EP char can be quickly generated under steam.

3.2. Morphology of the r-CF

The scanning electron microscopy (SEM) images of the surface morphology of Ar-CF and r-CF are shown in Figure 6. In steam pyrolysis at 500–600 °C, the EP of CFRP did not completely decompose. Therefore, some EP remained on the CF surface. However, an increase in the pyrolysis temperature affected the EP decomposition. Then, air pyrolysis induced the complete removal of EP on the CF surface within a short time. However, air pyrolysis at 500 °C did not completely remove the EP char on the r-CF and remained in the form of crumbs. On the other hand, air pyrolysis at 550 °C completely removed the EP char on the r-CF surface, and the degree of etching on the surface was small. Air pyrolysis at 600 °C also completely removed the EP char on the CF surface. However, the r-CF surface was etched, and a curve was formed due to excessive air treatment. Based on the results above, the conditions of the 50-55-10 sample were effective for EP/DDS-based CFRP recycling.

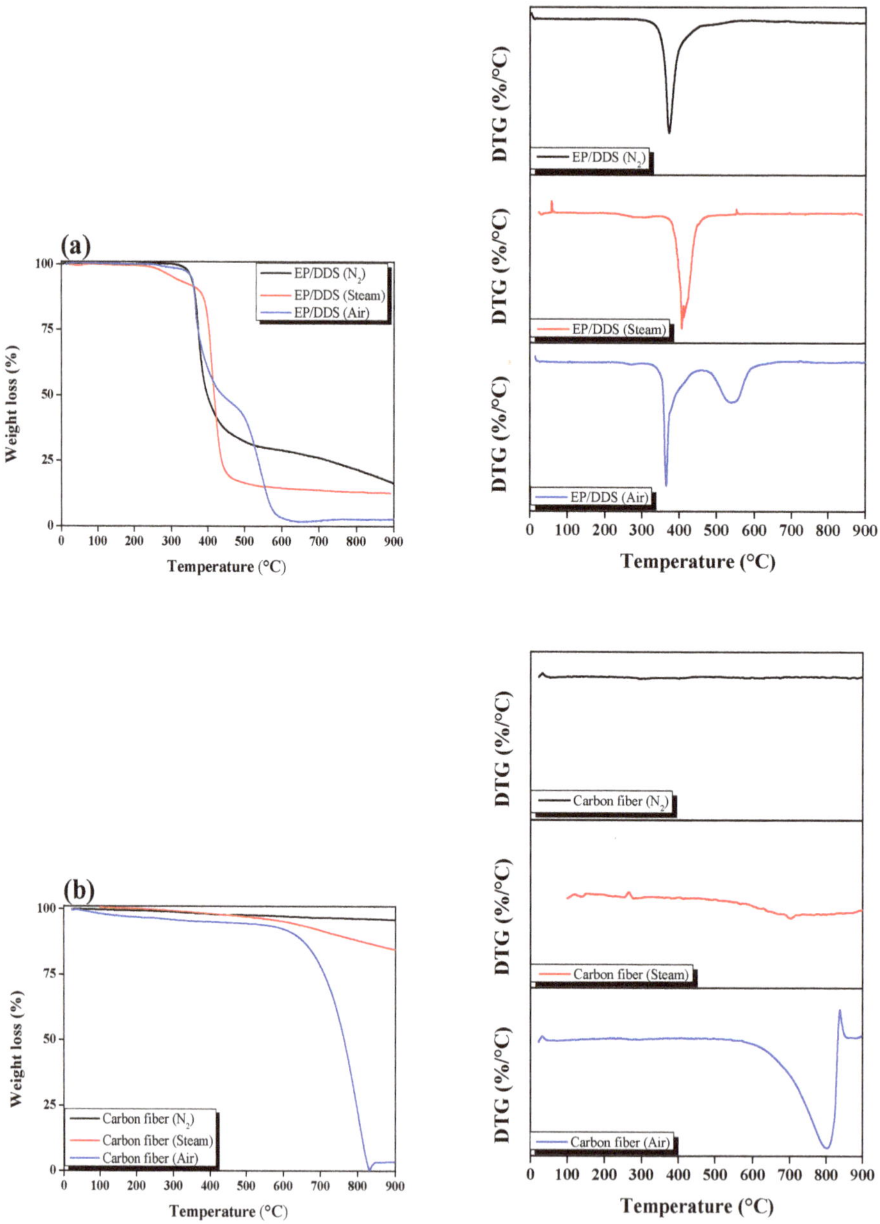

Figure 4. TGA and DTG graphs of as-received carbon fiber and epoxy resin pyrolysis under gas atmosphere: (**a**) cured epoxy/4,4′-diaminodiphenyl sulfone and (**b**) carbon fiber.

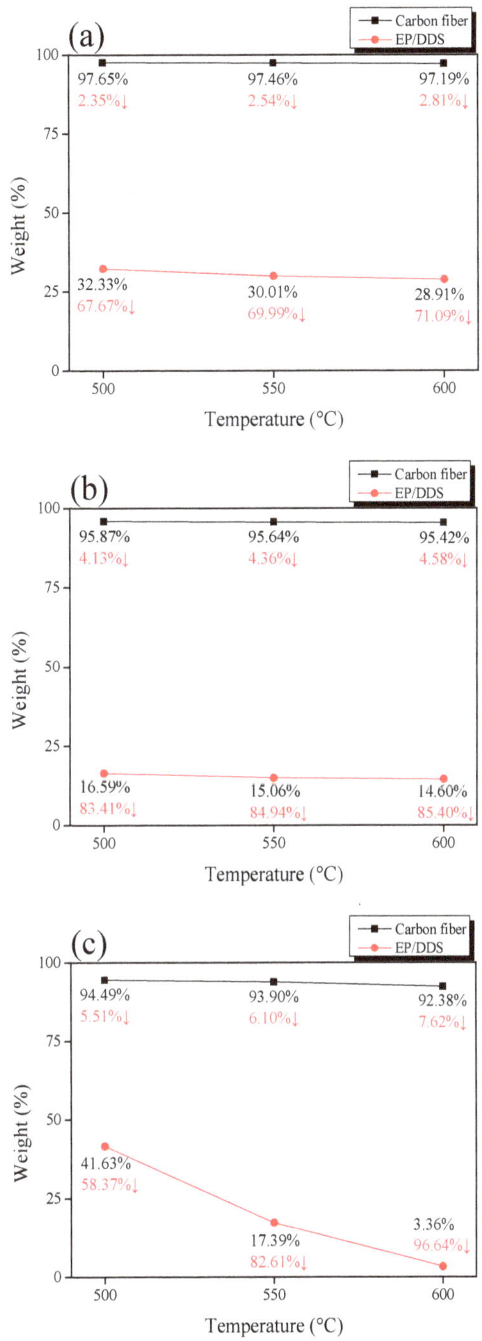

Figure 5. Weight loss of as-received carbon fiber and epoxy resin pyrolysis under gas atmosphere: (**a**) N_2, (**b**) steam, and (**c**) air.

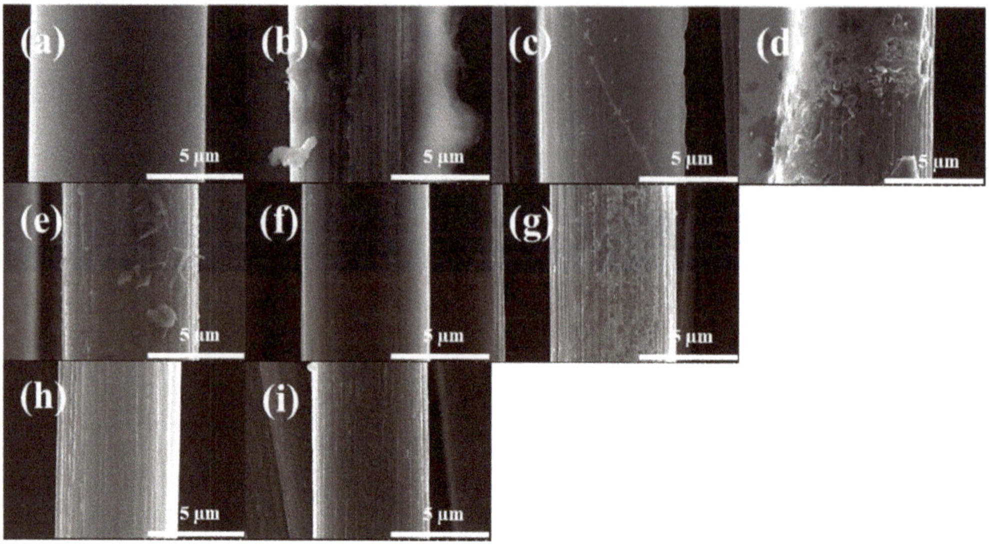

Figure 6. Scanning electron microscopy images of the as-received and recycled carbon fibers under varying pyrolysis conditions; (**a**) Ar-CF, (**b**) 50, (**c**) 55, (**d**) 60, (**e**) 50-50-10, (**f**) 50-55-10, (**g**) 50-60-10, (**h**) 55-60-10, and (**i**) 60-60-10.

3.3. Mechanical Properties of the r-CF

The tensile strength of the recovered r-CF is shown in Figure 7a. Equation (1) was used to calculate the tensile strength.

$$Strength = F/A \quad (1)$$

where F is the force at the break of the fiber, and A is the average cross-sectional area of the fiber.

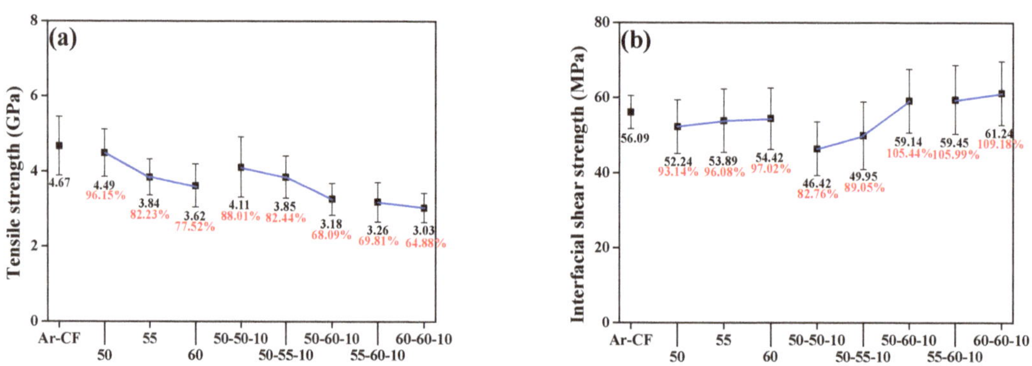

Figure 7. Physical properties of the as-received and recycled carbon fibers under varying pyrolysis conditions: (**a**) tensile strength and (**b**) interfacial shear strength.

The tensile strength of r-CF was decreased as the pyrolysis temperature increased. In the case of Ar-CF, the tensile strength decreased from 4.67 GPa to 3.03 GPa (60-60-10). As the temperature increased to 500–600 °C, the tensile strength of r-CF tended toward 22.48% (4.49 to 3.62 GPa) in the case of the sample that only underwent steam pyrolysis.

It is considered that the EP on the CF surface was decomposed by steam to form EP char, and thus the tensile strength was reduced as the CF was oxidized. The tensile strength of the 50-50-10 sample, which was subjected to air post-treatment after steam pyrolysis at 500 °C, was reduced by 11.99% (4.11 GPa), and the error rate was calculated to be large compared with that of the other samples. The EP char formed by steam was not completely decomposed in the air atmosphere and remained on the surface in the form of crumbs, which reduced the tensile strength and seemed to be the cause of the large error rate. The tensile strength of the sample subjected to secondary pyrolysis at 550–600 °C was reduced by up to 35.12% (3.85–3.03 GPa). Short air pyrolysis at temperatures above 550 °C induced a decrease in tensile strength due to oxidation and etching of the r-CF surface. However, the EP char on the CF surface was completely removed.

The IFSS of the recovered r-CF is shown in Figure 7b. Equation (2) was used to calculate the IFSS.

$$IFSS = F/\pi DL \tag{2}$$

where F is the equipment force when the fiber breaks, D is the droplet size, and L is the diameter of the carbon fiber.

The IFSS of Ar-CF was 56.09 MPa. The IFSS of r-CF recovered under steam pyrolysis conditions was lower than the IFSS of Ar-CF, regardless of temperature conditions. However, the IFSS of r-CF after air treatment at 500–550 °C was calculated as 46.42 MPa (82.76%) and 49.95 MPa (89.05%), respectively, compared to Ar-CF. As steam-air pyrolysis proceeded, EP char debris was formed on the surface of r-CF, and EP droplets did not completely adhere to CF but instead formed on the EP char. Meanwhile, the IFSS of the r-CF after air treatment at 600° C for 10 min was calculated as 59.14 MPa (105.44%), 59.45 MPa (105.99%), and 61.24 MPa (109.18%), respectively, compared with Ar-CF. As shown in the SEM images in Figure 5, the specific surface area increased due to the curvature of the r-CF surface. In addition, it is thought that the improvement in the IFSS was caused by the increase in functional groups containing oxygen atoms on r-CF by air.

3.4. Functional Group of r-CF

Figure 8 shows the FT-IR analysis that was done to identify the change in functional groups containing oxygen atoms in Ar-CF and r-CF. The hydroxyl (-OH) group on the CF was observed at 3431 cm^{-1} [43,44]. The peaks corresponding to the sp^2 C-H bond (=CH) and sp^3 C-H bond (-CH) were observed at 2920 cm^{-1} and 2851 cm^{-1} [43–45], respectively. In addition, the peak for the C=C bond was observed at 1575 cm^{-1} [43–45], and the peaks for the ketone (C=O) and C-O bond were observed at 1638 cm^{-1} and 1100 cm^{-1} [43–45], respectively.

The FT-IR spectra of r-CF were similar to those of Ar-CF. As the steam-air heat treatment proceeded, the peaks corresponding to the -OH, C=O, and C-O groups increased, and the peaks corresponding to the -CH, =CH, and C=C bonds decreased, suggesting that steam and air heat treatment converted -CH, =CH, and C=C bonds to functional groups containing oxygen atoms. Therefore, it is considered that the introduction of functional groups containing oxygen atoms improved the interfacial bonding interaction between CF and EP.

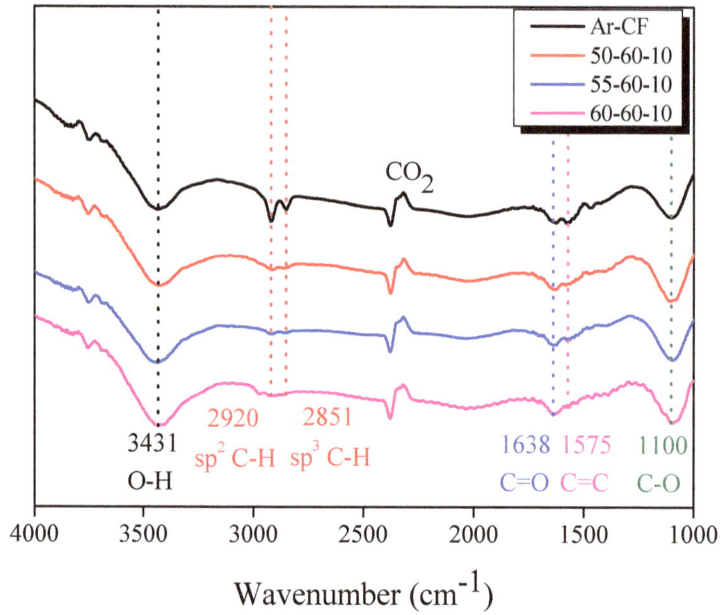

Figure 8. Fourier-transform infrared spectra of the as-received and recycled carbon fibers with varying pyrolysis conditions.

4. Conclusions

In this study, CFRPs were prepared using epoxy (EP) and 4,4′-diaminodiphenyl sulfone (DDS) as a curing agent. The prepared CFRPs were pyrolyzed under steam and air atmospheres to recover r-CF, and the structural and mechanical properties of the r-CF were evaluated. As the pyrolysis temperature increased, r-CF was oxidized, and the tensile strength was reduced by up to 64.88% relative to that of Ar-CF. As the thermal decomposition temperature increased, r-CF was oxidized and the tensile strength decreased by up to 64.88% compared to Ar-CF, but the IFSS tended to increase by 9.18% due to the increase in oxygen functional groups. In conclusion, a short-duration air post-treatment after steam pyrolysis had the effect of the complete removal of the residual EP char on the CF surface. Based on these results, clean r-CF was recovered from EP/DDS-based CFRP after air post-treatment. Although the tensile strength decreased, the IFSS improved. Therefore, the r-CF with improved IFSS has advantageous for manufacturing high-quality CFRP because of its excellent interfacial bonding with EP.

Author Contributions: Conceptualization, B.-J.K.; methodology, B.-J.K.; validation, Y.-M.L.; formal analysis, Y.-M.L.; investigation, Y.-M.L.; writing—original draft preparation, Y.-M.L.; writing—review and editing, K.-W.K.; visualization, Y.-M.L.; supervision, K.-W.K., B.-J.K.; project administration, K.-W.K. All authors have read and agreed to the published version of the manuscript.

Funding: This study was supported by the "Industrial Core Technology Development Program" (no. 20012817). The development of functional parts manufacturing technology, which reduced the cost by more than 15% by upcycling carbon fiber and intermediate materials, was funded by the Ministry of Trade, Industry, and Energy, Republic of Korea. This study was supported by the Material Parts Technology Development Project (Project No. 20015733) funded by the Ministry of Trade, Industry Energy (MOTIE), Republic of Korea.

Institutional Review Board Statement: Not applicable.

Informed Consent Statement: Not applicable.

Data Availability Statement: The data presented in this study are available on request from the corresponding author.

Conflicts of Interest: The authors declare no conflict of interest.

References

1. Hernandez, T.P.A.; Mills, A.R.; Yazdani Nezhad, H.Y. Shear driven deformation and damage mechanisms in High-performance carbon Fibre-reinforced thermoplastic and toughened thermoset composites subjected to high strain loading. *Compos. Struct.* **2021**, *261*, 113289. [CrossRef]
2. Zhang, J.; Chevali, V.S.; Wang, H.; Wang, C.H. Current status of carbon fibre and carbon fibre composites recycling. *Compos. B Eng.* **2020**, *193*, 108053. [CrossRef]
3. Rao, P.S.; Hardiman, M.; O'Dowd, N.P.; Sebaey, T.A. Comparison of progressive damage between thermoset and thermoplastic CFRP composites under in-situ tensile loading. *J. Compos. Mater.* **2020**, *55*, 1473–1484. [CrossRef]
4. EUR. L.Ex. 12 April 2022. Available online: eur-lex.europa.eu/legal-content/EN/TXT/?uri=CELEX%3A32022D0591 (accessed on 6 April 2022).
5. Kupski, J.; Teixeira de Freitas, S.T.D. Design of adhesively bonded lap joints with laminated CFRP adherends: Review, challenges and new opportunities for aerospace structures. *Compos. Struct.* **2021**, *268*, 113923. [CrossRef]
6. Morioka, K.; Tomita, Y.; Takigawa, K. High-temperature fracture properties of CFRP composite for aerospace applications. *Mater. Sci. Eng. A* **2001**, *319*, 675–678. [CrossRef]
7. Petrakli, F.; Gkika, A.; Bonou, A.; Karayannis, P.; Koumoulos, E.P.; Semitekolos, D.; Trompeta, A.-F.; Rocha, N.; Santos, R.M.; Simmonds, G.; et al. End-of-Life Recycling Options of (Nano)Enhanced CFRP Composite Prototypes Waste—A Life Cycle perspective. *Polymers* **2020**, *12*, 2129. [CrossRef]
8. Shi, J.; Bao, L. Optimum decomposition conditions for glass fiber reinforced plastic recycling by superheated steam. *Jpn. J. Appl. Phys.* **2011**, *50*, 01AJ01. [CrossRef]
9. Bere, P.; Dudescu, M.; Neamțu, C.; Cocian, C. Design, manufacturing and test of CFRP front hood concepts for a light-weight vehicle. *Polymers* **2021**, *13*, 1374. [CrossRef] [PubMed]
10. Fragassa, C.; Pavlovic, A.; Minak, G. On the structural behaviour of a CFRP safety cage in a solar powered electric vehicle. *Compos. Struct.* **2020**, *252*, 112698. [CrossRef]
11. Jeong, J.S.; Kim, K.W.; An, K.-H.; Kim, B.-J. Fast recovery process of carbon fibers from waste carbon fibers-reinforced thermoset plastics. *J. Environ. Manag.* **2019**, *247*, 816–821. [CrossRef]
12. Berger, D.; Brabandt, D.; Bakir, C.; Hornung, T.; Lanza, G.; Summa, J.; Schwarz, M.; Herrmann, H.-G.; Pohl, M.; Stommel, M. Effects of defects in series production of hybrid CFRP lightweight components—Detection and evaluation of quality critical characteristics. *Measurement* **2017**, *95*, 389–394. [CrossRef]
13. Zhang, J. Zhou, P.; Guan, C.; Liu, T.Q.; Kang, W.-H.; Feng, P.; Gao, Shuoqi. An ultra-lightweight CFRP beam-string structure. *Compos. Struct.* **2021**, *257*, 113149. [CrossRef]
14. Wong, D.W.Y.; Lin, L.; McGrail, P.T.; Peijs, T.; Hogg, P.J. Improved fracture toughness of carbon fibre/epoxy composite laminates using dissolvable thermoplastic fibres. *Compos. A* **2010**, *41*, 759–767. [CrossRef]
15. El-Tahan, M.; Galal, K.; Hoa, V.S. New thermoplastic CFRP bendable rebars for reinforcing structural concrete elements. *Compos. B Eng.* **2013**, *45*, 1207–1215. [CrossRef]
16. Miyano, Y.; Nakada, M.; Ichimura, J.; Hayakawa, E. Accelerated testing for long-term strength of innovative CFRP laminates for marine use. *Compos. B Eng.* **2008**, *39*, 5–12. [CrossRef]
17. Alam, P.; Robert, C.; Ó Brádaigh, C.M.Ó. Tidal turbine blade composites—A review on the effects of hygrothermal aging on the properties of CFRP. *Compos. B Eng.* **2018**, *149*, 248–259. [CrossRef]
18. Wang, W.-X.; Matsubara, T.; Hu, J.; Odahara, S.; Nagai, T.; Karasutani, T.; Ohya, Y. Experimental investigation into the influence of the flanged diffuser on the dynamic behavior of CFRP blade of a shrouded wind turbine. *Renew. Energy* **2015**, *78*, 386–397. [CrossRef]
19. Taniguchi, N.; Nishiwaki, T.; Kawada, H. Tensile strength of unidirectional CFRP laminate under high strain rate. *Adv. Compos. Mater.* **2007**, *16*, 167–180. [CrossRef]
20. Mechtcherine, V.; Michel, A.; Liebscher, M.; Schneider, K.; Großmann, C. Mineral-impregnated carbon fiber composites as novel reinforcement for concrete construction: Material and automation perspectives. *Autom. Constr.* **2020**, *110*, 103002. [CrossRef]
21. Firmo, J.P.; Roquette, M.G.; Correia, J.R.; Azevedo, A.S. Influence of elevated temperatures on epoxy adhesive used in CFRP strengthening systems for civil engineering applications. *Int. J. Adhes. Adhes.* **2019**, *93*, 102333. [CrossRef]
22. Danilov, A.I. Some aspects of CFRP steel structures reinforcement in civil engineering. *Procedia Eng.* **2016**, *153*, 124–130. [CrossRef]
23. Kim, K.W.; Lee, H.M.; An, J.H.; Chung, D.C.; An, K.-H.; Kim, B.-J. Recycling and characterization of carbon fibers from carbon fiber reinforced epoxy matrix composites by a novel super-heated-steam method. *J. Environ. Manag.* **2017**, *203*, 872–879. [CrossRef] [PubMed]
24. Vo Dong, P.A.; Azzaro-Pantel, C.; Cadene, A.-L. Economic and environmental assessment of recovery and disposal pathways for CFRP waste management. *Resour. Conserv. Recycl.* **2018**, *133*, 63–75. [CrossRef]

25. Liu, W.; Huang, H.; Zhu, L.; Liu, Z. Integrating carbon fiber reclamation and additive manufacturing for recycling CFRP waste. *Compos. B Eng.* **2021**, *215*, 108808. [CrossRef]
26. EUR. L.Ex. 21 October 2000. Available online: eur-lex.europa.eu/legal-content/EN/ALL/?uri=celex%3A32000L0053 (accessed on 18 September 2000).
27. Maaß, S. *Aircraft Recycling—A Literature Review*; Hamburg University of Applied Sciences: Hamburg, Germany, 2020; pp. 8–23.
28. Gopalraj, S.K. Impacts of Recycled Carbon Fibre and Glass Fibre as Sustainable Raw Materials for Thermosetting Composites. Ph.D. Thesis, Acta Universitatis Lappeenrantensis, Lappeenranta, Finland, 2022; pp. 15–17.
29. Meng, F.; McKechnie, J.; Turner, T.A.; Pickering, S.J. Energy and environmental assessment and reuse of fluidised bed recycled carbon fibres. *Compos. A* **2017**, *100*, 206–214. [CrossRef]
30. Khalil, Y.F. Comparative environmental and human health evaluations of thermolysis and solvolysis recycling technologies of carbon fiber reinforced polymer waste. *Waste Manag.* **2018**, *76*, 767–778. [CrossRef]
31. Tapper, R.J.; Longana, M.L.; Hamerton, I.; Potter, K.D. A closed-loop recycling process for discontinuous carbon fibre polyamide 6 composites. *Compos. B Eng.* **2019**, *179*, 107418. [CrossRef]
32. Thomas, C.; Borges, P.H.R.; Panzera, T.H.; Cimentada, A.; Lombillo, I. Epoxy composites containing CFRP powder wastes. *Compos. B Eng.* **2014**, *59*, 260–268. [CrossRef]
33. Obunai, K.; Fukuta, T.; Ozaki, K. Carbon fiber extraction from waste CFRP by microwave irradiation. *Compos. A* **2015**, *78*, 160–165. [CrossRef]
34. Zhao, Q.; An, L.; Li, C.; Zhang, L.; Jiang, J.; Li, Y. Environment-friendly recycling of CFRP composites via gentle solvent system at atmospheric pressure. *Compos. Sci. Technol.* **2022**, *224*, 109461. [CrossRef]
35. Hanaoka, T.; Arao, Y.; Kayaki, Y.; Kuwata, S.; Kubouchi, M. Analysis of nitric acid decomposition of epoxy resin network structures for chemical recycling. *Polym. Degrad. Stab.* **2021**, *186*, 109537. [CrossRef]
36. Hanaoka, T.; Ikematsu, H.; Takahashi, S.; Ito, N.; Ijuin, N.; Kawada, H.; Arao, Y.; Kubouchi, M. Recovery of carbon fiber from prepreg using nitric acid and evaluation of recycled CFRP. *Compos. B. Eng.* **2022**, *231*, 109560. [CrossRef]
37. Feraboli, P.; Kawakami, H.; Wade, B.; Gasco, F.; DeOto, L.; Masini, A. Recyclability and reutilization of carbon fiber fabric/epoxy composites. *J. Compos. Mater.* **2012**, *46*, 1459–1473. [CrossRef]
38. Xu, P.; Li, J.; Ding, J. Chemical recycling of carbon fibre/epoxy composites in a mixed solution of peroxide hydrogen and N,N-dimethylformamide. *Compos. Sci. Technol.* **2013**, *82*, 54–59. [CrossRef]
39. Okajima, I.; Sako, T. Recycling of carbon fiber-reinforced plastic using supercritical and subcritical fluids. *J. Mater. Cycles Waste Manag.* **2017**, *19*, 15–20. [CrossRef]
40. Okajima, I.; Hiramatsu, M.; Shimamura, Y.; Awaya, T.; Sako, T. Chemical recycling of carbon fiber reinforced plastic using supercritical methanol. *J. Supercrit. Fluids* **2014**, *91*, 68–76. [CrossRef]
41. Okajima, I.; Sako, T. Recycling fiber-reinforced plastic using supercritical acetone. *Polym. Degrad. Stab.* **2019**, *163*, 1–6. [CrossRef]
42. Meyer, L.O.; Schulte, K.; Grove-Nielsen, E. CFRP-recycling following a pyrolysis route: Process optimization and potentials. *J. Compos. Mater.* **2009**, *43*, 1121–1132. [CrossRef]
43. Deng, Z.; Xu, L.; Zhang, L.; Peng, J.; Guo, S.; Liu, J.; Koppala, S. Recycling of Carbon Fibers from CFRP Waste by Microwave Thermolysis. *Processes* **2019**, *7*, 207. [CrossRef]
44. Deng, Z.; Yue, J.; Huang, Z. Solvothermal degradation and reuse of carbon fiber reinforced boron phenolic resin composites. *Compos. B. Eng.* **2021**, *221*, 109011. [CrossRef]
45. Huang, Z.; Deng, Z.; Dong, C.; Fan, C.; Ren, Y. A close-loop recycling process for carbon fiber reinforced vinyl ester resin composite. *Chem. Eng. J.* **2022**, *446*, 137254. [CrossRef]

polymers

MDPI

Article

Fiber Orientation and Strain Rate-Dependent Tensile and Compressive Behavior of Injection Molded Polyamide-6 Reinforced with 20% Short Carbon Fiber

Joonhee Lee , Hyungyil Lee and Naksoo Kim *

Department of Mechanical Engineering, Sogang University, Seoul 04107, Republic of Korea
* Correspondence: nskim@sogang.ac.kr; Tel.: +82-010-2703-8635

Abstract: As the interest in short-fiber reinforced polymer (SFRP) composites manufactured by injection molding increases, predicting the failure of SFRP structures becomes important. This study aims to systemize the prediction of failure of SFRP through mechanical property evaluation considering the anisotropy and strain rate dependency. To characterize the mechanical properties of polyamide-6 reinforced with carbon fiber of a weight fraction of 20% (PA6-20CF), tensile and compressive experiments were conducted with different load-applying directions and strain rates. Additionally, the results were discussed in detail by SEM image analysis of the fracture faces of the specimen. FE simulations based on the experimental condition were constructed, and the numerical model coefficients were derived through comparison with experimental results. The coefficients obtained were verified by bending tests of the specimens manufactured from composite cross members fabricated by injection molding. Predicting under static and high strain rate conditions, small errors of about 9.6% and 9.3% were shown, respectively. As a result, it proves that explained procedures allow for better failure prediction and for contribution to the systematization of structural design.

Keywords: short fiber-reinforced composites; injection molding; thermoplastic material characterization; numerical simulation; complex geometry

Citation: Lee, J.; Lee, H.; Kim, N. Fiber Orientation and Strain Rate-Dependent Tensile and Compressive Behavior of Injection Molded Polyamide-6 Reinforced with 20% Short Carbon Fiber. *Polymers* **2023**, *15*, 738. https://doi.org/10.3390/polym15030738

Academic Editors: R.A. Ilyas, S.M. Sapuan, Emin Bayraktar, Shukur Abu Hassan, Nabil Hayeemasae and Khubab Shaker

Received: 1 December 2022
Revised: 25 January 2023
Accepted: 29 January 2023
Published: 31 January 2023

1. Introduction

Short fiber reinforced thermoplastic polymers are largely utilized in the industry due to their versatility, recyclability, and low cost and cycle time related to the injection molding process [1]. For these reasons, they have been largely employed in the automotive industry, whose fuel efficiency and emission reduction regulations have been strengthening in recent years. Among the recent regulations, the European one concerning CO2 emission poses a serious challenge to emission reduction (< 95 g/km within 2020). Thus, several automotive manufacturers are researching the direction of replacing aluminum and steel components with reinforced thermoplastic ones because of the considerable weight reduction [2]. In order to properly apply fiber-reinforced thermo-plastic materials to high-performance structural components, the local material properties resulting from the material flow during the thermoplastic injection molding (TIM) process should be properly accounted for. Hence, in the literature, several authors focused their attention on estimating the variation of the mechanical properties of short fiber reinforced (SFR) material, as summarized hereafter.

The effect of the anisotropy developing in injection-molded short glass fiber reinforced polybutylene terephthalate (PBT) polymer was investigated by means of cutting several specimens along 0°, 45°, and 90°, in comparison to the injection direction [3]. One of the most interesting aspects is related to the shell-core structure developing in the material during the injection molding process. This fact results in the fiber in the shell region, close to the surfaces of the specimen, being strongly aligned with the polymer flow direction, whereas, in the core of the specimen, they are transversally aligned with the polymer flow. This aspect was also investigated by means of numerical simulations [4–8]. In addition, much effort has also been focused on investigating the influence of the fiber length on the mechanical properties of SFR thermoplastics. Fu and Lauke presented an analytical model for the estimation of the probabilistic fiber length and orientation distribution in injection molded components and validated it by comparing its prediction to experimental results [9]. In their conclusions, they highlighted how the strength of the composite increases with the fiber length as well as with the reduction of the critical fiber length or, in other words, with the increase of the interfacial adhesion between fibers and matrix polymer. Similar conclusions have been also drawn by other researchers [10,11].

Many studies related to the experimental characterization of the mechanical behavior of SFRP under various loading conditions have also been performed [12–27]. Experiments were conducted by cutting the specimens with various angles from the plates made of glass fiber-reinforced polyamide-6 (PA6) and polyphenylene sulfide (PPS). Changes in mechanical properties according to glass fiber content and fiber orientation angle can be confirmed. Uniaxial tensile and compressive tests for $\theta = 0°,90°$ specimens were conducted, and the experimental results were verified with the simulation results and application for complex geometry. Similarly, the stress triaxiality-dependent failure behavior of the polyamide-6 with 20% carbon fiber (PA6-20CF) was investigated [22]. In addition, many studies have been conducted to investigate the mechanical behavior and failure mechanism of polymer composites reinforced with glass fiber and carbon fiber under compressive load [28–32]. The progress of the kink-band and the fracture mode were especially described in detail in terms of fractography [28,29]. Additionally, there were experimental results that confirmed that the mechanical behavior of fiber-reinforced composites can be different depending on the strain rate [6,24,28,29,33,34]. The deviation in stress–strain response was influenced by the interfacial properties because of the strain rate sensitivity of the fiber and matrix. In addition, multi-scale modeling and simulation for characterizing the material mechanical behavior have been utilized [8,35–37]. As mentioned earlier, in the cases of composites, since consideration of anisotropy greatly affects the accurate prediction of mechanical properties, attempts to numerically model the fiber orientation through experiments and simulation have been made steadily [38–43].

For the mechanical properties of polyamide-6 reinforced with carbon fiber studied in this paper, Karsli and Aytac investigated the thermal and morphological properties [44]. They focused on the influences of fiber length and fiber content. Additionally, the failures of fibers and matrix were analyzed through SEM image analysis of fracture surfaces of the specimens. In addition, Zhang investigated the strain rate and temperature-dependent mechanical properties of carbon fiber-reinforced thermoplastic composites fabricated by hot press molding [45]. This research presented that the modulus and strength of continuous carbon fiber reinforced thermoplastic (CCFRT) increase with a higher strain rate. Additionally, the relationship is non-linear, and it can be confirmed that it is due to the increase in the strength of the thermoplastic matrix and the fiber interface. In this study, changes in mechanical properties of PA6-20CF manufactured through injection molding according to fiber orientation distribution and strain rate are identified. Additionally, by appropriately considering local tensile-compressive mechanical properties caused by injection molding, contributed to the systematization of the structural design of complex structures. The research is articulated according to the following goals. First, experimental characterization of PA6-20CF through specimens of different thicknesses and angles fabricated by injection molding. Second, verification by injection molding simulation. Last, consideration of local mechanical properties and verification of tensile-compressive mechanical properties through the mapping process. Therefore, this paper presents descriptions of the material and the geometry of the specimen used in the experiment to obtain tensile-compressive mechanical properties according to the fiber orientation. Then, the experimental results are discussed and compared in detail with the analysis of the failure mechanism observed in SEM images. Based on the anisotropic elastoplastic mechanical behavior of the material, the numerical model is explained, and simulations were conducted. Considering the strain rate-dependent tensile-compressive mechanical properties, the numerical model coefficients according to the fiber orientation were derived by calibration of the load–displacement results of the experiment and simulation. As a final step, mechanical properties are input based on the obtained coefficients, and a bending test under static and high strain rate conditions was conducted which was machined from a cross member manufactured by injection molding of PA6-20CF (each jig speed: 1 mm/min, 20 mm/min). It allowed evaluation of how much the previously described mechanism can improve the failure prediction accuracy in the analysis of complex geometries. Under the static and high strain rate conditions, failure was predicted with small errors of about 9.6% and 9.3%, respectively. Therefore, the validity of the derived sets of properties was confirmed, suggesting that these research procedures can improve the reliability of the structural analysis of composite materials with complex shapes under complex stress states and contribute to design systematization.

2. Materials and Methods

2.1. PA6-20CF Plane Specimen

In this study, PA6-20CF material reinforced with KN111 polyamide-6 (Kolon Plastics Inc., Gimcheon, Republic of Korea) and Torayca T700S (Toray Carbon Fibers America Inc., Tacoma, USA.) short carbon fiber with a weight fraction of 20% was investigated. PA6-20CF plate specimens with a thickness of t = 1.8 mm and 4.0 mm were manufactured through the injection molding of the PA6-20CF pellet, and the specifications are shown in Figure 1a, b. Before the injection process of the plate specimen, the pellet was dried at 120 °C for 6 h to prevent deterioration of properties due to hygroscopicity and was injected under the following process conditions: the mold surface temperature of 85 °C, melt temperature of 285 °C, injection time with 2 sec, and cooling time with 20 sec. After that, specimens were manufactured within a small processing error range of 0.25 mm through wire cutting of the plates. Tensile and compression specimens were prepared for three orientation angles θ = 0°, 45°, and 90°, which are defined as the angle between flow direction with the direction of applied load during tests (see Figure 1b,c).

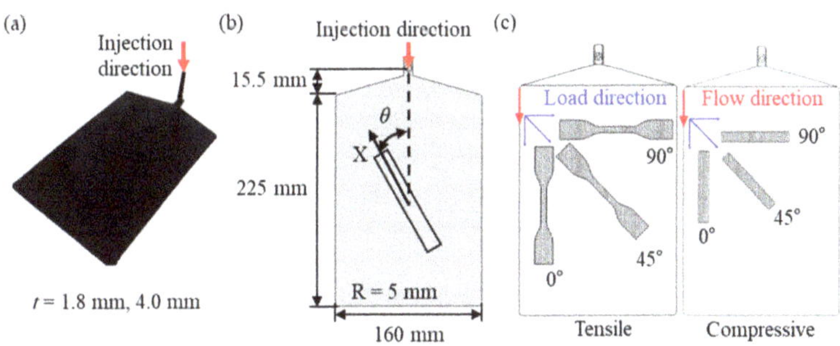

Figure 1. (**a**) PA6-20CF plate manufactured by injection molding, and injection direction, (**b**) Schematic illustration of different fiber orientation (θ) specimens obtained from the injection-molded plate (X = loading axis), (**c**) fiber alignment in specimen with fiber orientation (θ) = 0°, 45°, 90°.

In addition to that, the machining locations of the specimen from the plate were decided through finite volume method (FVM) analysis, and a detailed description will be presented in Section 2.2. After wire cutting, all the machined specimens were stored in a sealed plastic bag with silica gel to prevent moisture absorption. To prevent deviations in physical properties due to the hygroscopicity of all specimens, prior to the test, the HKOD-50 electric furnace (heat flow: 12 m^3/h) was used for drying at 120 °C for 6 h, and then the temperature was lowered for 1 h at room temperature. After that, the tests were conducted. The tensile tests were carried out according to ASTM-D638-02a-Type IV standard and the machined positions on the plate are the same at the thickness t = 1.8 and 4.0 mm of the plate as shown in Figure 2a [46]. An Epsilon 3542 extensometer with a gauge length of 25 mm was used. Additionally, the tests were performed with different jig speeds of 1.5, 15, and 150 mm/min for considering strain rate dependency (strain rates of 0.001, 0.01, and 0.1 s^{-1}, respectively).

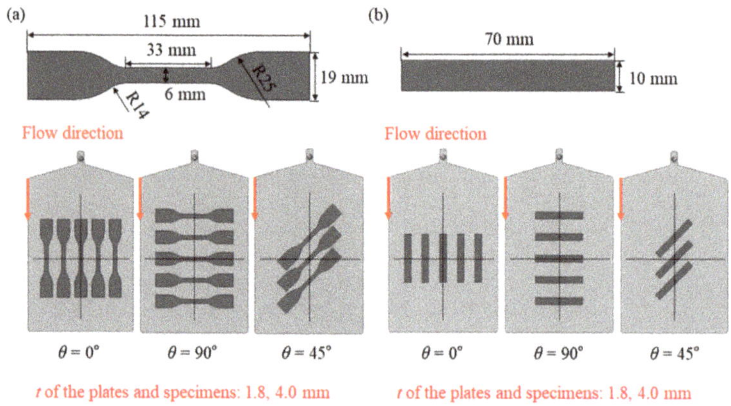

Figure 2. (**a**) ASTM-D 638-02a-Type IV specimen dimensions and cutting location of the tensile specimens from the plate, (**b**) ASTM-D3410M-16 specimen dimensions and cutting location of the compressive specimens from the plate.

The compressive tests were carried out according to ASTM D3341M-16 standard and the machined positions on the plate are the same at the thickness t = 1.8 and 4.0 mm of the plate as shown in Figure 2b [47]. The tests were performed with different jig speeds of 0.6, 6, and 60 mm/min for considering strain rate dependency (strain rates of 0.001, 0.01, and 0.1 s^{-1}, respectively). To prevent buckling that occurs in the specimen when a compressive load is applied, the specimen has a relatively short gauge length compared to the grip part. Additionally, it was confirmed that only acceptable failure characteristic results within the ASTM D3341M-16 standard were derived. For repeatability verification, $\theta = 0°, 90°$: 5 times, and $\theta = 45°$: 3 times or more were performed according to each test condition (fiber orientation and strain rate) both tensile and compressive.

2.2. Numerical Implementation

In the research, three different commercial software were utilized. Autodesk Moldflow Insight (AMI) 2019 (San Rafael, CA, USA) was employed for the estimation of the fiber orientation distributions in the injection-molded plates and specimens. The Autodesk Helius 2019 (San Rafael, CA, USA) allowed for the mapping of the local mechanical properties, as a result of the fiber length and orientation distributions. The mapping was carried out from the results thermoplastic injection molding (TIM) process simulation into the input file of the ABAQUS/Standard structural simulation. Finally, ABAQUS 2017 (Dassault Systemes, Velizy-Villacoublay, France) was employed for the estimation of local stress and strain distribution on specimens.

2.2.1. Thermoplastic Injection Molding Simulation

The thermoplastic injection molding simulation was developed by means of the commercial software Autodesk Moldflow Insight 2019. To properly locate the specimens on the plate used for machining the tensile and compressive specimens, a finite volume method simulation of injection molding of the plate was conducted as shown in Figure 3. By this action, cross-sections at the center of the specimens could have a consistent fiber orientation distribution.

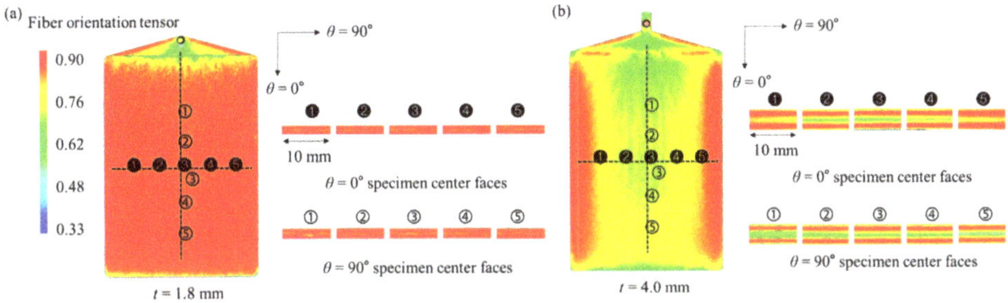

Figure 3. Results of the fiber orientation distribution from cut faces of the specimens from the plate of thickness t = (**a**) 1.8 mm, (**b**) 4.0 mm.

In the simulation, the same process parameters utilized in the real injection molding process have been utilized, as summarized previously. The fiber orientation distribution was derived through the ARD-RSC model presented in Equation (1) [48,49]. The ARD-RSC model considered the deformation rate tensor ($\dot{\gamma}$) as a polynomial rather than a constant in the original Folgar–Tucker formulation [50]. Thus, the overestimation of the rate of change of the fourth-order fiber orientation tensor was adjusted. As a result, it was improved to better describe the interactions between fibers with a length of 1 mm or longer.

$$\begin{aligned}\frac{\partial a_{ij}}{\partial t} &= -\frac{1}{2}\left(w_{ik}a_{kj} - a_{ik}w_{kj}\right) + \frac{1}{2}\lambda\left(\dot{\gamma}_{ik}a_{kj} + a_{ik}\dot{\gamma}_{kj} - 2\dot{\gamma}_{kl}\left[a_{ijkl} + (1-k)\left(L_{ijkl} - M_{ijkl}\cdot a_{mnkl}\right)\right]\right) \\ &+\dot{\gamma}\left[2\left(c_{ij} - (1-k)c_{kl}M_{ijkl}\right) - 2k\cdot c_{kk}a_{ij} - 5\left(c_{ik}a_{kj} + a_{ik}c_{kj}\right) + 10c_{kj}\left(a_{ijkl} + (1-k)\left(L_{ijkl} - M_{ijkl}\cdot a_{mnkl}\right)\right)\right]\end{aligned} \quad (1)$$

In Equation (1), the constant k is a dynamic scalar factor describing the ability to realign the fibers along the mold after injection. A low k means that the orientation ability of the fiber itself is low and, therefore, the degree of agreement between the fiber orientation distribution and the polymer resin flow is low. Conversely, when the value of k is large, the fibers are realigned to a high degree of agreement with the direction of resin flow in the mold between injections. In this research for the PA6-20CF, the orientation dynamic scalar factor k = 0.5 of ARD-RSC was utilized, which was utilized and verified through the comparison of fiber orientation distribution confirmed by SEM analysis and simulation in previous studies [22,23]. In the TIM simulation, 486,195 and 1,285,994 tetrahedral elements were utilized with t = 1.8 mm plates and for t = 4.0 mm plates, respectively. The mechanical properties of the PA6-20CF material were inputted in the Moldflow numerical simulation and the constants for the Ramberg–Osgood flow stress model, Equation (2), were calculated and are reported in the next section.

$$\sigma = E^{1/n}(K)^{(n-1)/n}\left(\varepsilon_{p,eff}\right)^{1/n} \quad (2)$$

This procedure allowed mapping the mechanical properties on the injection molding simulation mesh during the filling phase, sparing time during the mapping operation. Along with the flow stress model of Equation (2), a modified Hill '48 yield function in Equation (3) was utilized. It allowed considering the variation of the mechanical properties, for each element of the mesh, caused by the local fiber length and orientation distributions.

$$\begin{aligned}\sigma_{eff} = \Big[\tfrac{1}{2}\Big\{&(\alpha\sigma_{11} - \beta\sigma_{22})^2 + (\beta\sigma_{22} - \beta\sigma_{33})^2 + (\beta\sigma_{33} - \alpha\sigma_{11})^2 \\ &+6\left[(\sigma_{12})^2 + (\sigma_{23})^2 + (\sigma_{31})^2\right]\Big\}\Big]^{1/2}\end{aligned} \quad (3)$$

The α and β parameters in Equation (3) account for the improvement of the polymer matrix mechanical properties due to the presence of the carbon fibers, along the direction parallel and transversal to the local flow direction, respectively, and were calculated by Equation (4).

$$\begin{aligned} \alpha(\lambda_I) &= \theta + \left[\frac{(\alpha_m-\theta)}{(\lambda_{m,I}-1/2)}\right](\lambda_I - 1/2) \\ \beta(\lambda_I) &= \theta + \left[\frac{(\beta_m-\theta)}{(\lambda_{m,I}-1/2)}\right](\lambda_I - 1/2) \end{aligned} \tag{4}$$

For each element of the mesh, α and β are calculated based on λ_I, α_m, β_m and $\lambda_{m,I}$ constants. The λ_I defines the first eigenvalue of the fiber orientation tensor of the considered element, and the $\lambda_{m,I}$ define the first eigenvalue of the fiber orientation tensor in strong fiber alignment along the flow direction region.

2.2.2. Mapping Procedure

The mapping procedure was carried out in the Autodesk Helius software and allowed us to map the local mechanical properties, as a result of the TIM process, into the ABAQUS numerical simulation. This procedure is an important task in determining how the original fiber orientation distribution is reflected in the specimen when the specimen was machined at each location on the plate. After the mapping operation, the ABAQUS input file containing a specific set of mechanical properties, for each element of the mesh, was created and ought to be run from the Autodesk Helius command shell.

The Ramberg–Osgood model constants are based on isotropic behavior, and the respective σ-ε curves for tension and compression were considered. Once the stress state is determined, the program inputs the mechanical properties suitable for the stress state of each element. At each iteration step, both tensile and compressive solutions are calculated using a set of tensile and compressive mechanical properties. Then, the stress state is determined according to two verification procedures as shown in Figure 4, and a corresponding appropriate solution is used for each step.

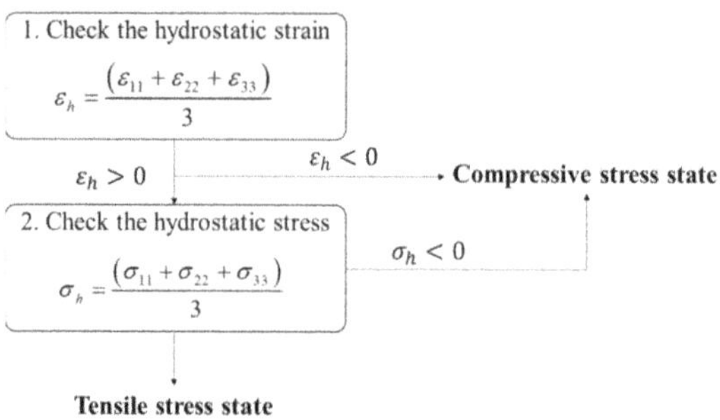

Figure 4. Schematic diagram of the procedure how to define the stress state.

2.2.3. Structural Simulation

The structural numerical simulations were constructed into ABAQUS/Standard utilizing the C3D10—quadratic hexahedral elements. Both tension and compression specimen geometries were modeled identically to the specifications of the specimen used in the test as previously shown in Figure 2. The region connected to the jig was given a rigid body condition, and in the case of the tensile specimen, displacement at the gauge distance node was used as output data. For t = 1.8 mm specimens, 9216, 22,400 elements were used for tensile and compressive specimens. Additionally, for t = 4.0 mm specimens, 18,432, 40,800 elements were used, respectively. The results of the ABAQUS numerical simulations are reported in detail in Sections 3.3 and 3.4 of the paper.

3. Results and Discussions

3.1. Experimental Results

In this section, the results of tensile and compressive tests performed at room temperature for specimen thicknesses t = 1.8, 4.0 mm, fiber orientation angles θ = 0°, 45°, 90°, and strain rates 0.001, 0.01, and 0.1 s^{-1} are presented. Furthermore, the mechanical properties and the behavior of the anisotropic elastoplastic material derived for each test condition are discussed. Afterward, it was analyzed based on the fractographic study of the SEM image, which is discussed in detail in Section 3.2.

3.1.1. Tensile Tests

Engineering stress-engineering strain curves obtained as a result of the tensile tests are shown in Figures 5 and 6. Additionally, the anisotropic elastoplastic material behavior (with hardening) of short fiber reinforced polymer (SFRP) according to the strain rate can be observed. As the fiber orientation angle θ decreased and strain rate increased at t = 1.8 mm and 4.0 mm, ultimate tensile strength σ_{UTS}, elastic modulus, and yield strength tended to increase. On the other hand, ultimate tensile strain ε_{UTS} tended to decrease as the fiber orientation angle θ decreased and the strain rate increased. Based on this, it shows that the material became brittle as the strain rate increased. When the thickness t of the plate is different (t = 1.8 and 4.0 mm), different material behavior is shown because of the deviation in the degree of anisotropy. For θ = 0°, the strength is greater at t = 1.8 mm, but for θ = 45° and 90°, the strength is greater at t = 4.0 mm. This is because the thinner the plate, the higher the shell ratio in the shell-core structure (higher average fiber orientation index), and the better the fibers are aligned in the main flow direction (Figure 3). Additionally, for the same reason, when t = 1.8 mm, it shows big differences in mechanical properties between θ = 0° and θ = 45°, 90°, but small at t = 4.0 mm.

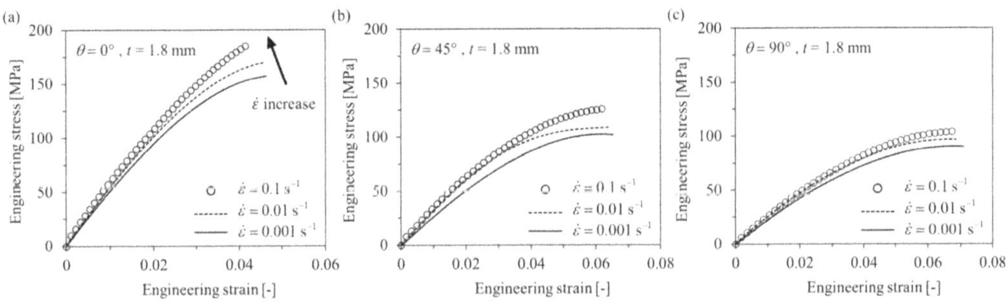

Figure 5. Tensile engineering stress–engineering strain curves of t = 1.8 mm specimens with different strain rates. (**a**) θ = 0°, (**b**) θ = 45°, (**c**) θ = 90°.

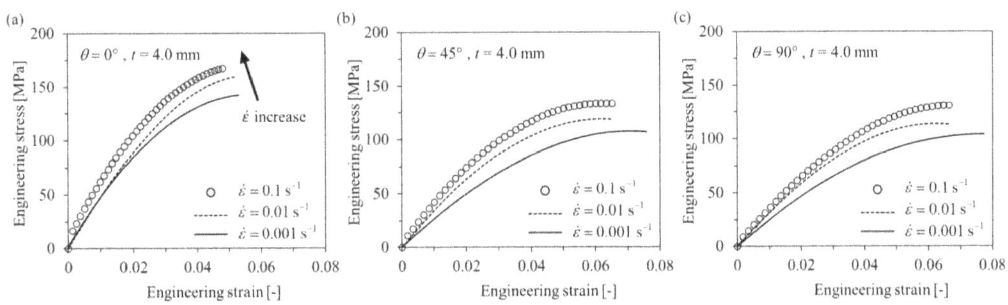

Figure 6. Tensile engineering stress–engineering strain curves of t = 4.0 mm specimens with different strain rates. (**a**) θ = 0°, (**b**) θ = 45°, (**c**) θ = 90°.

3.1.2. Compressive Tests

Engineering stress-engineering strain curves obtained as a result of the compressive tests are shown below in Figures 7 and 8. Similar to the tensile test results, the behavior of the anisotropic elastoplastic material of the short fiber reinforced polymer according to the strain rate could be observed, and different results were obtained when the thickness t was changed. At t = 1.8 mm, the strain rate sensitivity of $\theta = 0°$ is greater than that of $\theta = 45°, 90°$. At both t = 1.8 mm and 4.0 mm, the ultimate compressive strength σ_{UCS}, elastic modulus, and yield strength tended to increase as the strain rate increased. On the other hand, the ultimate compressive strain ε_{UCS} tended to decrease as the strain rate increased.

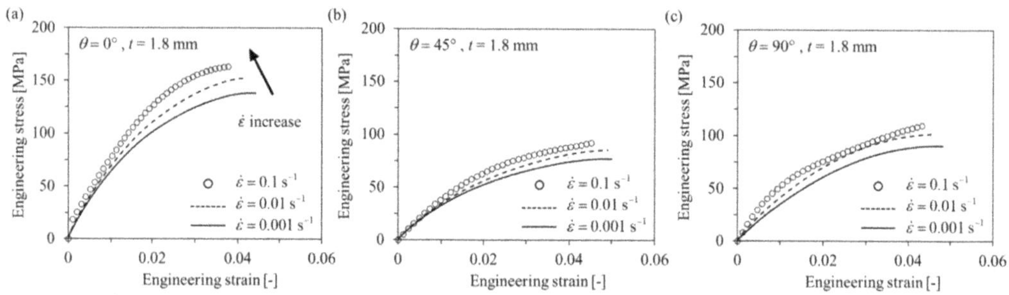

Figure 7. Compressive engineering stress–engineering strain curves of t = 1.8 mm specimens with different strain rates. (**a**) $\theta = 0°$, (**b**) $\theta = 45°$, (**c**) $\theta = 90°$.

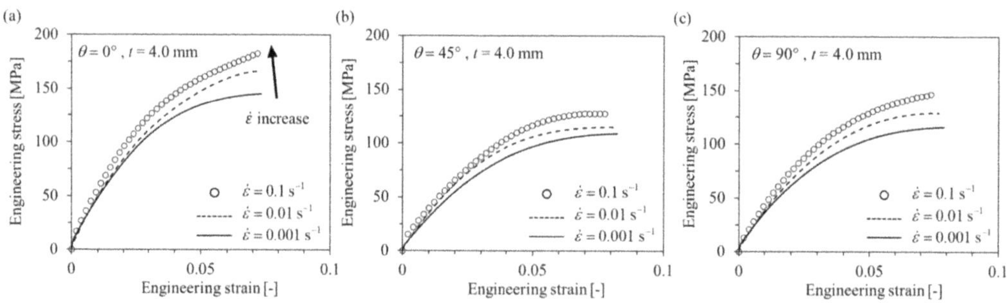

Figure 8. Compressive engineering stress–engineering strain curves of t = 4.0 mm specimens with different strain rates. (**a**) $\theta = 0°$, (**b**) $\theta = 45°$, (**c**) $\theta = 90°$.

Comparing the results of the compressive test with the tensile test, it can be observed that the short fiber-reinforced polymer exhibited completely different mechanical behavior under tensile and compressive loading conditions. First, under the compressive load condition at the same strain rate, it showed that the σ_{UCS} at $\theta = 45°$ was the lowest and the ε_{UCS} was the biggest. In addition, it can be observed that the difference in σ_{UTS} between t = 1.8 mm and t = 4.0 mm was relatively smaller than the difference in σ_{UCS}. Furthermore, in compression, when t = 4.0 mm, it showed that the strength was higher than that of t = 1.8 mm, at all θ. From this, it can be concluded that the mechanical properties of the short-fiber reinforced polymer are relatively strongly influenced by the matrix in the compressive state [31,32,45].

3.2. SEM Image Analysis

In this section, the experimental results explained in Section 3.1 are analyzed and verified through SEM (Scanning Electron Microscope) image analysis of the fracture surface of the specimen. For SEM analysis, a JSM-7100F FE-SEM machine (JEOL Ltd., Tokyo, Japan) with a resolution of 3.0 mm (at 15.0 kV) was utilized. The specimen fracture faces were coated with Pt, and SEM imaging was taken perpendicular to the fracture surface. The fiber orientation at the fracture surface and failure characteristics of the matrix and fiber were analyzed.

3.2.1. Tensile Specimens

Through SEM analysis of the tensile specimen, it can be confirmed that the alignments of the fibers observed at the fracture surface were different depending on the fiber orientation θ (see Figure 9). It can be observed that the fibers of the $\theta = 0°$ specimen are cut perpendicular to the fracture plane, and in the case of the $\theta = 45°$ and 90° specimens, the fibers are aligned parallel to the fracture plane. In addition to that, the length of the fibers observed at the fracture surface of the $\theta = 90°$ specimen is the longest, and the length of $\theta = 0°$ is the shortest [3,21].

Figure 9. Fracture faces observed from tensile specimens which were tested with a strain rate of 0.001 s^{-1} (magnification: ×35, scale: 100 µm).

In the case of $\theta = 0°$, since the fracture plane is perpendicular to the fiber direction, it can be expected that the fiber fracture occurs predominantly. Various microscopic damage mechanisms according to the fiber orientation θ under tensile load can be observed. It is possible to observe fiber failure of shortened fibers after breakage and fiber pull-outs in which fibers were pulled out from the matrix in Figure 10a. Additionally, as shown in Figure 10b, debonding occurred as traces in which fiber separated from the matrix [51]. The fracture face of the fiber in Figure 10c is consistent with the fracture characteristics of the carbon fiber under the tensile load studied previously [52–54].

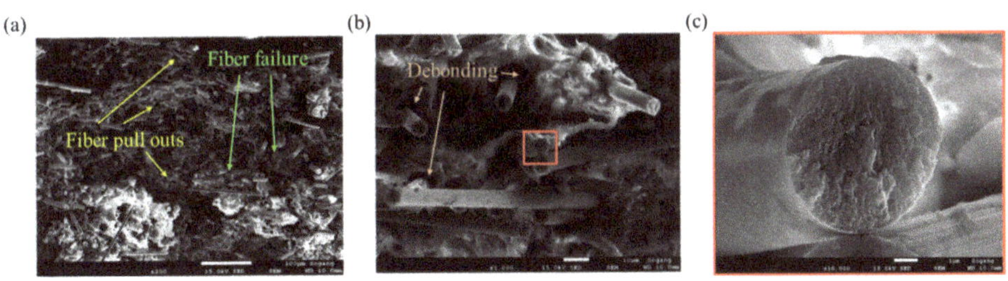

Figure 10. Fracture mechanisms observed for $\theta = 0°$ tensile specimen: (**a**) Fiber failure and fiber pull-outs (magnification: ×200, scale: 100 µm), (**b**) fiber-matrix debonding (magnification: ×1000, scale: 10 µm), (**c**) fracture face of the fiber after the tensile failure (magnification: ×10,000, scale: 1 µm).

At the failure surface of the $\theta = 90°$ specimen, completely different failure characteristics were observed, as shown in Figure 11. The average length of the fibers was long. Additionally, the fiber tracks left as traces after the fibers were separated from the matrix and the hackles distributed along them can be observed (Figure 11a,b). Moreover, the matrix that was bonded to the fibers was weakened and a smeared matrix appeared. Through the deformation of the matrix, the ductile behavior of $\theta = 90°$ can be explained. In the case of $\theta = 45°$, the behavior is intermediate between $\theta = 0°$ and $\theta = 90°$ (Figure 12). Although the average fiber length observed was long, a large number of fibers in which breakage has occurred can also be found. Therefore, it can be confirmed that the failure characteristics that can be seen at $\theta = 0°$ and $90°$ coexist. The fracture face of the fiber was the roughest at $\theta = 0°$, and conversely, the smoothest at $\theta = 90°$.

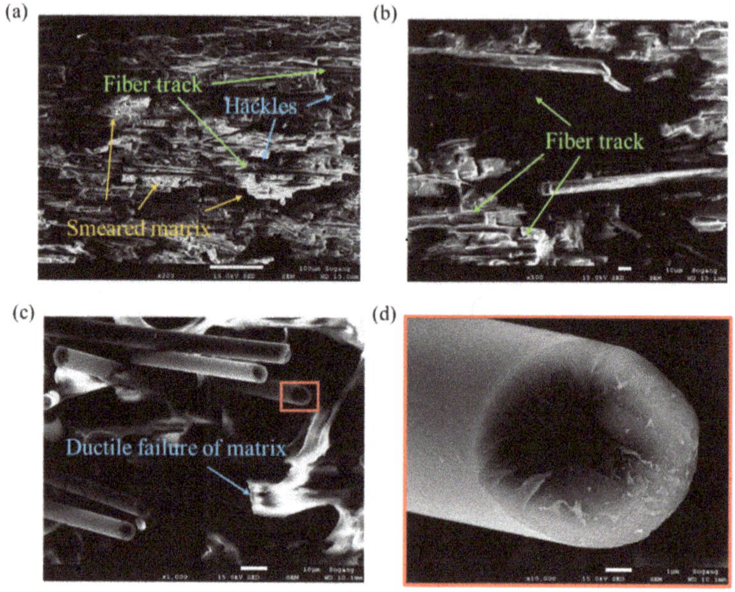

Figure 11. Fracture mechanisms observed for $\theta = 90°$ tensile specimen: (**a**) Fiber tracks, hackles, and smeared matrix (magnification: ×200, scale: 100 μm), (**b**) fiber tracks and broken long fiber (magnification: ×500, scale: 10 μm), (**c**) ductile failure of the matrix (magnification: ×1000, scale: 10 μm), (**d**) smooth fracture face of the fiber (magnification: ×10,000, scale: 1 μm).

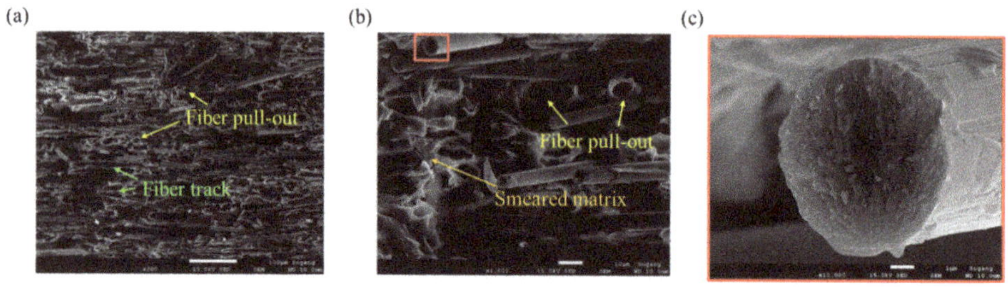

Figure 12. Fracture mechanism observed for $\theta = 45°$ tensile specimen, (**a**) fiber pull-outs and fiber tracks (magnification: ×200, scale: 100 μm), (**b**) smeared matrix and fiber pull-outs (magnification: ×1000, scale: 10 μm), (**c**) fracture surface of the fiber (magnification: ×10,000, scale: 1 μm).

3.2.2. Compressive Specimens

In compressive specimen SEM images in Figure 13, the direction of fibers aligned at each fiber orientation θ show similar results with tensile (see Figure 9). However, the average length of fibers observed at the fracture surface did not show a significant difference between $\theta = 45°$ and $\theta = 90°$. Moreover, the deformed matrix appears brighter at $\theta = 45°$ than at $\theta = 90°$. Through this, more deformation of the matrix can be grasped at $\theta = 45°$.

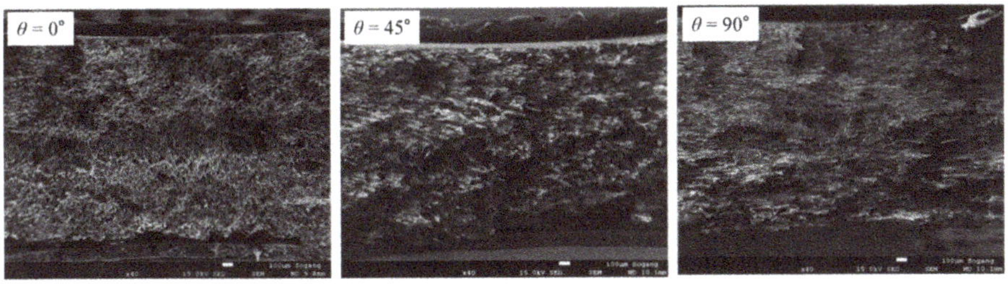

Figure 13. Fracture surfaces observed for compressive specimens with a strain rate of 0.001 s^{-1} (magnification: ×40, scale: 100 μm).

For θ = 0° compressive specimen, the failure characteristics were similar to those of the tensile specimen, but in compression, fiber breakage was promoted by the buckling of the fiber, and fibrillation induced by strong shear ductile damage occurred, as shown in Figure 14a,b. Additionally, fiber-matrix debonding can be observed in Figure 14b,c [51]. The fracture surface of the fiber in Figure 15c is comparable to the fracture characteristics of the carbon fiber under flexural load studied previously [52,53].

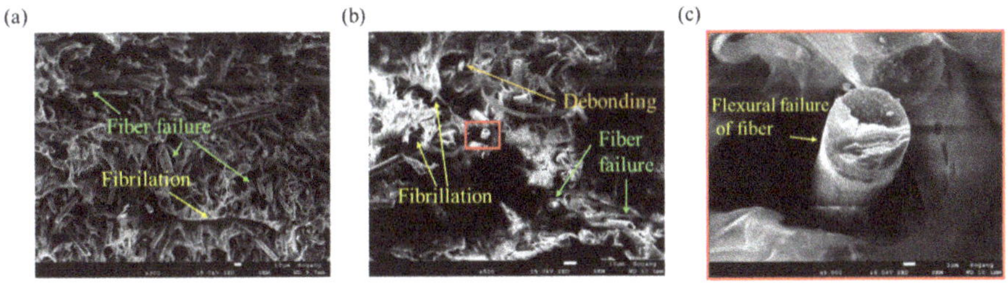

Figure 14. Fracture mechanism observed for θ = 0° compressive specimen, (**a**) fiber failure and fibrillation (magnification: ×300, scale: 10 μm), (**b**) fiber-matrix debonding (magnification: ×500, scale: 10 μm), (**c**) fracture face of the fiber after flexural failure (magnification: ×5000, scale: 1 μm).

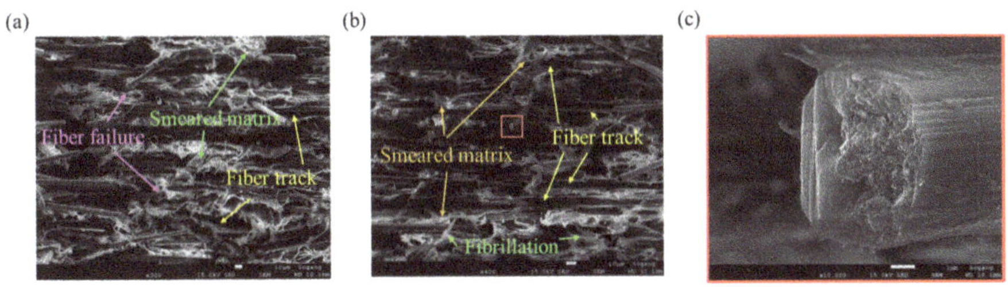

Figure 15. Fracture mechanisms observed for θ = 90° compressive specimen: (**a**) fiber failure, smeared matrix, and fiber tracks (magnification: ×300, scale: 10 μm), (**b**) fiber tracks and fibrillation (magnification: ×400, scale: 10 μm), (**c**) fracture face of the fiber after flexural failure (magnification: ×10,000, scale: 1 μm).

Both in θ = 90° and 45°, fiber tracks, the smeared matrix near the fiber, and fibrillation occurred frequently as shown in Figures 15 and 16. In compression, fibers commonly showed flexural failure due to buckling, and ductile shear damage was dominant [28–32,52–54]. Fiber failure was observed more frequently at θ = 90° than at θ = 45°, conversely, there was more deformation on the matrix in the case of θ = 45°. These facts support that the $\sigma_{UCS,\,90^\circ}$, is bigger than $\sigma_{UCS,\,45^\circ}$.

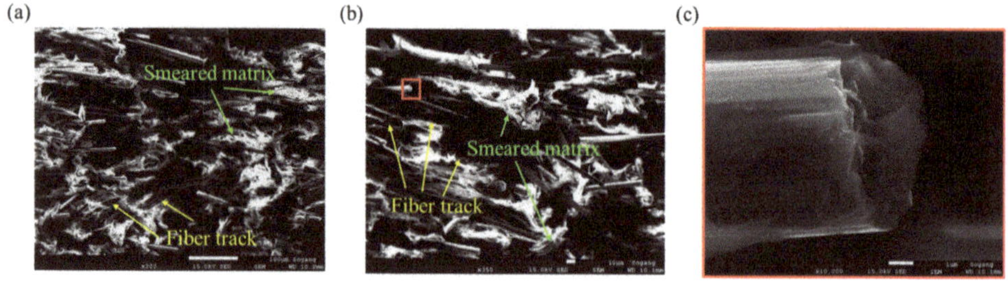

Figure 16. Fracture mechanism observed for $\theta = 45°$ compressive specimen: (**a**) smeared matrix and fiber tracks (magnification: $\times 200$, scale: 100 μm), (**b**) smeared matrix and fiber tracks (magnification: $\times 350$, scale: 10 μm), (**c**) fracture surface of fiber due to flexural failure (magnification: $\times 10{,}000$, scale: 1 μm).

3.3. Numerical Model Verification

To prove the reliability of the mapping procedure and the anisotropic elastoplastic behavior of PA6-20CF, EXP-FE load–displacement curves for tensile and compression specimens were compared. The same Ramberg–Osgood coefficients were applied to both $t = 1.8$ and 4.0 mm at the same load condition and strain rate. Additionally, the comparison result of the load–displacement curves of the EXP-FE under each test condition (load condition and strain rate) is shown in Figures 17 and 18. The input coefficient sets were as shown in Table 1. In both tension and compression, as the strain rate increased, K, n decreased, α, β, and the elastic modulus of matrix and fiber tended to increase. Moreover, in the case of compression, the change in the elastic modulus was more significant. As mentioned earlier, different fiber orientation distribution causes a difference in mechanical properties, and it is important to consider this properly. By the mapping procedure, local mechanical properties as a result of the injection molding process can be input for each element in FE simulation, so the reliability of the results can be improved. The possibility to consider the thickness-dependent (fiber orientation-dependent) mechanical properties with one set of model coefficients is a huge advantage for the simulation of complex geometries. The coefficients derived from the EXP-FE comparison were finally verified by being used for structural analysis of structures with complex shapes under a complex stress state in the next section—Section 3.4.

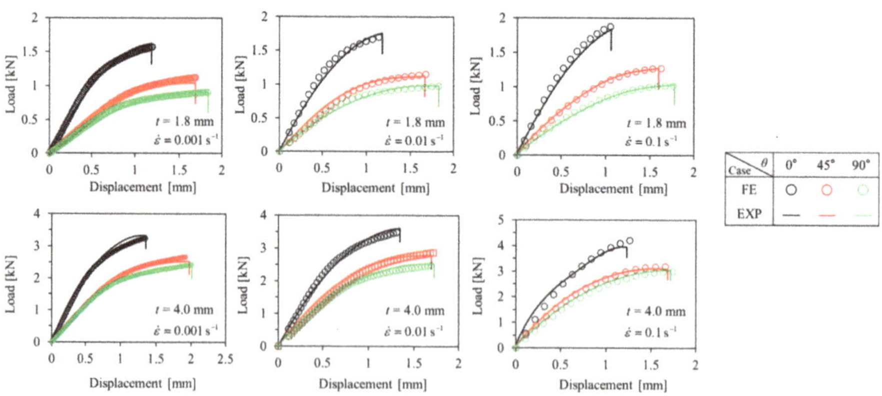

Figure 17. Comparison between experimental and numerical results for ASTM-D638 tensile specimens.

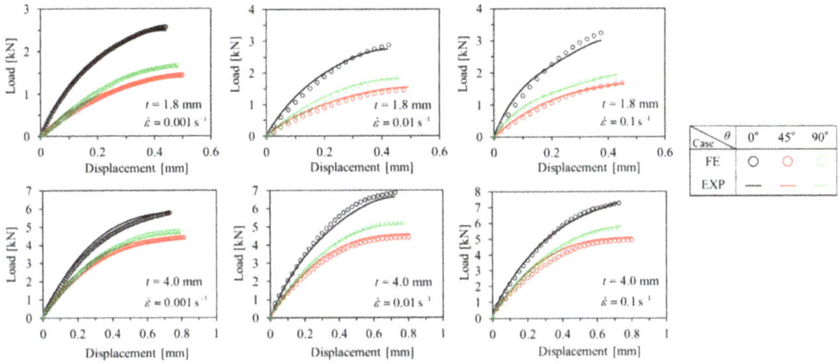

Figure 18. Comparison between experimental and numerical results for ASTM-D3410 compressive specimens.

Table 1. Model constants for the Ramberg–Osgood flow stress model and for the modified Hill'48 yield function.

Stress State	Strain Rate [s^{-1}]	K [MPa]	n	E_m [GPa]	E_f [GPa]	α	β	λ_i
Tension	0.001	83.4	10.4	1.3	50.1	2.795	1.574	0.85
	0.01	79.3	9.9	1.4	55.1	2.929	1.606	0.85
	0.1	77.9	9.2	2.1	63.7	3.031	1.612	0.85
Compression	0.001	42.6	10.6	1.6	50.7	1.662	0.435	0.85
	0.01	40.6	9.1	3.7	65.9	1.811	0.575	0.85
	0.1	38.2	8.8	4.0	70.2	2.222	0.613	0.85

3.4. Application: Composite Cross Member Rear Part Bending Specimen

In order to verify the effect of considering the tensile-compressive mechanical behavior of PA6-20CF according to the fiber orientation and load conditions, a bending test of a specimen machined from PA6-20CF composite cross member was conducted (see Figures 19 and 20). The cross member was manufactured by injection molding under the following process conditions (Table 2), and TIM analysis was performed by inputting the same process parameters in the simulation. In addition to that, the model coefficients set shown in Table 1 was input to evaluate the reliability of the derived PA6-20CF mechanical properties. By the mapping procedure, the local mechanical properties due to the injection molding were considered in the specimen machined by cutting the cross member. In order to secure the reliability of the mechanical properties at the static and high strain rates derived previously, the jig speed was varied, and EXP-FE comparison was performed for two cases (jig speed: 1 mm/min, 20 mm/min). The experiment setup is the same as shown in Figure 20, and for the bending experiments, a universal test-one machine was used.

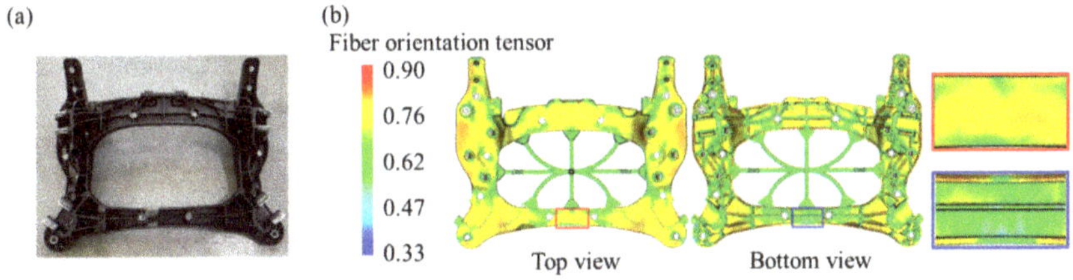

Figure 19. (**a**) PA6-20CF composite cross-member with aluminum insert fabricated by injection molding (bottom view), (**b**) fiber orientation distribution result of the cross member and detailed view at the machined location.

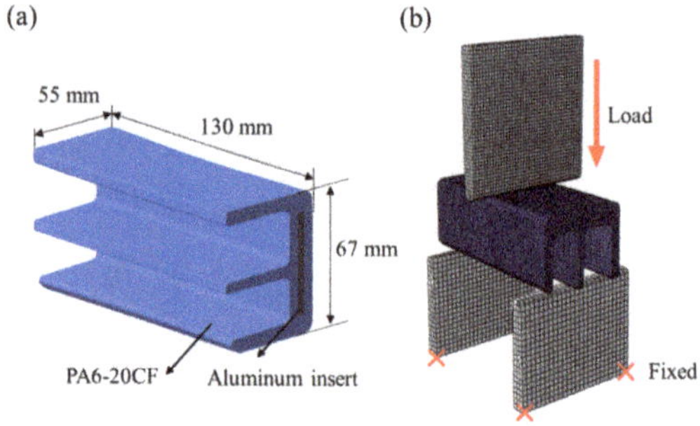

Figure 20. (**a**) Dimensions and configuration of the specimen after cutting, (**b**) FE model and boundary conditions.

Table 2. Injection molding process parameters of PA6-20CF composite cross member.

Process Parameter	Value
Mold surface temperature [°C]	85
Melt temperature [°C]	285
Injection time [s]	20
Cooling time [s]	140

In Figure 21, the bending test load–displacement curves for two different jig speeds were compared. The simulation results in which only tensile properties were input and results in which both tensile and compressive properties were entered were compared with the experimental results. In the case of the jig speed with 1 mm/min, as seen in Figure 21a, when only the tensile properties were considered, the result was 3.1 kN higher than the experimental result (relative error: 27.2%), whereas when tensile and compressive properties were simultaneously considered, the result was 1.1 kN lower (relative error: 9.6%). In addition to that, the result of the jig speed with 20 mm/min is shown in Figure 21b. When only the tensile properties were input, there was an error of 5.5 kN (relative error: 42.6%), and when both were considered, an error of 1.2 kN was obtained (relative error: 9.3%). By comparing the load at the failure displacement, it can be seen that the error was improved by about 26% at the jig speed of 1 mm/min and by about 13% at 20 mm/min. In the case of PA6-20CF, the tensile and compressive mechanical behaviors are different due to carbon fibers, and this reinforces the argument that these should be properly considered. Additionally, comparing Figure 21a,b, it can be confirmed that the prediction error was significantly reduced when the strain rate was properly considered. As a result, these show that investigated procedures can contribute to predicting the failure of the PA6-20CF structures.

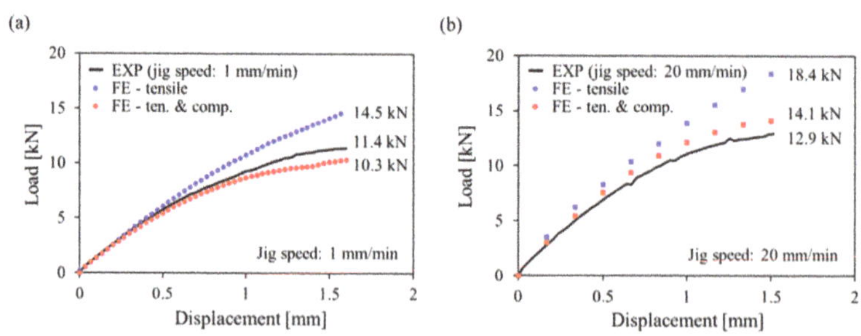

Figure 21. Bending test results comparison between EXP and FE with jig speed: (**a**) 1 mm/min, (**b**) 20 mm/min.

4. Conclusions

The main objectives of this paper were as follows: first, failure mechanism investigation through experimental characterization and SEM verification of injection molded PA6-20CF; second, defining the advantage of simultaneous consideration of tensile-compression properties of anisotropic elastoplastic material behavior in the complex stress state; third, design systemization of SFRP which is fabricated by injection molding considering mechanical properties according to fiber orientation distribution and strain rate.

By experiments and SEM image analysis, mechanical behavior of PA6-20CF with various load conditions and fiber orientation was identified. Additionally, it was found that the failure mechanism was very different depending on the fiber orientation and the loading conditions. Afterwards, utilizing the mapping procedure, inputting local mechanical properties induced by the injection molding process into the structural numerical simulation was carried out. By this method, it was possible to accurately consider the anisotropic mechanical behavior of the uniaxial tensile and compressive specimen, and by the comparison of EXP-FE results, the Ramberg–Osgood flow stress model coefficients were derived. A structural analysis model was constructed in which the mechanical properties at each load condition and strain rate were inputted and, to verify it, a bending test was conducted with a specimen manufactured by machining a composite crossmember. It demonstrated high EXP-FE correlation under static conditions and high strain rate conditions to prove the validity and applicability of modeling.

Author Contributions: Conceptualization, J.L.; methodology, J.L.; software, J.L.; validation, J.L.; formal analysis, J.L.; investigation, J.L.; resources, N.K.; writing—original draft preparation, J.L.; writing—review and editing, N.K.; visualization, J.L.; supervision, N.K.; project administration, H.L., N.K.; All authors have read and agreed to the published version of the manuscript.

Funding: This research received no external funding.

Institutional Review Board Statement: Not applicable.

Informed Consent Statement: Not applicable.

Data Availability Statement: The data presented in this study are available on request from the corresponding author.

Acknowledgments: This research was supported by the Carbon Industrial Cluster Development Program (1008 3609) funded by the Ministry of Trade, Industry & Energy (MOTIE, Republic of Korea) and by the Material Component Technology Development Program (2000 4983) funded by the Ministry of Trade, Industry & Energy (MOTIE, Republic of Korea).

Conflicts of Interest: The authors declare no conflict of interest.

References

1. Nie, W.Z.; Li, J.; Zhang, Y.F. Tensile properties of surface treated carbon fibre reinforced ABS/PA6 composites. *Plast. Rubber Compos.* **2010**, *39*, 16–20. [CrossRef]
2. Ishikawa, T.; Amaoka, K.; Masubuchi, Y.; Yamamoto, T.; Yamanaka, A.; Arai, M.; Takahashi, J. Overview of automotive structural composites technology developments in Japan. *Compos. Sci. Technol.* **2018**, *155*, 221–246. [CrossRef]
3. Mortazavian, S.; Fatemi, A. Effects of fiber orientation and anisotropy on tensile strength and elastic modulus of short fiber reinforced polymer composites. *Compos. Part B Eng.* **2015**, *72*, 116–129. [CrossRef]
4. Oumer, A.N.; Mamat, O. A study of fiber orientation in short fiber-reinforced composites with simultaneous mold filling and phase change effects. *Compos. Part B Eng.* **2012**, *43*, 1087–1094. [CrossRef]
5. Tseng, H.-C.; Chang, Y.-J.; Wang, T.-C.; Hsu, C.-H.; Chang, R.-Y. Three Dimensional Predictions of Fiber Orientation for Injection Molding of Long Fiber Reinforced Thermoplastics. In Proceedings of the SPE ACCE Conference, Michigan, AT, USA, 11–13 September 2013.
6. Lienhard, J.; Schulenberg, L. Strain rate dependent multiaxial characterization of long fiber reinforced plastic. *Compos. Part B Eng.* **2018**, *141*, 164–173. [CrossRef]
7. Kim, E.; Park, J.; Jo, S. A study on fiber orientation during the injection molding of fiber-reinforced polymeric composites: (Comparison between image processing results and numerical simulation). *J. Mater. Process. Technol.* **2001**, *111*, 225–232. [CrossRef]

8. Li, Z.; Lu, J.; Qiu, R.; Liu, Z.; Zhu, P. Multiscale modeling based failure criterion of injection molded SFRP composites considering skin-core-skin layered microstructure and variable parameters. *Compos. Struct.* **2022**, *286*, 115277. [CrossRef]
9. Fu, S.-Y.; Lauke, B. Effects of fiber length and fiber orientation distributions on the tensile strength of short-fiber-reinforced polymers. *Compos. Sci. Technol.* **1996**, *56*, 1179–1190. [CrossRef]
10. Capela, C.; Oliveira, S.; Pestana, J.; Ferreira, J. Effect of fiber length on the mechanical properties of high dosage carbon reinforced. *Procedia Struct. Integr.* **2017**, *5*, 539–546. [CrossRef]
11. Rezaei, F.; Yunus, R.; Ibrahim, N.A. Effect of fiber length on thermomechanical properties of short carbon fiber reinforced polypropylene composites. *Mater. Des.* **2009**, *30*, 260–263. [CrossRef]
12. Belmonte, E.; De Monte, M.; Hoffmann, C.-J.; Quaresimin, M. Damage initiation and evolution in short fiber reinforced polyamide under fatigue loading: Influence of fiber volume fraction. *Compos. Part B Eng.* **2017**, *113*, 331–341. [CrossRef]
13. Botelho, E.; Figiel, Ł.; Rezende, M.; Lauke, B. Mechanical behavior of carbon fiber reinforced polyamide composites. *Compos. Sci. Technol.* **2003**, *63*, 1843–1855. [CrossRef]
14. Arif, M.; Saintier, N.; Meraghni, F.; Fitoussi, J.; Chemisky, Y.; Robert, G. Multiscale fatigue damage characterization in short glass fiber reinforced polyamide-66. *Compos. Part B Eng.* **2014**, *61*, 55–65. [CrossRef]
15. Launay, A.; Maitournam, M.; Marco, Y.; Raoult, I.; Szmytka, F. Cyclic behaviour of short glass fibre reinforced polyamide: Experimental study and constitutive equations. *Int. J. Plast.* **2011**, *27*, 1267–1293. [CrossRef]
16. Horst, J.J.; Spoormaker, J.L. Mechanisms of fatigue in short glass fiber reinforced polyamide 6. *Polym. Eng. Sci.* **1996**, *36*, 2718–2726. [CrossRef]
17. Vieille, B.; Taleb, L. About the influence of temperature and matrix ductility on the behavior of carbon woven-ply PPS or epoxy laminates: Notched and unnotched laminates. *Compos. Sci. Technol.* **2011**, *71*, 998–1007. [CrossRef]
18. Vieille, B.; Chabchoub, M.; Bouscarrat, D.; Keller, C. Prediction of the notched strength of woven-ply PolyPhenylene Sulfide thermoplastic composites at a constant high temperature by a physically-based model. *Compos. Struct.* **2016**, *153*, 529–537. [CrossRef]
19. Crupi, V.; Guglielmino, E.; Risitano, G.; Tavilla, F. Experimental analyses of SFRP material under static and fatigue loading by means of thermographic and DIC techniques. *Compos. Part B Eng.* **2015**, *77*, 268–277. [CrossRef]
20. Dean, A.; Grbic, N.; Rolfes, R.; Behrens, B. Macro-mechanical modeling and experimental validation of anisotropic, pressure- and temperature-dependent behavior of short fiber composites. *Compos. Struct.* **2019**, *211*, 630–643. [CrossRef]
21. Ricotta, M.; Sorgato, M.; Zappalorto, M. Tensile and compressive quasi-static behaviour of 40% short glass fibre-PPS reinforced composites with and without geometrical variations. *Theor. Appl. Fract. Mech.* **2021**, *114*, 102990. [CrossRef]
22. Quagliato, L.; Lee, J.; Fonseca, J.H.; Han, D.; Lee, H.; Kim, N. Influences of stress triaxiality and local fiber orientation on the failure strain for injection-molded carbon fiber reinforced polyamide-6. *Eng. Fract. Mech.* **2021**, *250*, 107784. [CrossRef]
23. Quagliato, L.; Kim, Y.; Fonseca, J.H.; Han, D.; Yun, S.; Lee, H.; Park, N.; Lee, H.; Kim, N. The influence of fiber orientation and geometry-induced strain concentration on the fatigue life of short carbon fibers reinforced polyamide-6. *Mater. Des.* **2020**, *190*, 108569. [CrossRef]
24. Mouhmid, B.; Imad, A.; Benseddiq, N.; Benmedakhène, S.; Maazouz, A. A study of the mechanical behaviour of a glass fibre reinforced polyamide 6,6: Experimental investigation. *Polym. Test.* **2006**, *25*, 544–552. [CrossRef]
25. Sato, N.; Kurauchi, T.; Sato, S.; Kamigaito, O. Microfailure behaviour of randomly dispersed short fibre reinforced thermoplastic composites obtained by direct SEM observation. *J. Mater. Sci.* **1991**, *26*, 3891–3898. [CrossRef]
26. Fu, S.-Y.; Lauke, B. The elastic modulus of misaligned short-fiber-reinforced polymers. Compos. *Sci. Technol.* **1998**, *58*, 389–400. [CrossRef]
27. Tanaka, K.; Kitano, T.; Egami, N. Effect of fiber orientation on fatigue crack propagation in short-fiber reinforced plastics. *Eng. Fract. Mech.* **2014**, *123*, 44–58. [CrossRef]
28. Fan, J.; Wang, C. Dynamic compressive response of a developed polymer composite at different strain rates. *Compos. Part B Eng.* **2018**, *152*, 96–101. [CrossRef]
29. Fan, J.; Weerheijm, J.; Sluys, L. Compressive response of multiple-particles-polymer systems at various strain rates. *Polymer* **2016**, *91*, 62–73. [CrossRef]
30. Yuan, Y.; Yao, X.; Niu, K.; Liu, B.; Wuyun, Q. Compressive failure of fiber reinforced polymer composites by imperfection. *Compos. Part A Appl. Sci. Manuf.* **2019**, *118*, 106–116. [CrossRef]
31. Opelt, C.; Cândido, G.; Rezende, M. Compressive failure of fiber reinforced polymer composites—A fractographic study of the compression failure modes. *Mater. Today Commun.* **2018**, *15*, 218–227. [CrossRef]
32. Li, A.-J.; Zhang, J.-J.; Zhang, F.-Z.; Li, L.; Zhu, S.-P.; Yang, Y.-H. Effects of fiber and matrix properties on the compression strength of carbon fiber reinforced polymer composites. *New Carbon Mater.* **2020**, *35*, 752–761. [CrossRef]
33. Ma, L.; Liu, F.; Liu, D.; Liu, Y. Review of Strain Rate Effects of Fiber-Reinforced Polymer Composites. *Polymers* **2021**, *13*, 2839. [CrossRef]
34. Cui, J.; Wang, S.; Wang, S.; Li, G.; Wang, P.; Liang, C. The Effects of Strain Rates on Mechanical Properties and Failure Behavior of Long Glass Fiber Reinforced Thermoplastic Composites. *Polymers* **2019**, *11*, 2019. [CrossRef]
35. Kanouté, P.; Boso, D.; Chaboche, J.-L.; Schrefler, B. Multiscale methods for composites: A review. *Arch. Comput. Methods Eng.* **2009**, *16*, 31–75. [CrossRef]

36. Feyel, F.; Chaboche, J.-L. FE2 multiscale approach for modelling the elastoviscoplastic behaviour of long fibre SiC/Ti composite materials. *Comput. Methods Appl. Mech. Eng.* **2000**, *183*, 309–330. [CrossRef]
37. Spahn, J.; Andrä, H.; Kabel, M.; Müller, R. A multiscale approach for modeling progressive damage of composite materials using fast Fourier transforms. *Comput. Methods Appl. Mech. Eng.* **2014**, *268*, 871–883. [CrossRef]
38. Chung, S.T.; Kwon, T.H. Numerical simulation of fiber orientation in injection molding of short-fiber-reinforced thermoplastics. *Polym. Eng. Sci.* **1995**, *35*, 604–618. [CrossRef]
39. Gupta, M.; Wang, K.K. Fiber orientation and mechanical properties of short-fiber-reinforced injection-molded composites: Simulated and experimental results. *Polym. Compos.* **1993**, *14*, 367–382. [CrossRef]
40. Bay, R.S.; Tucker, C.L., III. Fiber orientation in simple injection moldings. Part I: Theory and numerical methods. *Polym. Compos.* **1992**, *13*, 317–331. [CrossRef]
41. VerWeyst, B.E.; Tucker, C.; Foss, P.H.; O'Gara, J.F. Fiber Orientation in 3-D Injection Molded Features. *Int. Polym. Process.* **1999**, *14*, 409–420. [CrossRef]
42. Hu, J.; Zhang, X.; Chen, Z.; Guo, W.; Li, H.; Deng, X. Experimental and Numerical Research on Open-Hole Strength and Damage Mechanism of Regularly Arrayed Short Fiber Reinforced Polymer Composite. *Polymers* **2020**, *12*, 1622. [CrossRef] [PubMed]
43. Wang, J.; Chen, L.; Shen, W.; Zhu, L. Research on Tensile Properties of Carbon Fiber Composite Laminates. *Polymers* **2022**, *14*, 2318. [CrossRef] [PubMed]
44. Karsli, N.G.; Aytac, A. Tensile and thermomechanical properties of short carbon fiber reinforced polyamide 6 composites. *Compos. Part B Eng.* **2013**, *51*, 270–275. [CrossRef]
45. Zhang, Y.; Sun, L.; Li, L.; Wei, J. Effects of strain rate and high temperature environment on the mechanical performance of carbon fiber reinforced thermoplastic composites fabricated by hot press molding. *Compos. Part A Appl. Sci. Manuf.* **2020**, *134*, 105905. [CrossRef]
46. *ASTM D638-10*; Standard Test Method for Tensile Properties of Plastic. ASTM Compass: West Conshohocken, PA, USA, 2015.
47. *ASTM D3410/D3410M-16e1*; Standard Test Method for Compressive Properties of Polymer Matrix Composite Materials with Unsupported Gage Section by Shear Loading. ASTM Compass: West Conshohocken, PA, USA, 2016.
48. Phelps, J.H.; Tucker, C.L., III. An anisotropic rotary diffusion model for fiber orientation in short-and long-fiber thermoplastics. *J. Non-Newton. Fluid Mech.* **2009**, *156*, 165–176. [CrossRef]
49. Favaloro, A.J.; Tucker, C.L., III. Analysis of anisotropic rotary diffusion models for fiber orientation. *Compos. Part A Appl. Sci. Manuf.* **2019**, *126*, 105605. [CrossRef]
50. Folgar, F.; Tucker, C.L., III. Orientation behavior of fibers in concentrated suspensions. *J. Reinf. Plast. Compos.* **1984**, *3*, 98–119. [CrossRef]
51. Velasco, M.; Correa, E.; París, F. Interaction between fibres in the transverse damage in composites. *Eng. Fract. Mech.* **2020**, *239*, 107273. [CrossRef]
52. Naito, K.; Tanaka, Y.; Yang, J.; Kagawa, Y. Tensile and flexural properties of single carbon fibres. In Proceedings of the ICCM-17-17th International Conference on Composite Materials, Edinburgh, UK, 27–31 July 2009.
53. Da Silva, J.L.G.; Johnson, D.J. Flexural studies of carbon fibres. *J. Mater. Sci.* **1984**, *19*, 3201–3210. [CrossRef]
54. Kumar, S.; Anderson, D.P.; Crasto, A.S. Carbon fibre compressive strength and its dependence on structure and morphology. *J. Mater. Sci.* **1993**, *28*, 423–439. [CrossRef]

polymers

MDPI

Article

Thermal Effects on Mechanical Strength of Additive Manufactured CFRP Composites at Stable and Cyclic Temperature

Isyna Izzal Muna [1,*], Magdalena Mieloszyk [1], Ruta Rimasauskiene [2], Nabeel Maqsood [2] and Marius Rimasauskas [2]

[1] Institute of Fluid Flow Machinery, Polish Academy of Sciences, Fiszera 14, 80-231 Gdansk, Poland
[2] Department of Production Engineering, Faculty of Mechanical Engineering and Design, Kaunas University of Technology, Studentu 56, 51424 Kaunas, Lithuania
* Correspondence: imuna@imp.gda.pl

Abstract: Additive manufacturing (AM) techniques can be applied to produce carbon-fiber-reinforced polymer (CFRP) elements. Such elements can be exposed to different environmental factors, e.g., temperature, moisture, and UV radiation, related to their operational conditions. From a variety of environmental factors, the temperature is one of the most typical. Temperature strongly influences matrix material joining together CFRP components, resulting in material strength reduction. Therefore, it is important to understand processes in the composite material caused by temperature. This experimental work investigated the thermal effects on the performances of AM CFRP composites. Specimens with unidirectional (UD) alignments of the fiber reinforcement were printed using the fused deposition modeling (FDM) technique. The printed specimens were subjected to two different thermal conditions: stable continuous at 65 °C and cyclic temperature between 50 and 70 °C. Tensile testing was performed to study the mechanical strength and Young's modulus of AM UD-CFRPs. In order to investigate the morphological structure on the surface of AM specimens, an optical microscope, scanning electron microscope (SEM), and digital microscope were utilized. Untreated (intact) samples attained the highest average tensile strength value of 226.14 MPa and Young's modulus of 28.65 GPa. The ultimate tensile strength of the sample group subjected to stable heat treatment decreased to 217.99 MPa, while the thermal cycling group reduced to 204.41 MPa. The Young's modulus of the sample group subjected to stable thermal exposure was decreased to 25.39 GPa, while for the thermal cycling group, it was reduced to 20.75 GPa. The visual investigations revealed that the intact or untreated specimen group exhibited lateral damage in top failure mode (LAT), the thermally stable group underwent edge delamination in the middle (DGM) as the nominated failure mode, and the explosive breakage at gauge in the middle (XGM) failure mode occurred in the sample from the thermal cycling group. Based on morphological observations at the microscale, the delamination, fiber pull-out, and matrix cracking were the dominant damages in the 3D-printed tensile-tested specimens. The molecular chains of the polymer changed their structure into an amorphous one, and only local motions of stretching occurred when the specimens were exposed to stable heating (prolonged). In the case of thermal cycling, the strain gradients were accumulated in the matrix material, and the local stresses increased as a result of the reheating and re-cooling exposure of the polymeric composites; the molecular motion of the long-range polymer structure was reactivated several times. Micro-cracking occurred as a result of internal stresses, which led to material failure and a reduction of the mechanical properties.

Keywords: thermal cycle; thermally stable; 3D-printing technology; fused deposition modeling; CFRP composites; mechanical degradation; morphological structure

Citation: Muna, I.I.; Mieloszyk, M.; Rimasauskiene, R.; Maqsood, N.; Rimasauskas, M. Thermal Effects on Mechanical Strength of Additive Manufactured CFRP Composites at Stable and Cyclic Temperature. *Polymers* **2022**, *14*, 4680. https://doi.org/10.3390/polym14214680

Academic Editor: R.A. Ilyas

Received: 29 September 2022
Accepted: 31 October 2022
Published: 2 November 2022

1. Introduction

The technological advancement in additive manufacturing (AM) over the last decade has been a major driver for the development of composite materials. It is widely known that AM technologies, specifically the fused deposition modeling (FDM) method, have numerous benefits owing to the low price, flexibility, and simplicity with small amounts of waste material, which indeed open up endless possibilities in their utilization in engineering and industrial fields [1–3]. Carbon-fiber-reinforced polymer (CFRP) has potential applications in the industrial field such as automotive bumper brackets [4] and the development of more durable wind turbines [5]. The printed composite materials must have a long service life in real applications. However, these materials are more likely exposed to harsh environmental conditions specifically related to temperature. Depending on its different magnitude and period time, the exposure to temperature of 3D-printed composites could affect the mechanical behavior of the materials differently. Researchers usually use different thermal cycling profiles to study the behavior of aerospace vehicles such as satellite components, which depend on their case study or thermal cycling test apparatus [6].

There have been a large number of papers that experimentally studied the effect of thermal treatment of the mechanical behavior of polymeric composites manufactured with the FDM method at a continuous temperature [7–12]. Several results reported that the heat treatment leads to the better mechanical properties of 3D-polymer-based composites [10–12]. Handweker et al. investigated the influence of a heat treatment (annealing) process on the mechanical performance of continuous- and chopped-fiber-reinforced polyamide 6 in the build-up direction. It was found that Young's modulus increased by a factor of three, while the ultimate tensile strength (UTS) increased by 50% for the chopped-carbon-fiber-reinforced material and 186% for the continuous-glass-fiber-reinforced material [11]. In Wang et al. [12], heat treatment was performed on three printed continuous-carbon-fiber-reinforced polymer (CCFRP) composite specimens with different layer distributions and heating conditions. The dimensional change of CCFRCs during heat treatment was shown to be much more closely related to the microstructure change than the material crystallization. Furthermore, heat treatment could improve the mechanical properties of CCFRCs by decreasing the porosity while also strengthening the interface. Heat treatment, specifically, delayed the initiation of cracks in CCFRCs during bending tests by changing the failure mode of the matrix layers and improving the bonding between fiber and bundles/matrix. Another experimental work by Nassar et al. revealed that heat treatment was able to enhance the bonding of filament layers and reduce the porosity content of CFRP composite printed using the FDM method, which resulted in an improvement of its tensile modulus. However, there was a limited improvement in the tensile strength and modulus of elasticity values for the samples treated at low heat treatment temperature [10]. Despite the benefits of using FDM to generate durable materials with high mechanical properties, this method is also known to have some limitations, such as nozzle clogging [13,14] and the appearance of satellite droplets [15].

However, some other experiments on CFRP composites showed a decreased mechanical behavior after thermal treatment [7,16]. Zhang et al. investigated the effect of high temperature on pure epoxy resin and 3D carbon/epoxy braided composites [7]. It was revealed that a constant decline in compressive properties occurred as a result of the combination of matrix degradation and fiber/resin interface cracking. At temperatures lower than the glass transition temperature (Tg) of the pure epoxy resin, the matrix aging was the main aging process, although fiber/matrix interface debonding can be detected at temperatures greater than the Tg, such as 180 °C. After thermal aging, their micro-morphologies and compressive behaviors were reported. Jia et al. experimentally investigated the influence of heat treatment performed at a stable temperature on CFRP composites generated conventionally, ranging from −100 °C to 100 °C [16]. The results showed that, under static and dynamic three-point bending tests, CFRP composites had improved flexural strength, maximum deflection, and energy absorption at lower temperatures (−60 °C, −100 °C), but performed poorly at higher temperatures (100 °C). At various temperatures, experimental photos from post-mortem photographs, scanning electron microscopy, and high-speed films were used to investigate various failure behaviors such as micro-buckling, kinking, and fiber breakage. In more recent experimental research by [6], 3D-printed CF/PEEK was exposed to thermal cycling and then evaluated using tensile and arc heating tests. It was reported that the thermal cycle resulted in decreased tensile strength and the length of the samples increased after the heating test.

Until recently, the number of experiments of thermal treatment conducted in another thermal mode, e.g. cyclic temperature, remains very limited since most heat treatment experiments on 3D-printed composites were mostly performed at a stable continuous temperature. However, some papers investigated the effect of thermal cycling at low temperatures on composites produced conventionally. Ghasemi et al. studied the residual stress during thermal cycling on a graphite/epoxy composite [17]. The cycle was performed one time and used as a change in temperature from 70 °C to 100 °C and back to 70 °C. This profile has a heating/cooling (transition) rate of 3–5 °C per minute, while the duration of the dwelling time at the temperature extremes was about 15 min. It was reported that the mechanical properties of this composite were degraded when the material was exposed to thermal cycling conditions. Up to a certain period of exposure, the composite may retain its strength and stiffness above its allowable limits. However, as time passes, the strength and stiffness may become very low.

Another thermal cycling treatment of a carbon fiber/epoxy composite generated conventionally was carried out by Gonzalez et al. [18]. On CFRP samples, 500 thermal cycles at 15 °C/min and a 5 min dwell time were applied during the experimental investigation. Two different atmospheres were used in the thermal cycling test. The first represented an inert atmosphere (nitrogen gas) for isolating any oxidative agents' effects. For the second atmosphere, dry air was used to simulate material oxidation effects. At high temperatures, matrix oxidation was expected, but matrix cracking occurred due to a mismatch in the thermal expansion coefficient. The combination of both effects resulted in CFRP aging acceleration. Lafarie et al. studied the characterization of the damage processes of carbon/epoxy laminates under isothermal aging and thermal cycling conditions in a vacuum and oxidative atmosphere [19]. The thermal cycling experiments were 500 thermal cycles with maximum and minimum temperatures of 180 and −50 °C, respectively, and constant cooling and heating rates of 4 °C/min. It was found that, during a thermal cycling test in an oxidative atmosphere such as air, there was a coupling effect between matrix oxidation, which occurs at the highest temperatures of the cycle, and matrix cracking caused by thermo-mechanical ply stresses caused by prohibited differential expansions of the plies. Previously, heating treatment of unidirectional (UD) continuous CFRP composite at an elevated temperature was experimentally and numerically investigated by Muna et al. [20]. The coupled thermal–displacement analysis was used during the numerical calculations. The strain in the sample due to its exposure to elevated temperature was measured using fiber Bragg grating (FBG) sensors.

Thermal degradation of additive manufactured polymer composite has not been explored by researchers comprehensively, yet it is an important phenomenon that will be useful in the structural health prediction of lightweight materials. Up to the present, the majority of research studies on the thermal degradation of 3D-printed composites have only been investigated on specimens at a stable temperature for a certain duration, while for the thermal cycling treatment, most of the research work has been performed on the specimens produced not with 3D-printing technology, but rather, conventionally. Furthermore, few explorations are available on the work performed on 3D-printed polymeric composites under both thermal conditions at continuous and cyclic temperatures and the effect on the mechanical degradation. This paper aims to investigate the effect of thermal exposure on the mechanical and morphological behavior of 3D continuous-carbon-fiber-reinforced polymers at stable continuous and cyclic temperatures.

2. Experimental Work

2.1. Additive Manufacturing of CFRP Composite Samples

Additive manufacturing (AM) is a well-known processing technique for adding materials to manufacture parts based on a 3D computer model in a layer-by-layer manner, which can also be referred to as 3D printing or rapid prototyping (RP). In this experimental work, the FDM method was chosen due to its simplicity and ability to print complex geometries with low cost and flexibility to use different materials. Unidirectional (UD) continuous CFRP composites were printed as experimental specimens. The benefit of employing continuous carbon fibers (CCFs) as a reinforcement agent in polymer-based composites is that their weight remains light with the increased strength. Thermoplastic polylactic acid (PLA) was chosen as the matrix agent.

The CFRP composite specimens were additively manufactured with the fused deposition modeling (FDM) method at Kaunas University of Technology. A modified 3D printer MeCreator2 with dual inputs at the extrusion printing head, which was designed and developed by Rimasauskas et al., was employed [14]. The pre-impregnation process was applied for the CCF reinforcement for better printing quality and adhesion. For the impregnation process, PLA pellets were dissolved in dichloromethane solution from Eurochemicals with the ratio concentration of 90 g/10 g using the magnetic LBX H01 mini-stirrer at 600 rpm. Later, the standard non-impregnated CCF was passed through this resin and simultaneously dried with a hot air gun temperature of 220 °C. After the impregnation process, the CCF was further printed with the PLA matrix filament to form a composite specimen. The continuous carbon fiber used was T300D-3000, consisting of 3000 fibers in an untwisted tow (Toray, France). The materials' properties are shown in Table 1. The information of all materials concerning the mechanical properties was provided by the material suppliers. The 3 K CCF was used as the reinforcement material and was kept the same throughout the experimental work. An extrusion multiplier was used to control the content, but it was also kept the same at 0.5 for all the manufactured composite specimens. The prepared materials were then ready to be used for additive manufacturing process with the FDM method. The 3D model of the specimens was created in a CAD package, which was then exported as a stereo lithography (STL) file and subsequently loaded into 3D printer slicing software (Simplify3D, Cincinnati, OH, USA) to generate the G-code, which the 3D printer used to print out each test specimen. The CAD files of the specimens were used to extrude and deposit molten thermoplastic, which was built up in layers from a horizontal base. Two types of FDM printers were used during the experiments, a modified one with two inputs at the extruder and another 3D printer with one input from Prusa company (for printing the gripping tabs). The printing parameters are shown in Table 2.

Table 1. Material properties of continuous carbon fiber and polylactic acid.

	Elastic Strength (MPa)	Elastic Modulus (GPa)	Strain at Failure (%)	Density (g/cm^3)
Fiber	3530	230	1.5	1.76
Matrix	46.6	2.636	1.9	1.17

Table 2. Printing parameters.

Parameter	Value
Nozzle diameter	1.5 mm
Extrusion multiplier	0.7
Primary layer height	0.5 mm
Interior infill	100%
Infill pattern	rectilinear
Nozzle temperature	220 °C
Bed temperature	90 °C
Printing speed	3 mm/s

The extrusion printing was updated in order to include two inputs (one for the polymer filament and the other for the carbon fiber tow) and one output that allowed the polymer to be fused with the CF for printing. As was discussed previously in the experimental research by [14], the working principle of PLA and CF mixing consisted of several steps. Firstly, PLA filament was fed through a PTFE tube by using a standard filament feeding system. When the filament reached the printing head, which was heated up to 200 °C, the polymer liquefied. Subsequently, the impregnated carbon fiber tow was inserted through the PTFE tube to the printing head, directly to the printing nozzle. The molten polymer came into contact with the carbon fiber in the mixing zone, and secondary carbon fiber impregnation was performed. The molten polymer was constantly pushed together with the carbon fiber through the nozzle. It should be emphasized that the carbon fiber was fed solely by the molten polymer and did not require any additional feeding equipment. Later, the carbon fiber coated with polymer was extruded through a printing nozzle onto borosilicate glass mounted on an aluminum plate.

Generally, the manufacturing process of CFRP composites with the FDM method requires a long period of time (e.g., in this experimental case, about 1 h for printing 1 specimen), and the 3D printer is not designed to handle errors. Printing mistakes are frequently expected, and continuing the printing process when an error appears may endanger the 3D printer, and the final part of the printed sample typically cannot be used. One common printing error is nozzle clogging [13,14]. In order to minimize this detrimental issue, numerous efforts have been made by researchers to overcome nozzle clogging during the printing process [13,21,22]. Gutierrez et al. proposed a method to decrease the clogging deposition rate of alumina inclusions in continuous casting nozzles through three simultaneous measures: flow modification, the use of raw materials with a low impurity content, and smoothed internal surfaces [13]. Internal flow regulation entails avoiding dead zones and generating symmetric patterns. The feasibility of these methods was tested using a mathematical model. Using this approach, the adherence of inclusions to the nozzle wall applied a boundary condition based on the thickness of the sub-laminar boundary, rather than the typical "trap" boundary condition. The generic boundary condition produces deposition rates that are unaffected by the inclusion size. The proposed boundary condition uses particle sizes to distinguish against the clogging deposition rate. Plant studies with these nozzles, together with water modeling showed that the current strategy may significantly reduce clogging occurrence. More recently, as a response to the problem of nozzle clogging, Sampedro et al. created a real-time process monitoring system capable of accurately forecasting abnormal behaviors in the FDM printing method [21]. A network of collaborative sensors was utilized to collect time series data, which were then processed by the suggested machine learning method. The multi-head encoder–decoder temporal convolutional network (MH-ED-TCN) gathers features from the data, assesses their impact on the many processes that occur throughout an operational printing cycle, and distinguishes between normal and problematic manufacturing operations. The studies conducted yielded a 97.2% accuracy in predicting the future behavior of a 3D printer. More recent experimental work has been conducted to improve the quality of the 3D-printed composite specimen using a controlled cooling effect [23]. In regard to achieving this printing quality, the effect of printing parameters (extrusion temperature, extrusion width, and extrusion multiplier) on the tensile and flexural properties were examined. The highest Young's and flexural modulus of 31.50 GPa and 15.68 GPa, respectively, were reached when the composite structure was manufactured using printing parameters E.W. −1.3 mm and E.M. −0.5.

Next, thermoplastic PLA reinforced with CF composite was extruded through the printing nozzle, which was printed on borosilicate glass mounted on the aluminum plate bed surface at a certain temperature according to the optimized printing parameters in the literature (i.e., 90 °C). Before starting the printing process, adhesive spray should be spread out on top of the borosilicate glass to avoid the dragging of printed filaments. Once the printing of the sample completed, a cutter was used to separate the filament spool and printed part. Subsequently, the borosilicate glass should be removed from the aluminum plate bed and to allowed cool down, then the 3D-printed sample can be removed from the glass by the aid of blades. The dimension of unidirectional (UD) CFRP composites was 150 × 13 × 2 mm with 6 layers of the same fiber alignment in all layers. The volume fraction of the composite was kept constant throughout the experimental work due to having the same printing parameters. The carbon fiber reinforcement content was measuring using the tool length path, and the estimated carbon fiber content was 18.2% (wt.). The total number of specimens printed was 15 samples in which 5 samples for each group of thermal treatment were required in accordance with the D3039 ASTM standard used for tensile testing. The specimens are depicted in Figure 1. After the manufacturing stage completed, the weight and size dimensions of the composite parts were measured using a digital scale and a caliper, respectively. The weight and dimension measurement was necessary in order to compare the changes in size and mass before and after thermal loading. Furthermore, it was important for the calculation of the mechanical properties after the mechanical testing.

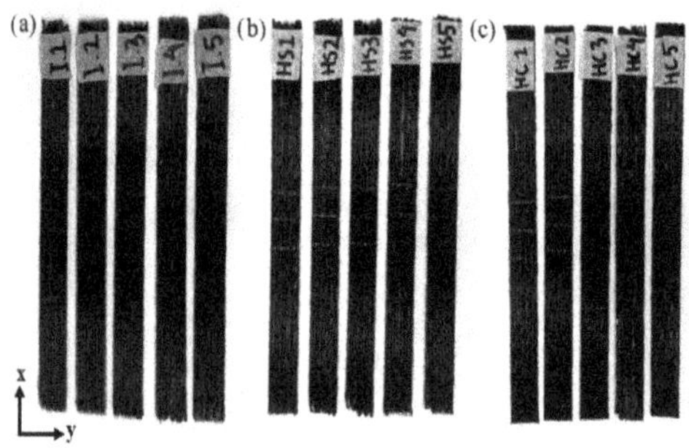

Figure 1. The 15 CFRP composite specimens printed with the FDM method for (**a**) intact/no treatment, (**b**) continuous temperature, and (**c**) cyclic temperature.

2.2. Thermal Treatment on Printed Specimens

In an attempt to investigate the effect of temperature exposures on the printed CFRP composites via FDM, the specimens were thermally treated under two different thermal conditions: stable continuous and cyclic at low temperature around the glass transition (Tg) of the PLA polymers used in this work. An air-circulated environmental oven "Memmert" Model UFB-400 was utilized for stable and cyclic heating, which supplied a continuous renewal of oxygen in a certain percentage of ambient air, as presented in Figure 2. For the stable continuous heat treatment, the specimens were placed inside the pre-heated oven at 65 °C and kept for 6 h, then naturally cooled at room temperature. As for the thermal cycling, the oven temperature was manually adjusted from 50 to 70 °C according to the cyclic plan (see Figure 3) for 6 cycles. One cycle consisted of two extreme temperatures with a dwell time of 10 min, a heating rate of 1 °C/min, and a cooling rate of 2.5 °C/min. These temperatures were chosen according to the glass transition temperature of the PLA used in this work, which lies around 65 °C. It was expected that the specimens would undergo cross-linking and a crystallization change of the polymer up to the state where its structural shape would particularly deform.

Dwell time means how long of an exposure period is appropriate at each temperature limit. If a prolonged dwell time is employed, the duration of the test will lengthen until the number of cycles decreases. Then, again, in some instances, a long dwell period can potentially speed up the testing process if it generates structural changes that increase the stresses during temperature fluctuations. At high temperatures, polymer materials, for instance, relax or creep—the polymer chains within the material move to alleviate the stresses imposed by the high temperature. When the temperature is decreased, these alterations may result in a large increase in the stresses imposed on buildings. However, sufficient dwell time is necessary for these modifications to take place.

Figure 2. An air-circulated oven for hot temperature treatment.

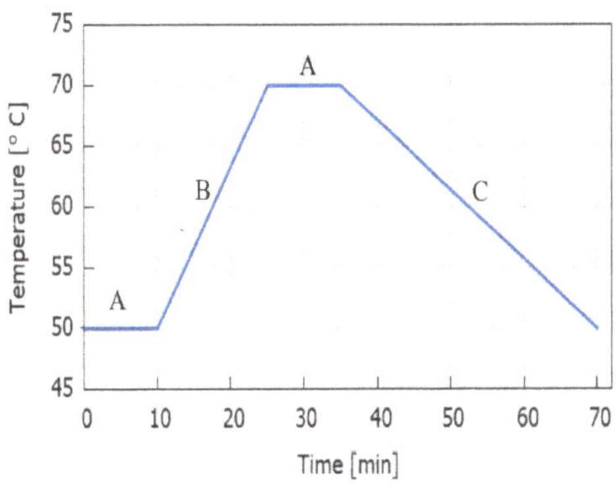

Figure 3. Thermal cycling profile for one cycle. A = dwell time; B = heating rate; C = cooling rate.

2.3. Static Tensile Testing

The tensile testing was performed after thermal treatment to examine the mechanical strength and Young's modulus of the treated specimens, as well as the untreated samples (intact). A Tilnius Olsen H25KT (capacity 25 kN) universal testing machine was utilized to perform tensile testing with the standard head displacement speed of 2 mm/min and at room temperature. The tensile testing setup is shown in Figure 4. The ASTM standard D3039 was used for the in-plane tensile testing, which requires each tested specimen to have four tabs in gripping position at the top and bottom. Prior to the tensile testing, PLA tabs having dimensions 50 × 12.5 × 2 mm were printed separately, and four points were marked 15 mm from the center of the specimens to measure the elastic strain. The printed tabs then were adhered to the specimens with a universal structural bonder (adhesive glue) by applying pressure via clamps to hold and secure the parts tightly. In this study, the ASTM D3039 standard was used to perform the tensile testing of the specimens, in which five samples were required for each group of thermal treatment to determine the average material properties.

2.4. Morphological Investigation

The micro-morphological analyses on the surface were performed before and after performing the cyclic and stable thermal loadings by utilizing an optical microscope device (Nikon Eclipse LV100ND) equipped with a high-definition color camera (Nikon DS-Ri2). The imaging software (NIS Elements 4.5.1.00, Nikon Europe B.V., Amstelveen, The Netherlands) was used to prepare and process the data at 5× magnification. The maximum sample size observed with the optical microscope was 150 × 150 mm. A scanning electron microscope device (FE-SEM SU5000, Hitachi Co., Tokyo, Japan) was employed to investigate the micro-structure damage of different specimen groups after tensile testing. The maximum specimen size observed with the SEM was 200 mm in diameter and 80 mm in height. A digital microscope (Levenhuk, DTX 500 LCD, Warsaw, Poland) was utilized to capture the macroscopic images of the tensile-fractured samples. The maximum specimen size observed with the digital microscope was about 150 × 100 mm. The microscopes are presented in Figure 5.

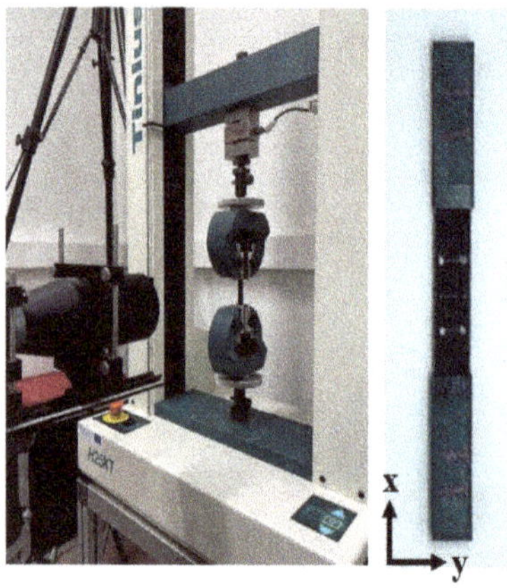

Figure 4. Experimental setup of tensile testing.

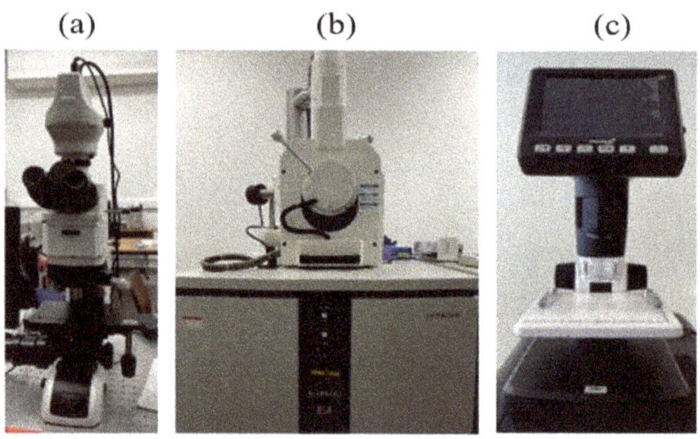

Figure 5. (**a**) Optical microscope; (**b**) scanning electron microscope; (**c**) digital microscope.

3. Results

3.1. Dimensional Change

The samples were measured before and after the heat treatment in order to determine the amount of geometrical changes, as shown in Table 3. Any thermal treatment is expected to have an effect on the dimensions and shape of a plastic part. For the test samples, the length, as well as the height and width were measured using a digital vernier caliper dimensional measurement tool. Table 1 shows the change of the specimens' length, width, and thickness before heat treatment (after 3D printing) and after heat treatment. The difference between the nominal design from the CAD software and the printed dimension in practice is referred as variations of the relative values of the length, width, and thickness. These values were then acquired and averaged. In comparison to the nominal model, the dimensions of printed CFRP samples in the xy-plane (length and width) were undersized around (0.62 ± 0.93)% and (0.07 ± 0.27)%, respectively. However, in comparison with the 3D model, the dimension in the z direction (thickness) was oversized, and a mean error of 13.44 ± 0.05% was obtained. Overall, these findings are consistent with previous research on the dimensional correctness of 3D-printed composite parts [8,24]. The noticeable difference in the dimension of the specimens can be hypothetically attributed to the difficulty of depositing impregnated continuous carbon fiber filament. During the printing process, the clogging of the printer nozzle could possibly occur due to the improper proportion of polymer matrix filament mounted into the impregnated CCF material. In addition to that, the issue with nozzle clogging might lead to the breakage of the CCF filament throughout the deposition process. The material deposition issue was encountered by the authors in this experimental work. In accordance with previous experimental research, a similar problem of nozzle clogging occurred when the amount of carbon fiber filament was higher than the polymer matrix, which caused a greater tension at the nozzle tip and resulted in damaged carbon fiber filament [14]. It is worth noting that carbon fiber filament was printed according to the pre-defined path where the carbon content was calculated by evaluating the path traveled by the printing head and carbon weight per unit length. Therefore, in an attempt to tackle this issue during the materials' deposition, the cooling fan shall be adjusted to prevent either a lack or excessive polymer extruded onto the printing bed.

Table 3. Dimensions of 3D-printed specimens before and after heating treatment.

Test Group	Before Treatment			After Treatment		
	L_o (mm)	W_o (mm)	T_o (mm)	L_t (mm)	W_t (mm)	T_t (mm)
Untreated	148.32 ± 1.04	12.65 ± 0.21	2.14 ± 0.02		-	
Stable	149.88 ± 0.28	13.19 ± 0.25	2.18 ± 0.05		no variation	
Cyclic	149.97 ± 0.42	12.95 ± 0.12	2.08 ± 0.02	150.37 ± 0.42	12.1 ± 0.21	1.74 ± 0.07

3.2. Static Tensile Results

The tensile properties of the 3D-printed specimens of each thermal group (untreated, continuous, and cyclic) are shown in Figure 6. From the bar graph plot, it can be seen that the untreated (intact) samples attained the highest average tensile strength value of 226.14 MPa and Young's modulus of 28.65 GPa. The ultimate tensile strength of the sample group subjected to stable heat treatment decreased to 217.99 MPa, while for the thermal cycling group, it reduced to 204.41 MPa. The Young's modulus of the sample group subjected to stable thermal exposure decreased to be 25.39 GPa, while for the thermal cycling group, it reduced to 20.75 GPa. The mean values in the bar plots were used to depict the trends of each printed group specimen and the range of their tensile characteristics effects. The degradation in the mechanical strength and elastic modulus after thermally stable and cyclic loading was attributed to the difference in the coefficient of thermal expansion (CTE) between the matrix and fiber, which was caused by the reduced cross-linking of the polymers [25]. This CTE discrepancy resulted in thermal stress, and it can cause the fibers to pull out due to fiber–matrix debonding, which then leads to the mechanical deterioration of composite specimens [18,26]. ANOVA analysis was performed to identify statistical differences and significance between the unheated group and the stable continuous and cyclic temperature on the tensile strength with a significance level of 5%. A one-way ANOVA revealed that there was a statistically significant difference in tensile strength between the three tested groups with a p-value of 0.0108 and test statistic F of 6.764 ($F(2, 27) = [6.764]$, $p = 0.0108$).

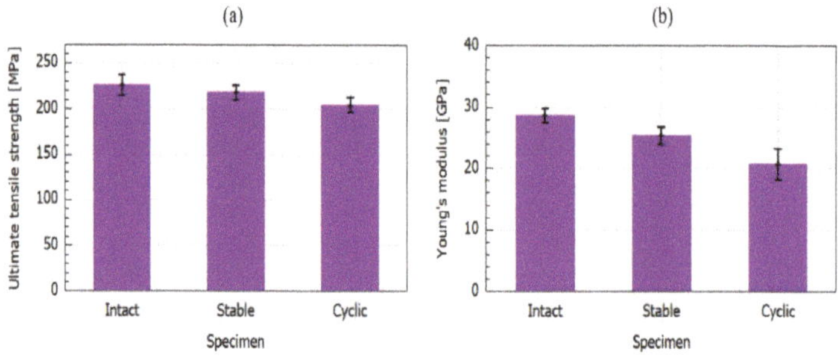

Figure 6. Experimental results of the tensile properties of 3D-printed samples. (**a**) Tensile strength; (**b**) Young's modulus.

Figure 7 depicts the representative stress–strain curves of unidirectional CFRP composites during the static tensile tests at various thermal conditions. All of the stress–strain curves showed a linear elastic regime, followed by a stress drop. From the tensile stress–strain curves, it can be observed that the untreated/intact composite specimen reached the maximum stress level, followed by the stable continuous and cyclic composite specimens. The stable continuous 3D-fabricated composite specimen was slightly less compared to the untreated specimen with almost similar behavior in strain, while the cyclic composite specimen attained the lowest stress level among the composite group and reached beyond the strain level of an untreated and stable continuous composite specimen, which indicated its greater elasticity.

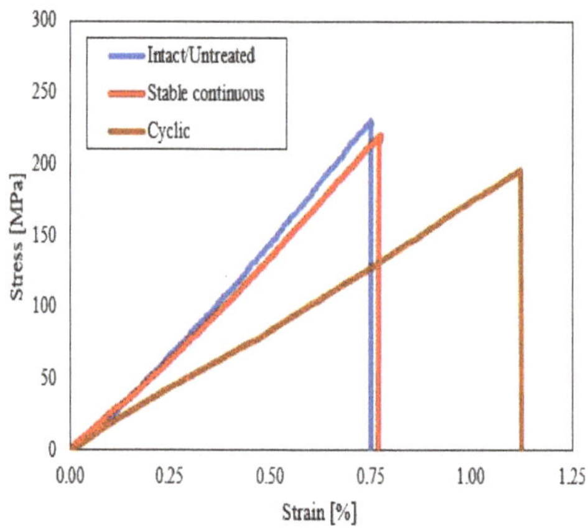

Figure 7. Average stress–strain curve for the tensile samples.

The visual investigation of the failure area of the 15 specimens was studied in accordance with ASTM standard D3039. The failure mode of each thermal group is presented in Figure 8. It was indicated that the intact or untreated specimen group exhibited the lateral-at tab-top (LAT) failure mode and well agreed with the result reported by researchers [27]. The intact sample after longitudinal tension loading exhibited lateral fiber breakage and fiber splitting completely at the top region near the tab, which is gripped by the gauge. Edge delamination in the middle (DGM) was the nominated failure mode in the stable continuous heated group's specimen. It can be seen that there was debonding between layers in the middle due to the lower strength of the interfacial adhesion formed after the continuous thermal loading for 6 h at 65 °C. The explosive breakage at gauge in the middle (XGM) failure mode occurred in the sample from the treated group after thermal exposure with cycling mode between 50 °C and 70 °C for six cycles, and this result was similar to what was studied by Ghasemi et al. [28]. In this damage mode, a large number of fibers pull out near the gauge, and this was presumed due to reduce the chain scission and cross-linking of the polymer matrix caused by thermal cycling. Moreover, it can be observed from Figure 8c that layer debonding also occurred after thermal cycling.

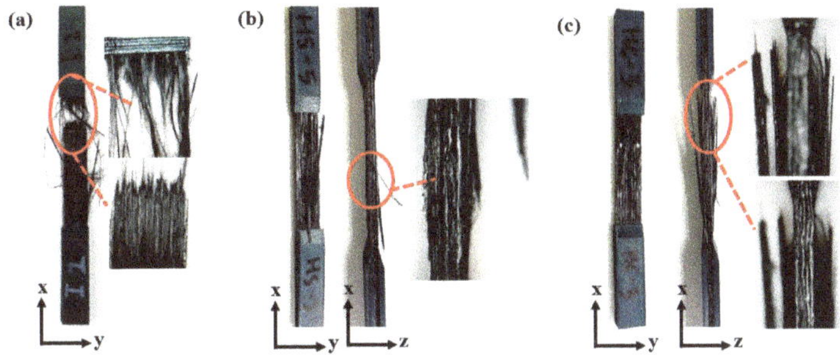

Figure 8. Fractured samples after performing tensile testing. (**a**) Intact; (**b**) continuous; (**c**) cyclic.

3.3. Morphological Investigation

Morphological analyses were performed before and after performing the thermal loading exposures by utilizing an optical microscope device, while a scanning electron microscope was employed to investigate the micro-structure of the different specimen groups after tensile testing. The microscopes are shown in Figure 5.

In order to represent the difference on the structural surface, one specimen of each heating group before and after thermal loading was compared. It can be seen from Figure 9 that there was a slight change in the morphological surface before and after the thermal performance in this experiment. The polymer PLA as a matrix material had a slightly smoother and finer appearance on the continuous and cyclic treatment samples. However, these specimens did not show some deformed shapes (some wrinkle shapes along the longitudinal direction) since it did not undergo sufficiently high-temperature exposure.

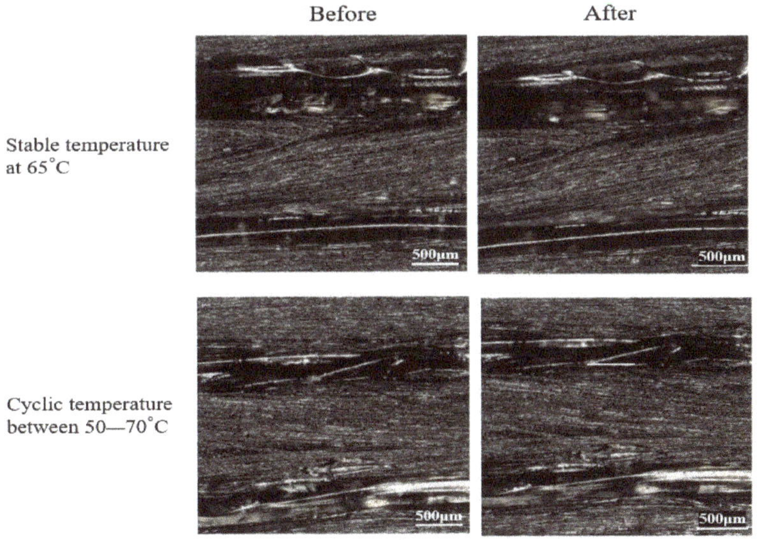

Figure 9. Optical micrographs of specimen before and after stable temperature at 65 °C and cyclic temperature between 50 and 70 °C.

From the SEM micrographs shown in Figure 10, it can be observed that the effect of thermal cycling and continuous heating on the 3D-printed CFRP specimens caused different microstructural fracture damages after tensile testing in both the polymers and carbon fibers compared to the untreated (intact) specimens. The PLA polymers from the thermally treated specimens were split up with a relatively micro and massive cracked hole located in the middle for continuous and cyclic heating, respectively. As for the intact sample, its polymer structure remained morphologically unseparated, and fracture did not exist. The presented micro-crack in the matrix material was one of the main damage mechanisms caused by the thermal cycles.

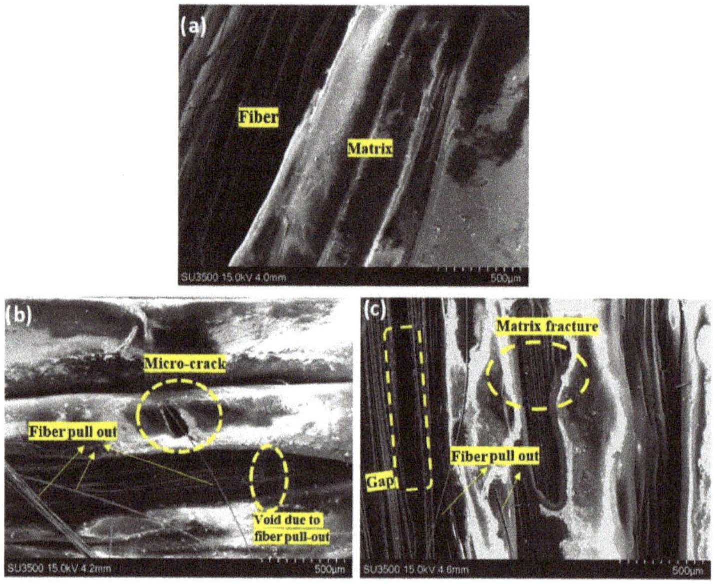

Figure 10. Scanning electron microscope images of the specimen after destructive tensile testing: (**a**) intact (untreated); (**b**) stable continuous at 65 °C; (**c**) cyclic between 50 and 70 °C in a 60°, 0°, and 90° view angle, respectively.

4. Discussion

Based on the experimental results of the tensile test, the untreated group displayed superior strength results when compared to the heated group subjected to stable continuous and cyclic temperatures. The mechanical response after stable temperature exposure showed a moderate loss in tensile strength and Young's modulus. The behavior trends of Young's modulus and the tensile strength were almost similar with a higher detrimental effect under thermal cycling. The mechanical strength of 3D CFRP specimens subjected to the continuous temperature and the cyclic temperature was reduced by 3.6% and 9.6%, respectively. The Young's modulus of the specimens subjected to the continuous temperature and the cyclic temperature was decreased by 11.4% and 26.5%, respectively. These values obtained from the tensile testing indicated the degradation of the mechanical properties after the thermal exposures.

It was decided to investigate the fracture interface of the 3D-printed samples in order to monitor the deformation behavior and how the fracture happened prior to and after mechanical testing. To study such occurrences, an optical microscope was employed to investigate the interfacial adhesion between the continuous carbon fiber (CCF) and PLA thermoplastic matrix prior to destructive tensile testing. The SEM microscope was used to examine the fracture interface of the 3D CFRP specimens and changes in the matrix microstructure from each group. To investigate such outcomes, one specimen from each group was chosen that best reflected the failure mode.

The thermal cycling performed in this work was with an oxidative condition, where environmental air was used during the cooling process by opening the ventilator to the maximum scale to allow the air to enter the oven chamber. This oxidative atmosphere led to accelerated aging of the CFRP composites due to a mismatch in the thermal expansion coefficient (CTE), which caused the polymer matrix's oxidation and cracking [18,19]. While microscopic-scale damage such as micro-cracking occurred as a result of the impact and internal stresses, micro-cracking was the leading cause of material failure due to its undetectable nature, as well as the induced structure fragmentation, which led to a reduction in the mechanical properties such as strength, stiffness, and dimensional stability [29].

The visual investigation of the failure area of the 15 specimens was studied in accordance with ASTM standard D3039. The failure mode of each thermal group is presented in Figure 8. It was indicated that the intact or untreated specimen group exhibited the lateral-at tab-top (LAT) failure mode and well-agreed with the result reported by researchers [27]. The intact sample after longitudinal tension loading exhibited lateral fiber breakage and fiber splitting completely at the top region near the tab, which was gripped by the gauge. Edge delamination in the middle (DGM) was the nominated failure mode in the stable continuous heated group's specimen. It can be seen that there was debonding between layers in the middle due to the lower strength of interfacial adhesion formed after the continuous thermal loading for 6 h at 65 °C. The explosive breakage at gauge in the middle (XGM) failure mode occurred in the sample from the treated group after thermal exposure with cycling mode between 50 °C and 70 °C for six cycles, and this result was similar to what was studied by Ghasemi et al. [28]. In this damage mode, a large amount of fibers pulled out near the gauge, and it was presumed to be due to the reduced chain scission and cross-linking of the polymer matrix caused by the thermal cycling. Moreover, it can be observed from Figure 8c that layer debonding also occurred after the thermal cycling.

The appearance of breakage phenomena can be also triggered by the interfacial forces exerted between the fiber and the matrix, as presented in Figure 8. It has been recently reported in research work where functionalized surfaces with pure cellulose nanocrystals [30] are faced with composites made by matrices with a polar nature and cellulose-rich hemp/flax fibers [31]. These studies concluded that the breakage of bundles and the single hemp/flax fibers was due to these high interfacial forces. In addition, the early stages of cycling were also linked to the production of microvoids, followed by interfacial sliding, which caused composite quality degradation [32].

The mechanical properties of polymeric composites are more matrix-dependent, making them more susceptible to thermal aging. However, it is worth noting that the type of thermal loading can also influence composite behavior. As mentioned previously, the detrimental effect on the mechanical behavior of the 3D CFRP composite was attributed to the difference of the coefficient of thermal expansion (CTE) between the matrix and fiber, and it was caused by the reduced cross-linking of the polymers [25]. This CTE disparity caused local thermal stress, which might cause fiber pull-out due to fiber–matrix debonding. The debonding of fiber and matrix occurs in an oxidative environment due to the significant strain gradients present in matrix locations near highly stiff fibers, causing high local stresses, which lead to crack initiation and mechanical deterioration of composite specimens [18,19,26].

The visual inspection of the morphological surface shown in Figure 9 showed that there was a slight change to the morphological surface before and after the thermal treatment in this experiment. On stable (prolonged) and cyclic treatment samples, the polymer PLA as the matrix material had a slightly smoother and finer appearance. However, because they were not exposed to sufficiently high temperatures, these specimens did not exhibit visible warpage or deformed shapes (some wrinkle shapes along the longitudinal direction). Moreover, the warpage of the 3D CFRP specimen was governed by the stress level released in the macroscopic dimension [33]. It can be observed from Figure 10 that matrix cracks occurred after the thermally stable treatment, and in the cycled group, the crack was larger. When the specimens were exposed to stable heating (prolonged) for 6 h at 65 °C, which is a few degrees above the glass transition temperature of the PLA polymer used, the molecular chains of the polymer changed their structure into an amorphous one, and only local motions of stretching occurred [20]. In addition to that, the prolonged heating allowed the formation of adsorptive inter-layer bonds and volatile groups. As a result, the polymer structure changed, and inner stress relaxation occurred as a result of this thermo-process, which is known as thermo-relaxation [34]. In the case of thermal cycling, the strain gradients were accumulated in the matrix material, and the local stresses increased as a result of the reheating and re-cooling exposure of the polymeric composites; the molecular motion of the long-range polymer structure was reactivated several times.

5. Conclusions

In summary, we investigated the effect of stable and cyclic temperature on the mechanical properties of unidirectional 3D CFRP composites. Static tensile testing was performed to obtain the mechanical strength, Young's modulus, and failure damage of these composites. The results indicated that unidirectional CFRP composites undergo thermal exposures, revealing a degraded mechanical strength and Young's modulus under stable temperature at 65 °C and cyclic temperature between 50 and 70 °C. The specimens under thermal cyclic possessed lower mechanical performance compared to the untreated group by having a 9.6% reduction in tensile strength and 26.5% in the elastic modulus. Meanwhile, the 3D CFRP composite group under continuous thermal exposure exhibited a lower decrease of the mechanical properties by having a 2.4% reduction in tensile strength and 11.4% in the elastic modulus. The degradation in the mechanical strength and elastic modulus after thermally stable and cyclic loading was attributed to the difference in the coefficient of thermal expansion (CTE) between the matrix and fiber, which was caused by the reduced cross-linking of the polymers.

Based on morphological investigations using an optical microscope, the failure mode of 15 specimens was studied in accordance with ASTM standard D3039. It was indicated that the intact or untreated specimen group exhibited lateral damage at top failure mode (LAT), the thermally stable group underwent edge delamination in the middle (DGM) as the nominated failure mode, and the explosive breakage at gauge in the middle (XGM) failure mode occurred in the sample from the thermal cycling group.

Based on the SEM observations, the delamination, fiber pull-out, and matrix cracking were the dominant damages in the 3D-printed tensile-tested specimens. The molecular chains of the polymer changed their structure into an amorphous one, and only local motions of stretching occurred when the specimens were exposed to stable heating (prolonged). In the case of thermal cycling, the strain gradients were accumulated in the matrix material, and the local stresses increased as a result of the reheating and re-cooling exposure of the polymeric composites; the molecular motion of the long-range polymer structure was reactivated several times. Micro-cracking in the matrix occurred as a result of internal stresses, which led to material failure and a reduction in the mechanical properties. The temperature had a considerable influence on the matrix material, and connecting together the CFRP components reduced the material's strength. As a result, it is critical to comprehend temperature-induced reactions in composite materials.

The findings presented in this work offer pathways to investigate further the degradation of 3D composites under several temperature conditions to study the integrity of the structure and the damage mechanism. In future work, similar thermal exposure modes (thermally stable and thermal cycle) will be performed at different magnitudes to provide an extensive horizon of thermal effects on the mechanical characteristics of 3D-printed CFRP composites. Furthermore, investigating the effect of different thermal cycling parameters such as heating and cooling rates and dwell time on the materials' properties would be beneficial for industrial applications. The thermal degradation of the polymer material used in the additive manufacturing of a composite should be considered to be investigated through rheological and thermogravimetric measurements to understand the thermal behavior of the material during the heating conditions.

Ultimately, understanding the thermal influence at different conditions on the morphological structure of FDM printed materials and their mechanical properties can show the first step to inspecting and eliminating them. These thermal effects may incite researchers to develop techniques of thermal treatment of printed specimens that withstand their mechanical characteristics to be potential applications for the operation of space under various thermal conditions.

Author Contributions: Conceptualization, I.I.M. and M.M.; methodology, I.I.M., M.M., R.R. and M.R.; software, I.I.M. and M.M.; validation, M.M.; formal analysis, I.I.M. and M.M.; investigation, I.I.M., M.M., R.R., N.M. and M.R.; resources, M.M., R.R. and M.R.; writing—original draft preparation, I.I.M., M.M. and N.M.; visualization, I.I.M., M.M. and N.M.; supervision, M.M. and R.R.; project administration, M.M.; funding acquisition, M.M. All authors have read and agreed to the published version of the manuscript.

Funding: The research was financed by National Science Centre, Poland, within the project entitled Thermal degradation processes of additive manufactured structures (2019/35/O/ST8/00757), and the Polish National Agency for Academic Exchange within the Foreign doctoral fellowship program (PPN/STA/2021/1/00006).

Institutional Review Board Statement: Not applicable.

Data Availability Statement: Not applicable.

Acknowledgments: The authors of this work would like to thank to researchers from the Kaunas University of Technology, Lithuania, for the sample manufacturing.

Conflicts of Interest: The authors declare that they have no competing financial interest that could have influenced the work reported in this study.

References

1. Hou, Z.; Kobayashi, H.; Tanaka, K.; Takarada, W.; Kikutani, T.; Takasaki, M. Laser-Assisted Melt Electrospinning of Poly (L-lactide-co-ε-caprolactone): Analyses on Processing Behavior and Characteristics of Prepared Fibers. *Polymers* **2022**, *14*, 2511. [CrossRef] [PubMed]
2. Jiang, C.; Wang, K.; Jiang, X.; Zhang, C.; Wang, B. Quantitative investigation of the process parameters of electrohydrodynamic direct-writing and their effects on fiber surface roughness and cell adhesion. *Polymers* **2020**, *12*, 2475. [CrossRef] [PubMed]
3. Habibi, M.; Shili, C.N.; Sutton, J.; Goodarzi, P.; Pezeshki, A. Dietary branched-chain amino acids modulate the dynamics of calcium absorption and reabsorption in protein-restricted pigs. *J. Anim. Sci. Biotechnol.* **2022**, *13*, 1–16. [CrossRef] [PubMed]
4. Böhm, R.; Hornig, A.; Weber, T.; Grüber, B.; Gude, M. Experimental and Numerical Impact Analysis of Automotive Bumper Brackets Made of 2D Triaxially Braided CFRP Composites. *Materials* **2020**, *13*, 3554. [CrossRef] [PubMed]
5. Rahman, A.S. Design of cost-effective and efficient fiber-reinforced composite blades for wind turbines. *Reinf. Plast.* **2019**, *63*, 21–25. [CrossRef]
6. Abdullah, F.; Okuyama, K.i.; Morimitsu, A.; Yamagata, N. Effects of thermal cycle and ultraviolet radiation on 3D-printed carbon fiber/polyether ether Ketone Ablator. *Aerospace* **2020**, *7*, 95. [CrossRef]
7. Zhang, M.; Sun, B.; Gu, B. Accelerated thermal ageing of epoxy resin and 3-D carbon fiber/epoxy braided composites. *Compos. Part A Appl. Sci.* **2016**, *85*, 163–171. [CrossRef]
8. Pascual-González, C.; San Martín, P.; Lizarralde, I.; Fernández, A.; León, A.; Lopes, C.; Fernández-Blázquez, J. Post-processing effects on microstructure, interlaminar and thermal properties of 3D-printed continuous carbon fiber composites. *Compos. B Eng.* **2021**, *210*, 108652. [CrossRef]
9. Arjun, P.; Bidhun, V.; Lenin, U.; Amritha, V.; Pazhamannil, R.V.; Govindan, P. Effects of process parameters and annealing on the tensile strength of 3D-printed carbon fiber reinforced polylactic acid. *Mater. Today Proc.* **2022**, *62*, 7379–7384. [CrossRef]
10. Nassar, A.; Younis, M.; Elzareef, M.; Nassar, E. Effects of Heat-Treatment on Tensile Behavior and Dimension Stability of 3D Printed Carbon Fiber Reinforced Composites. *Polymers* **2021**, *13*, 4305. [CrossRef]
11. Handwerker, M.; Wellnitz, J.; Marzbani, H.; Tetzlaff, U. Annealing of chopped and continuous fiber reinforced polyamide 6 produced by fused filament fabrication. *Compos. B Eng.* **2021**, *223*, 109119. [CrossRef]
12. Wang, K.; Long, H.; Chen, Y.; Baniassadi, M.; Rao, Y.; Peng, Y. Heat-treatment effects on dimensional stability and mechanical properties of 3D-printed continuous carbon-fiber-reinforced composites. *Compos. Part A Appl. Sci.* **2021**, *147*, 106460. [CrossRef]
13. Gutiérrez, E.; Barreto, J.d.J.; Garcia-Hernandez, S.; Morales, R.; González-Solorzano, M.G. Decrease of nozzle clogging through fluid flow control. *Metals* **2020**, *10*, 1420. [CrossRef]
14. Rimašauskas, M.; Kuncius, T.; Rimašauskienė, R. Processing of carbon fiber for 3D-printed continuous composite structures. *Mater. Manuf.* **2019**, *34*, 1528–1536. [CrossRef]

15. Tofan, T.; Borodinas, S.; Kačianauskas, R.; Jasevičius, R. Modeling 3D Droplet Movement Using a Drop-on-Demand Inkjet Printhead Model. *Processes* **2022**, *10*, 1467. [CrossRef]
16. Jia, Z.; Li, T.; Chiang, F.p.; Wang, L. An experimental investigation of the temperature effect on the mechanics of carbon fiber reinforced polymer composites. *Compos. Sci. Technol.* **2018**, *154*, 53–63. [CrossRef]
17. Ghasemi, A.; Tabatabaeian, A.; Moradi, M. Residual stress and failure analyses of polymer matrix composites considering thermal cycling and temperature effects based on classical laminate plate theory. *J. Compos. Mater.* **2019**, *53*, 3021–3032. [CrossRef]
18. González, S.; Laera, G.; Koussios, S.; Domínguez, J.; Lasagni, F.A. Simulation of thermal cycle aging process on fiber-reinforced polymers by extended finite element method. *J. Compos. Mater.* **2018**, *52*, 1947–1958. [CrossRef]
19. Lafarie-Frenot, M.; Rouquie, S.; Ho, N.; Bellenger, V. Comparison of damage development in C/epoxy laminates during isothermal ageing or thermal cycling. *Compos. Part A Appl. Sci.* **2006**, *37*, 662–671. [CrossRef]
20. Muna, I.I.; Mieloszyk, M. Temperature Influence on Additive Manufactured Carbon Fiber Reinforced Polymer Composites. *Materials* **2021**, *14*, 6413. [CrossRef]
21. Sampedro, G.A.R.; Agron, D.J.S.; Amaizu, G.C.; Kim, D.S.; Lee, J.M. Design of an In-Process Quality Monitoring Strategy for FDM-Type 3D Printer Using Deep Learning. *Appl. Sci.* **2022**, *12*, 8753. [CrossRef]
22. Agron, D.J.S.; Lee, J.M.; Kim, D.S. Nozzle thermal estimation for fused filament fabricating 3d printer using temporal convolutional neural networks. *Appl. Sci.* **2021**, *11*, 6424. [CrossRef]
23. Maqsood, N.; Rimašauskas, M. Influence of printing process parameters and controlled cooling effect on the quality and mechanical properties of additively manufactured CCFRPC. *Compos. Commun.* **2022**, *35*, 101338. [CrossRef]
24. Al Abadi, H.; Thai, H.T.; Paton-Cole, V.; Patel, V. Elastic properties of 3D-printed fiber-reinforced structures. *Compos. Struct.* **2018**, *193*, 8–18. [CrossRef]
25. Azimpour-Shishevan, F.; Akbulut, H.; Mohtadi-Bonab, M. Thermal shock behavior of twill woven carbon fiber reinforced polymer composites. *J. Compos. Sci.* **2021**, *5*, 33. [CrossRef]
26. Chung, D.D. 4-Polymer-Matrix Composites: Mechanical Properties and Thermal Performance. *Carbon Compos.* **2017**, 218–255.
27. Maqsood, N.; Rimašauskas, M. Characterization of carbon fiber reinforced PLA composites manufactured by fused deposition modeling. *JCOMC* **2021**, *4*, 100112. [CrossRef]
28. Ghasemi, A.; Moradi, M. Low thermal cycling effects on mechanical properties of laminated composite materials. *Mech. Mater.* **2016**, *96*, 126–137. [CrossRef]
29. Razak, Z.; Sulong, A.B.; Muhamad, N.; Haron, C.H.C.; Radzi, M.K.F.M.; Ismail, N.F.; Tholibon, D.; Tharazi, I. Effects of thermal cycling on physical and tensile properties of injection moulded kenaf/carbon nanotubes/polypropylene hybrid composites. *Compos. B Eng.* **2019**, *168*, 159–165. [CrossRef]
30. Marcuello, C.; Foulon, L.; Chabbert, B.; Molinari, M.; Aguié-Béghin, V. Langmuir–Blodgett Procedure to Precisely Control the Coverage of Functionalized AFM Cantilevers for SMFS Measurements: Application with Cellulose Nanocrystals. *Langmuir* **2018**, *34*, 9376–9386. [CrossRef]
31. Berzin, F.; Lemkhanter, L.; Marcuello, C.; Chabbert, B.; Aguié-Béghin, V.; Molinari, M.; Castellani, R.; Vergnes, B. Influence of the polarity of the matrix on the breakage mechanisms of lignocellulosic fibers during twin-screw extrusion. *Polym. Compos.* **2020**, *41*, 1106–1117. [CrossRef]
32. Russell-Stevens, M.; Todd, R.; Papakyriacou, M. The effect of thermal cycling on the properties of a carbon fiber reinforced magnesium composite. *Mater. Sci. Eng. A* **2005**, *397*, 249–256. [CrossRef]
33. Jensen, M.; Jakobsen, J. Effect of cure cycle on enthalpy relaxation and post shrinkage in neat epoxy and epoxy composites. *J. Non. Cryst. Solids* **2016**, *452*, 109–113. [CrossRef]
34. Korolev, A.; Mishnev, M.; Vatin, N.I.; Ignatova, A. Prolonged thermal relaxation of the thermosetting polymers. *Polymers* **2021**, *13*, 4104. [CrossRef] [PubMed]

polymers

MDPI

Article

Ballistic Limit of UHMWPE Composite Armor under Impact of Ogive-Nose Projectile

Li Ding [1,2,*], Xiaohui Gu [1], Peihui Shen [1] and Xiangsheng Kong [2]

[1] School of Mechanical Engineering, Nanjing University of Science and Technology, Nanjing 210094, China
[2] Nanjing Changjiang Electronics Group Co., Ltd., Nanjing 210037, China
* Correspondence: dingli@njust.edu.cn

Abstract: The ballistic response of armor has been widely used to evaluate its feasibility and advantages as a protective structure. To obtain the ballistic performance and ballistic limitations of composite armor, a type of ultra–high molecular weight polyethylene (UHMWPE) composite armor is designed, which is composed of UHMWPE laminates and steel face sheets of Q235. The total thickness of the armor is 53 mm, with an in–plane dimension of 300 mm × 300 mm. Then, an experimental study of the ballistic impact response of composite armor subject to a typical ogive–nose projectile was carried out. In the velocity range of 501.1 to 1026.1 m/s, the 14.5 mm caliber armor–piercing projectile could penetrate through the composite armor. At the velocity of 433.3 m/s, the A–P projectile was embedded in the armor, leaving a bulge mark on the back sheet. Therefore, 467.2 m/s is taken as the ballistic limit of the armor under the impact of the ogive–nose projectile. In addition, a corresponding numerical simulation model is also established to predict the ballistic limit of the projectile. The numerical predictions are consistent with the experimental results. The ballistic limit obtained from the numerical simulation results is 500 m/s, which is acceptable with a relative error of 7.02%. The failure mechanism of the composite armor is also obtained. Petaling is the main dominant failure mode for both face sheets, while delamination and shear failure dominate the penetration process of UHMWPE laminates. Finally, the perforation mechanism of composite armor under the impact of an A–P projectile is analyzed with theoretical models to predict the residual velocity, the work performed during the perforation, and the resisting stress of σ_s in the cavity–expansion model. The experimental and numerical simulation results can provide necessary data in the analysis of the composite structure's dynamic response under the impact of sharp head penetrators. The research results present the ballistic performance, failure mechanism, and ballistic limit of the composite armor under the impact of a typical ogive–nose projectile, which can be significant in the design of composite armor in the areas of ship shield, fortifications protection, and bulletproof structures against threats from sharp head penetrators.

Keywords: ballistic limit; composite armors; ogive–nose projectile; theoretical analysis; ultra–high molecular weight polyethylene

Citation: Ding, L.; Gu, X.; Shen, P.; Kong, X. Ballistic Limit of UHMWPE Composite Armor under Impact of Ogive-Nose Projectile. *Polymers* **2022**, *14*, 4866. https://doi.org/10.3390/polym14224866

Academic Editors: R. A. Ilyas, S. M. Sapuan, Emin Bayraktar, Shukur Abu Hassan, Nabil Hayeemasae and Khubab Shaker

Received: 13 October 2022
Accepted: 8 November 2022
Published: 11 November 2022

1. Introduction

Multilayered armor systems have been widely used in numerous ballistic and bulletproof applications, including bulletproof helmets, vests, and other armor parts, providing an acceptable range of protection for soldiers and structures [1–6]. Due to the advantages of high stiffness and low density, composites reinforced with ultra–high molecular weight polyethylene (UHMWPE) fibers are increasingly being used in the field of national defense as parts of lightweight armor systems to protect fortifications and structures from ballistic impacts [7–9]. The ballistic performance of monolithic UHMWPE composite under the impact of a blunt projectile has been studied [10–14], and the failure mechanisms analyzed. A common design of composite armor is a sandwich structure, which is made of thin face sheets and low–density non–metal cores. In addition, the ballistic impact response of a sandwich structure consisting of UHMWPE under a blunt projectile has also been investigated [15–19]. Deflection and bulging [4] consisting of shear plugging, formation of a transition plane, and bulging are the failure modes of UHMWPE [11], which allow it to have an excellent ability to resist the penetration of blunt projectiles.

It is reported that the bulletproof properties of composite materials are greatly discounted under the ballistic impact of projectiles with a sharp head [17,20,21]. However, little is reported on the ballistic performance of the UHMWPE or composite armors containing UHMWPE laminates under the impact of sharp head projectiles. Whether the ballistic performance of the composite armor can be predicted using numerical simulation has not been discussed yet. In addition, to evaluate the resistance of composite armors, especially with the sandwich structure containing UHMWPE laminates, the failure mode and failure mechanism of UHMWPE should also be further studied. An experimental study is one of the basic methods of scientific research, which provides the most direct evidence to help understand physical phenomena. However, the penetration process of the ballistic impact can hardly be observed in experiments. Therefore, due to the limits of testing conditions and the costs of the investment, it is difficult to carry out the relevant research that depends on experiments. Numerical simulations are similar to experimental research that is based on theoretical research [10,14,22]; they are efficient but irrelevant in external conditions. With the help of computer realization, the laws of penetration and dynamic response of the target can be obtained with accurate simulation models. The intermediate process in the penetration process could also be observed, and the crucial parameters could be extracted in the post–processing section to help understand the mechanism of the physical experiment better.

In this paper, a sandwich structure of UHMWPE composite armor is designed, which is made of two pieces of UHMWPE laminates in the middle and Q235 steel face sheets. The ballistic performance of the composite armor is studied systematically both from experiment and numerical simulation. The ballistic limit of UHMWPE composite armor under the impact of an ogive–nose penetrator is obtained, and the failure mechanism of the armor is analyzed. Finally, typical theoretical models are used to predict the residual velocity, the work performed during the perforation, and the resisting stress of σ_s in the cavity–expansion model to help better understand the penetration process of the composite armor under the impact of sharp head projectiles.

2. Configuration of the Armor and Projectile

2.1. Design of the Armor

As shown in Figure 1, the UHMWPE composite armor is made up of two pieces of UHMWPE laminates in the middle and Q235 steel face sheets. Typical UHMWPE laminate with a material grade of FDB4-HW-S1 is selected. Each piece of UHMWPE laminate has a thickness of 20 mm, and each Q235 steel face sheet has a thickness of 6 mm. Each layer of armor has the same in–plane dimensions of 300 mm × 300 mm, with a total thickness of 52 mm. The thin binder layer is replaced in each panel, then followed by pressing to obtain the overall panel structure of the composite armor. Due to the existence of a binder layer, the total thickness of the composite armor may increase from 1 mm to 53 mm.

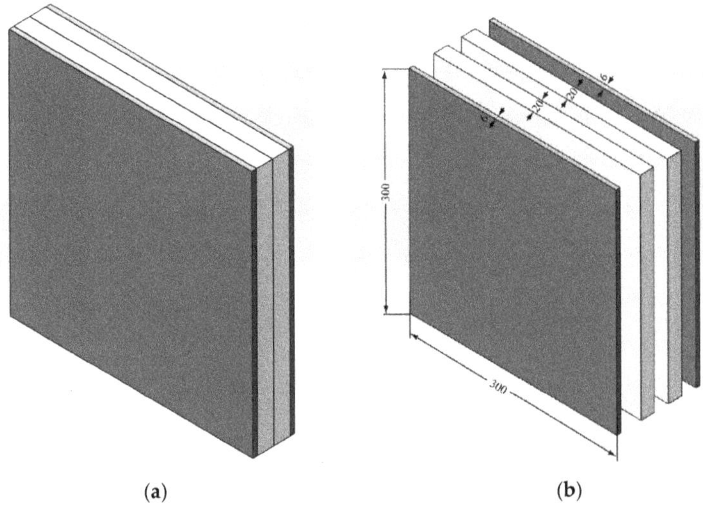

Figure 1. Structures of UHMWPE composite armor. (a) Assembly drawing. (b) Component drawing.

The material properties of Q235 steel are presented in Table 1, which is provided by the manufacturer of Wuhan Iron & Steel Co., Ltd., Wuhan, China.

Table 1. Material properties of Q235.

Steel	Yield Strength (MPa)	Tensile Strength (MPa)	Elongation after Break (%)	Poisson's Ratio (%)	Impact Energy Aku (J)
Q235	305	426	30	0.33	≥27

2.2. Structure of the Projectile

The structure of the standard Chinese projectile of 14.5 mm caliber armor–piercing (A–P) projectile is shown in Figure 2. The main geometric parameters of the projectile are presented in Figure 2a, with a total diameter of 14.93 mm and a length of 66.7 mm. The projectile is mainly composed of the brass jacket, steel core, and lead filler. The steel core with an ogive–nose head is made of a non–deformable hardened steel core, which will mainly contribute to the penetration performance of the projectile. The A–P core has a diameter of 12.48 mm and a length of 53.4 mm, with a mass of 40.2 to 41 g. Generally, the jacket is made of gliding brass, and the filler is made of lead, which serves to protect the barrel from the core, enhance the sealing effect and provide an optimized shape for flight in the air. They have a relatively small effect during the penetration, and they are not modeled in the following numerical simulation [23–25].

The ballistic limit or limit velocity is the velocity required for a particular projectile to reliably penetrate a particular piece of material. In other words, a given projectile will not pierce a given target when the projectile velocity is lower than the ballistic limit [26]. In addition, it is also important to evaluate the resistance of the armor. Due to the unknown perforation mechanism and the unknown ballistic limit of composite armor under the ballistic impact, especially the sharply pointed projectile, an experiment of the UHMWPE composite armor under the ballistic impact of the A–P projectile was carried out first.

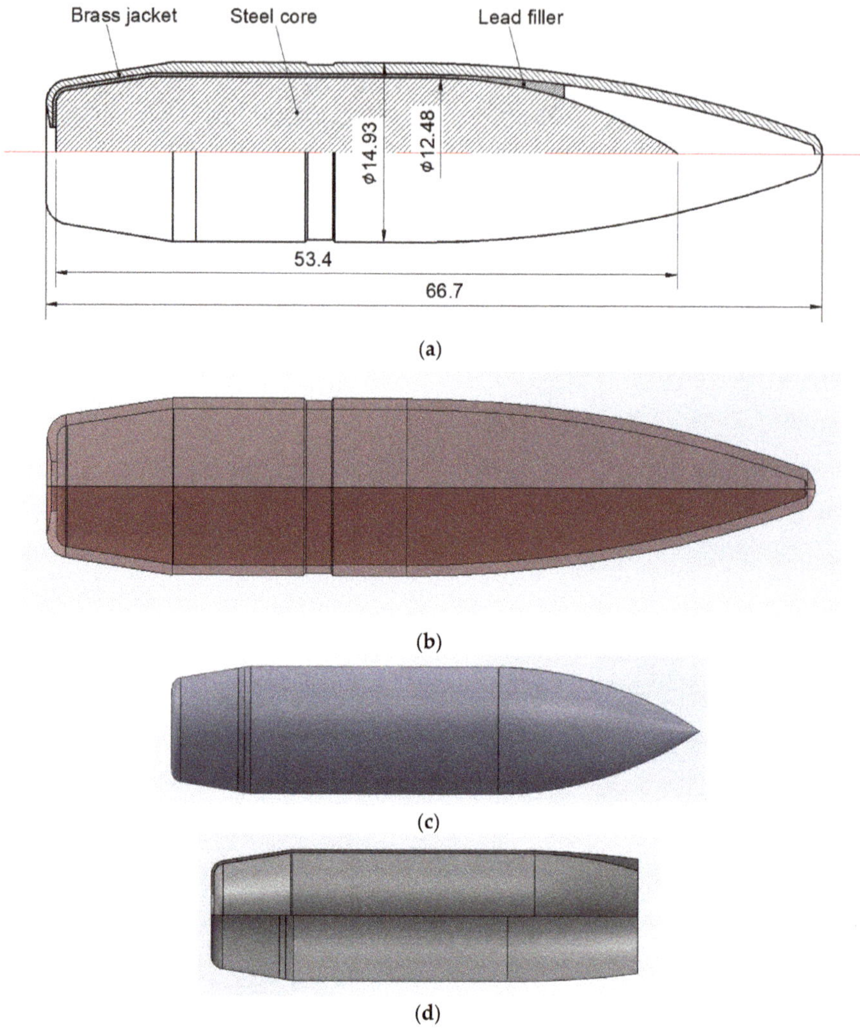

Figure 2. Structure of a 14.5 mm caliber armor–piercing projectile. (**a**) Geometric structure. (**b**) Brass jacket. (**c**) Steel core. (**d**) Lead filler.

3. Experimental Details and Results

3.1. Design of the Experiment

Figure 3 shows the 14.5 mm caliber A–P projectile and the state of the assembly of the projectile in the cartridge. By adjusting the quantity of the propellent in the cartridge, the pre–set velocities of the A–P projectile can be acquired.

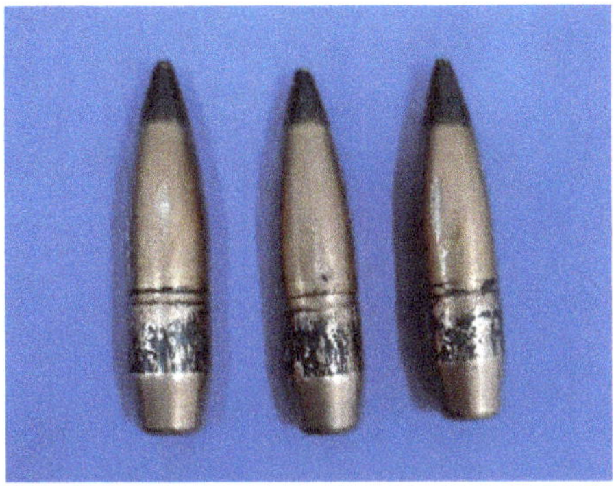

(a)

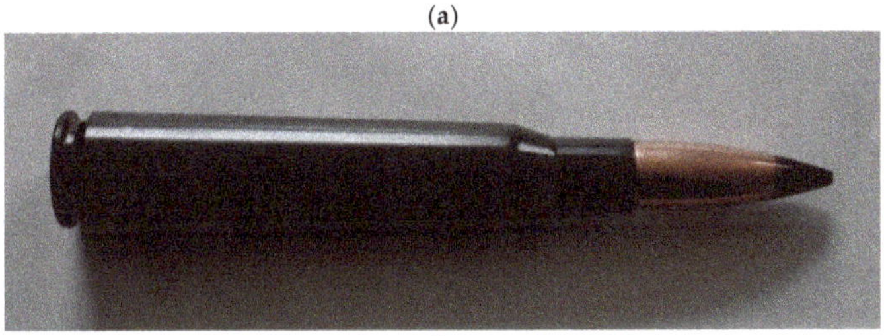

(b)

Figure 3. State of the projectile in the experiment. (**a**) 14.5 mm caliber A–P projectile. (**b**) Assembly of the projectile in the cartridge.

Figure 4 shows the states of the composite armors used in the experiment. The armors were clamped to the rear base on the steel shelf. The rear base was a triangle–shaped bracket, which was fixed to the steel shelf with bundles of iron wire. Figure 5 presents the electronic time–measuring instrument with six channels to capture the signal when the penetrator perforates through the tinfoil target. The electronic time–measuring instrument is of high sensitivity and is able to record the minimum time difference of 1 μs. The layout of the ballistic impact experiment is presented in Figure 6. The 14.5 mm caliber smooth–bore gun was mounted on a rigid platform. The distance between the gun muzzle and armor is about 4 m. To measure the impact velocity of the projectile, two tinfoil targets were placed in front of the armor. The composite armor was placed on a steel shelf at the same height as the ballistic gun.

(a)

(b)

Figure 4. Photograph of the armor in the experiment. (**a**) front view. (**b**) side view.

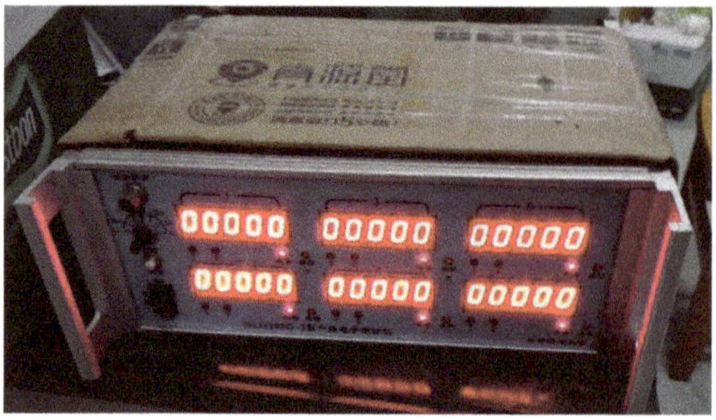

Figure 5. Photograph of the electronic time–measuring instrument.

(**a**)

(**b**)

Figure 6. Configuration of the ballistic impact experiment. (**a**) 14.5 mm caliber smooth–bore gun. (**b**) Layout of the targets.

3.2. Experimental Results

The perforation results of composite armor by the A–P projectile are listed in Table 2. For the unknown perforation mechanism and the unknown ballistic limit of composite armor under the ballistic impact of the sharply pointed projectile especially, the impact velocities v_i of the ogive–nose projectile were set at 1026 m/s initially to much lower velocities. In the velocity range of 501.1 to 1026.1 m/s, the A–P projectile could penetrate through the composite armor. While at the velocity of 433.3 m/s, the A–P projectile was embedded in the armor, leaving a bulge mark on the back sheet. The diameter of the entrance hole on the front sheet was about 14–16 mm, while the diameter of the outlet on the back sheet was about 13–24 mm. The entrance dimension discrepancies are small on the front sheet, and the divergence increased with the increase in impact velocity.

Table 2. Perforation results in the experiment.

v_i (m/s)	Perforation State in the Front and Back	
1026.1	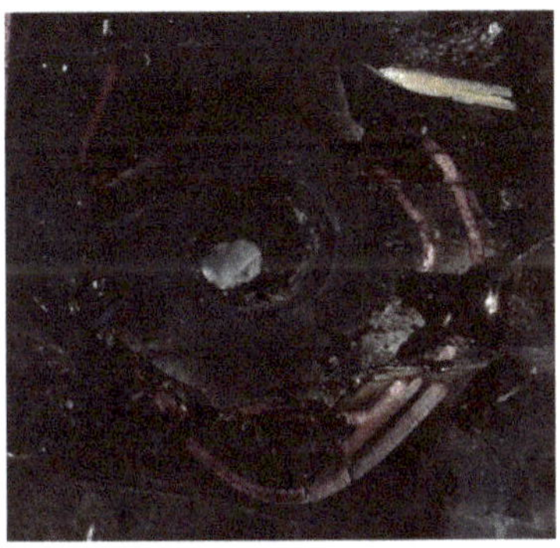	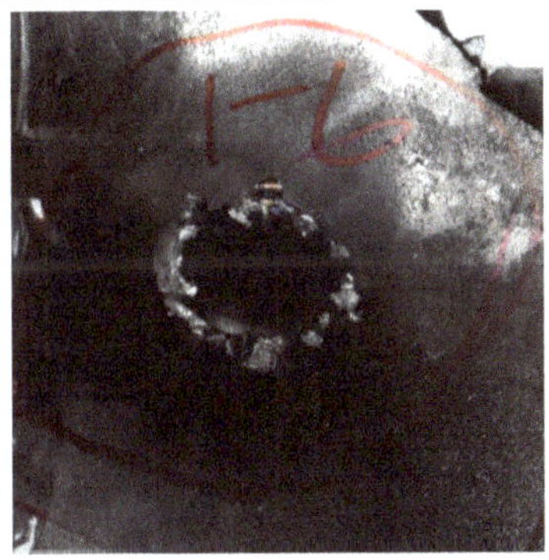
966.6	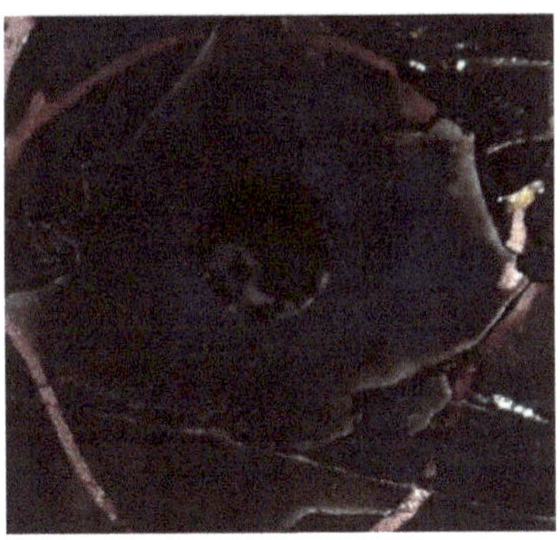	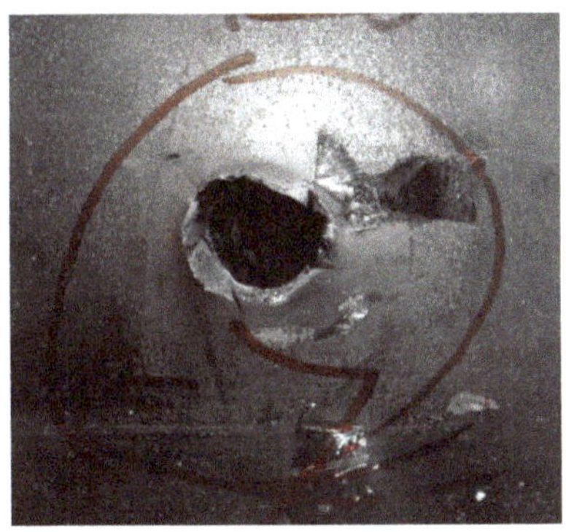

Table 2. *Cont.*

v_i (m/s)	Perforation State in the Front and Back	
768.7	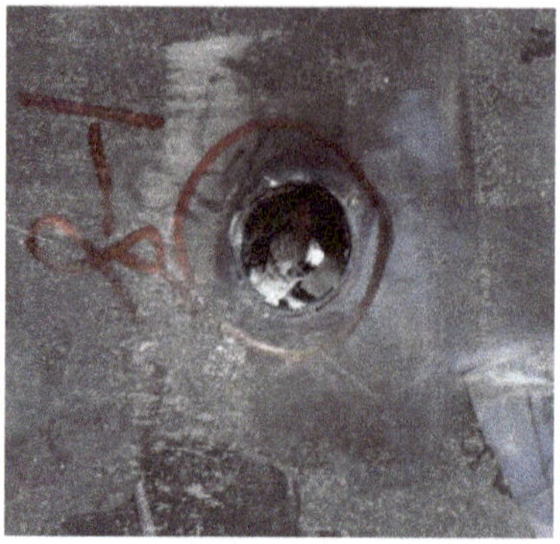	
688.4		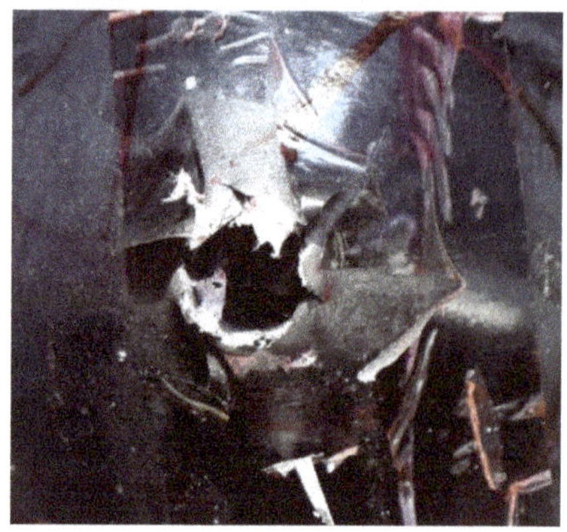
616.8		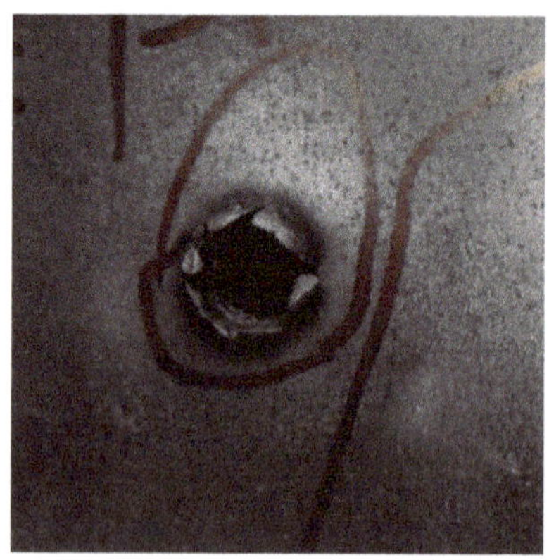

Table 2. *Cont.*

v_i (m/s)	Perforation State in the Front and Back	
595.4	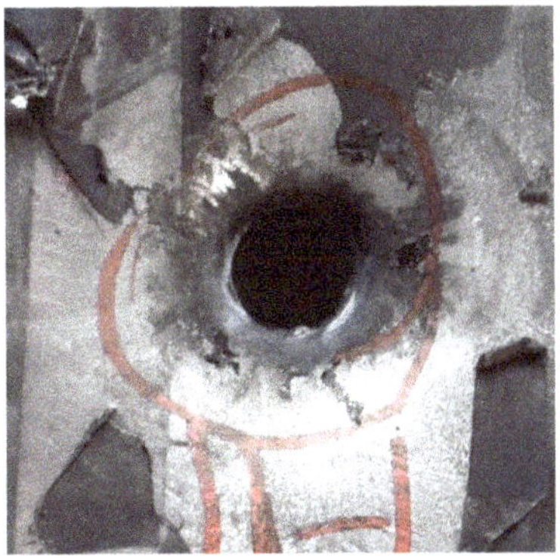	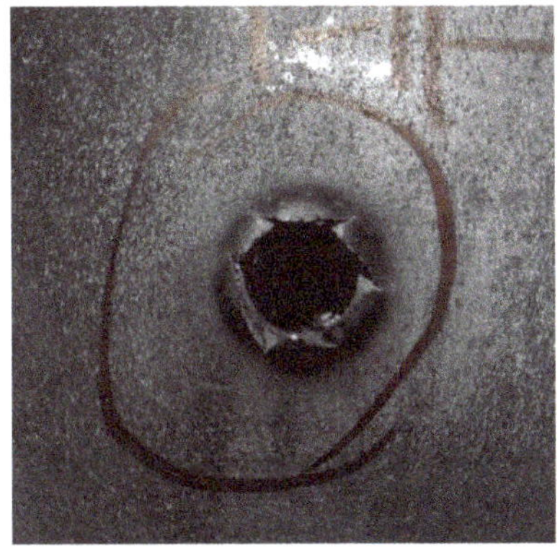
524.8	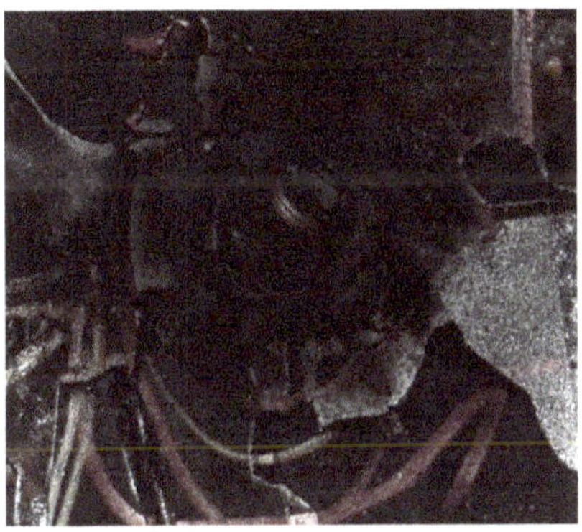	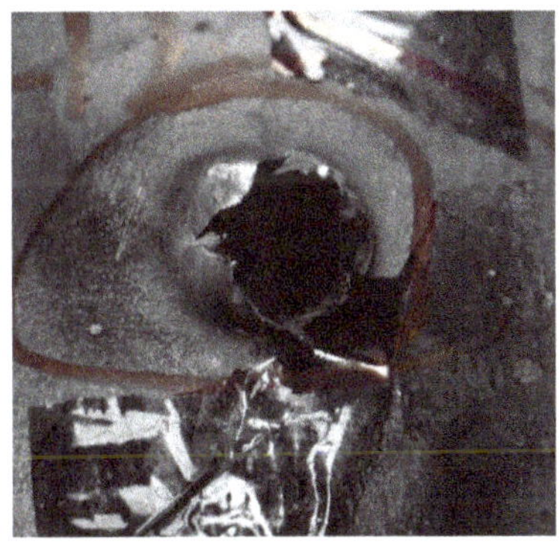
501.1	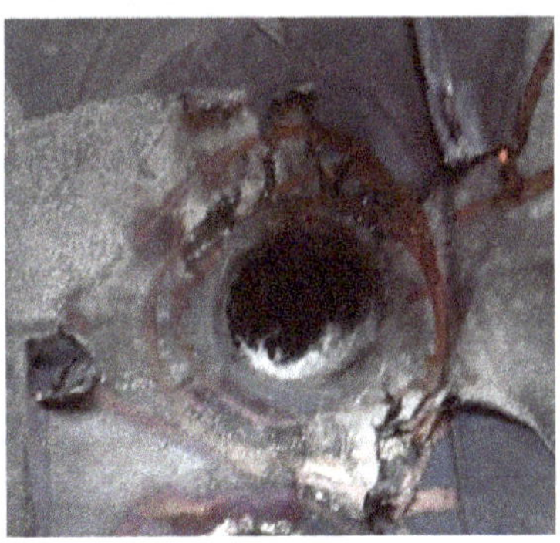	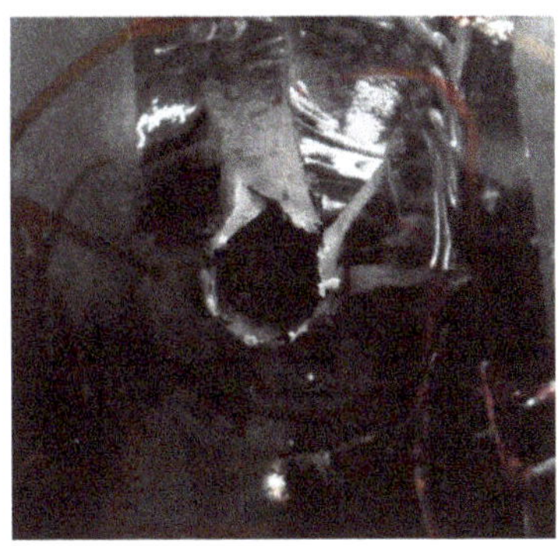

Table 2. *Cont.*

v_i (m/s)	Perforation State in the Front and Back	
433.3	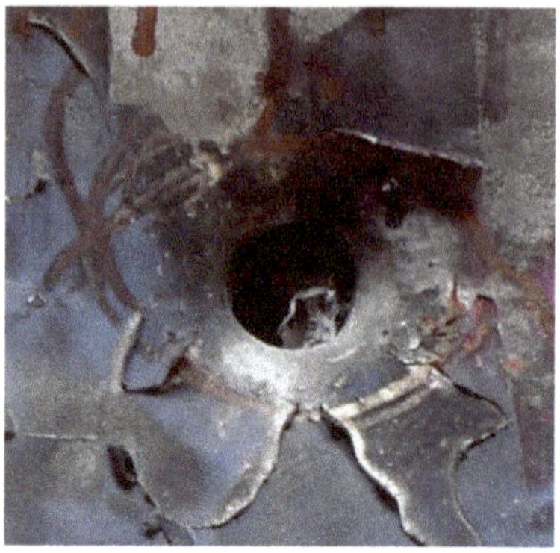	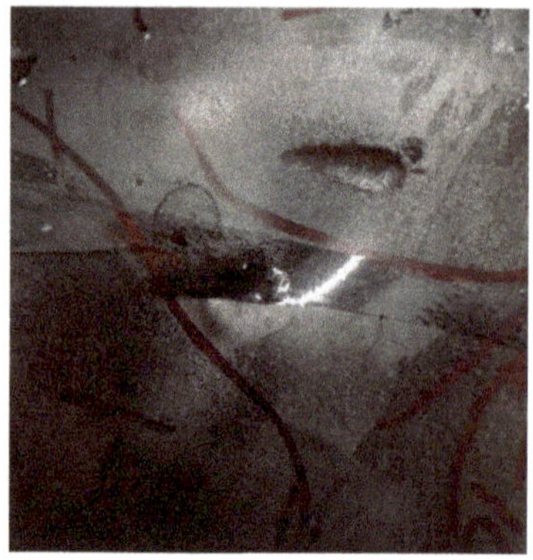

Figure 7 shows the state of a stripped brass jacket and lead filler. Figure 7a,c shows the front and back views of the perforation on the front sheet at the velocity of 524.8 m/s, and Figure 7b,d shows the front and back views of the perforation at the velocity of 501.1 m/s. The ablative phenomenon can be observed, especially on the back view of the front sheet, resulting from a severe interaction during the penetration.

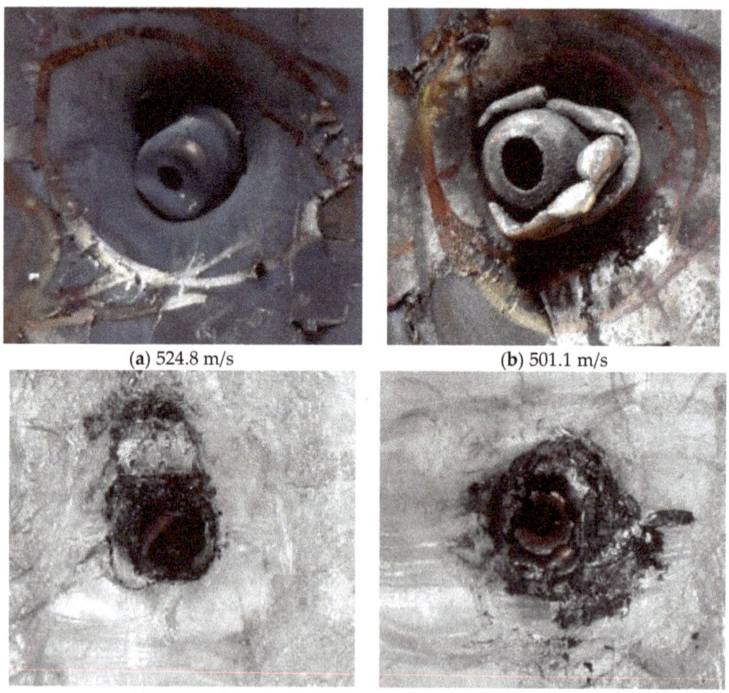

(**a**) 524.8 m/s (**b**) 501.1 m/s (**c**) 524.8 m/s (**d**) 501.1 m/s

Figure 7. The striped brass jacket and lead filler in the perforation of the front sheet at different impact velocities ((**a**,**b**) for the front view, (**c**,**d**) for the back view).

Figure 8 shows the failure states of the PE laminates. As shown in Figure 8a, in the side view of the PE laminates, the obvious phenomenon of delamination can be observed. As presented in Figure 8b, a penetration cavity was formed by the ogive–nose penetrator with relatively neat cutting edges accompanied by a charring layer. In addition, the PE laminate exhibited an extent of fibrillation, and the bare bunches of fibers can be observed around the penetration hole.

(a) (b)

Figure 8. Failure states of the PE laminates. (**a**) Side view of PE laminates. (**b**) Outlet of PE laminate.

Figure 9 presents additional detail on the perforation results at the impact velocity of 433.3 m/s, with the penetrator embedded in the armor. An obvious indentation was formed on the back of the back sheet, as shown in Figure 9a. After the back sheet was removed, the head of the steel core of the A–P projectile can be seen in Figure 9b, accompanied by an extent of fibrillation.

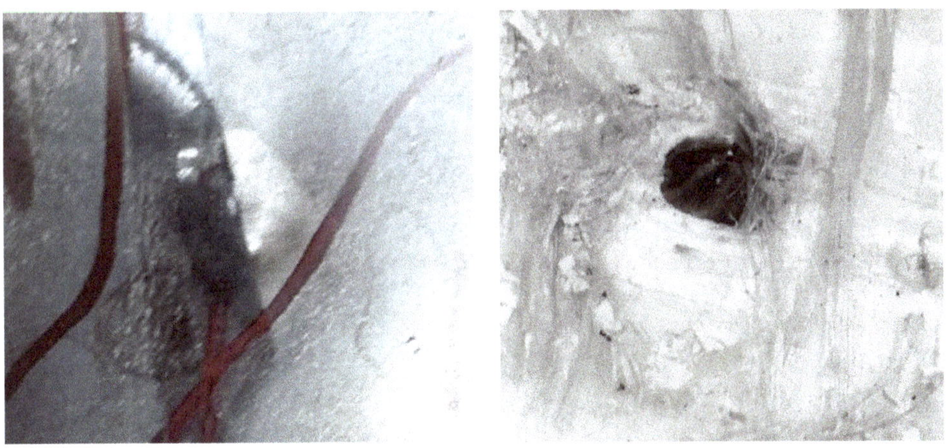

(a) (b)

Figure 9. Additional details for the perforation results at the impact velocity of 433.3 m/s. (**a**) Indentation of back plate by penetration. (**b**) Embedded penetrator.

Regarding the results of Table 2 and Figures 7–9, it can be summarized from the perforation results that: (1) The average velocities of 501.1 m/s and 433.3 m/s can be taken as the ballistic limits of the UHMWPE composite armor under the impact of the ogive–nose penetrator at 467.2 m/s. (2) Petaling, as the main dominant failure mode for both face sheets, can be observed within the range of impact velocity of 501 m/s to 1026 m/s. The surface of both face sheets stays relatively flat, with small overall deformation except for the protruding petal–shaped holes. Small pieces of petals accompanied by gaped rifts formed the perforation. (3) Delamination and shear failure dominate the penetration process of UHMWPE laminates. Due to the low interlaminar stiffness and strength in the PE laminate, delamination is prevalent through the panel's thickness, as can be seen in Figure 10a. (4) The charring layer on the front steel plate can be observed, and more severe ablation could be noticed at the impact velocity of around 1000 m/s.

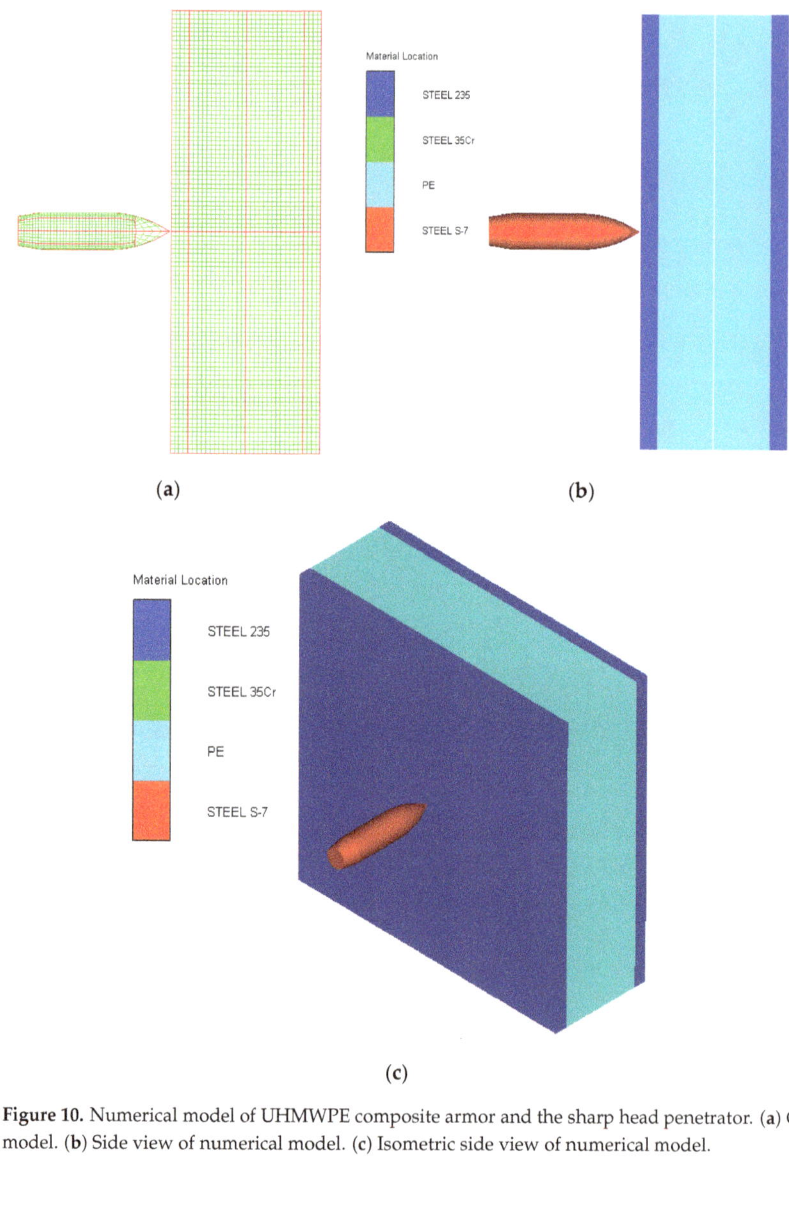

Figure 10. Numerical model of UHMWPE composite armor and the sharp head penetrator. (**a**) Grid model. (**b**) Side view of numerical model. (**c**) Isometric side view of numerical model.

4. Numerical Simulation and Analysis

4.1. Setup of Numerical Model

To predict the dynamic response and obtain the ballistic limit of UHMWPE composite armor under the ballistic impact of the A–P core, three–dimensional numerical models are carried out using the AUTODYN nonlinear software. The version of AUTODYN is v11.0 in the software of ANSYS 11.0, located in Nanjing, China.

As shown in Figure 10, the 3D Lagrange algorithm is adopted for all of the components in numerical simulation. The half 3D model is carried out with a mesh size of about 1.2 mm per grid. A hexahedral structured grid is used to model both the projectile and the composite armor. The numerical simulation model is composed of about 810 thousand nodes and 800 elements. On the edge of the target, fixed boundaries are used to constrain the movement of the armor. The boundary conditions are applied on the edges of both the face and back sheets. Different initial velocities are applied to the ogive–nose head penetrator to simulate the dynamic penetration behavior with different impact velocities. The material models and the parameters will be described below.

As presented in Table 3, the material models for the penetrator, face sheet, and UHMWPE laminate are listed. For steel, the shock equation of state, also called Grüneisen, is employed in conjunction with the Johnson–Cook constitutive model to simulate the dynamic response under ballistic impact. The Grüneisen EOS [27] can be used to describe how the materials interact with the shock wave and are based on Hugoniot's relation between the v_s. and the v_p, as $v_s = c_0 + sv_p$, where v_s. is the shock wave velocity, v_p is the material particle velocity, c_0 is the wave speed, and s is a material–related coefficient. The expression of the equation of state of Grüneisen for the compressed state is:

$$p = \frac{\rho_0 C^2 \mu \left[1 + \left(1 - \frac{\gamma_0}{2}\right)\mu - \frac{a}{2}\mu^2\right]}{\left[1 - (S_1 - 1)\mu - S_2\frac{\mu^2}{\mu+1} - S_3\frac{\mu^3}{(\mu+1)^2}\right]} + (\gamma_0 + a\mu)E. \quad (1)$$

Table 3. Material models used in numerical simulation.

Components	Material	ρ (g/cm^3)	Equation of State	Constitutive Model
Penetrator	Steel S-7	7.83	Shock	Johonson-Cook
Face sheet	Steel Q235	7.896	Shock	Johonson-Cook
PE laminates	UHMWPE	0.98	Ortho	Orthotropic Yield

In the expanded state,

$$p = \rho_0 C^2 \mu + (\gamma_0 + a\mu)E \quad (2)$$

where C is the intercept of the velocity curve between the shock wave and particle; S_1, S_2, and S_3 represent the slope of the $v_s - v_p$ curve; γ_0 is the coefficient of the Grüneisen; a is the one–order correction of γ_0. $\mu = \rho/\rho_0 - 1$ is a non–dimensional coefficient based on initial and instantaneous material densities. The parameters of the Grüneisen equation of state are listed in Table 4.

Table 4. EOS parameters of S-7 and Q235.

Material	Grüneisen Coefficient	C (m/s)	S_1	S_2	a
S-7	2.17	4569	1.330	0	0.47
Q235	2.17	4569	1.490	0	0.46

The Johnson–Cook model [28,29] incorporates the effect of strain rate–dependent work hardening and thermal softening, which is given by:

$$\sigma = (A + B\varepsilon^n)\left(1 + C\ln\frac{\dot{\varepsilon}}{\dot{\varepsilon}_0}\right)(1 - T^{*m}) \quad (3)$$

where ε is the plastic strain, and the temperature factor is expressed as:

$$T^* = \frac{T - T_r}{T_m - T_r} \tag{4}$$

where T_r is the room temperature, and T_m is the melt temperature of the material. A, B, n, C, and m are material–related parameters. The material parameters of S-7 tool steel and Q235 steel are presented in Table 5.

Table 5. Material constants for S-7 and Q235.

Steel	ρ (g/cm^3)	A (MPa)	B (MPa)	n	C	m	$\dot{\varepsilon}_0$ (s^{-1})	T_r (K)	T_m (K)
S-7	7.850	1540	477	0.16	0.016	1.0	1	293	1763
Q235	7.896	350	275	0.36	0.022	1.0	1	293	1793

The orthotropic material model proposed by Long H. Nguyen et al. [14] was used for modeling the dynamic behavior of the UHMWPE layer subjected to ballistic impact. The material models consist of a nonlinear equation of the state of orthotropic, a strength model, and a failure model. The constitutive response of the material in the elastic regime is described as the orthotropic EOS composed of volumetric and deviatoric components. The pressure is defined by:

$$\begin{aligned} P = P(\varepsilon_{vol}, e) - \tfrac{1}{3}(C_{11} + C_{21} + C_{31})\varepsilon_{11}^d - \tfrac{1}{3}(C_{12} + C_{22} + C_{32})\varepsilon_{22}^d \\ -\tfrac{1}{3}(C_{13} + C_{23} + C_{33})\varepsilon_{33}^d \end{aligned} \tag{5}$$

where C_{ij} are the coefficients of the stiffness matrix, ε_{ij}^d refers to the deviatoric strains in the principal directions, and the volumetric component $P(\varepsilon_{vol}, e)$ is defined by the Mie–Grüneisen EOS:

$$P(\varepsilon_{vol}, e) = P_r(v) + \frac{\Gamma(v)}{v}[e - e_r(v)] \tag{6}$$

where v, e, and $\Gamma(v)$ represent the volume, internal energy, and the Grüneisen coefficient, respectively. $P_r(v)$ is the reference pressure, and $e_r(v)$ is the reference internal energy. The quadratic yield surface was adopted as the material strength model to describe the nonlinear, irreversible hardening behavior of the composite laminate:

$$\begin{aligned} f(\sigma_{ij}) &= a_{11}\sigma_{11}^2 + a_{22}\sigma_{22}^2 + a_{33}\sigma_{33}^2 + 2a_{12}\sigma_{11}\sigma_{22} + 2a_{23}\sigma_{22}\sigma_{33} \\ &+2a_{13}\sigma_{11}\sigma_{33} + 2a_{44}\sigma_{23}^2 + 2a_{55}\sigma_{31}^2 + 2a_{66}\sigma_{12}^2 = k \end{aligned} \tag{7}$$

where a_{ij} are the plasticity coefficients, and σ_{ij} represent the stresses in the principal directions of the material. In addition, the state variable, k, is used to define the border of the yield surface. It is described with a master and stress–effective plastic strain curve defined by ten piecewise points to consider the effect of strain hardening.

In the numerical models, the failure model of the orthotropic material is based on a combined stress criterion given as follows:

$$\left(\frac{\sigma_{ii}}{S_{ii}(1 - D_{ii})}\right)^2 + \left(\frac{\sigma_{ij}}{S_{ij}(1 - D_{ij})}\right)^2 + \left(\frac{\sigma_{ki}}{S_{ki}(1 - D_{ki})}\right)^2 \geq 1 \text{ for } i, j, k = 1, 2, 3 \tag{8}$$

where S is the failure strength in the respective directions of the material, and D is the damage parameter following a linear relationship with stress and strain, as shown below:

$$D_{ii} = \frac{L\sigma_{ii,f}\varepsilon_{cr}}{2G_{ii,f}} \tag{9}$$

where L is the characteristic cell length, ε_{cr} refers to the crack strain, and $G_{ii,f}$ presents the fracture energy in the direction of damage.

The constants for the orthotropic equation of state are presented in Table 6, and the parameters for orthotropic yield strength are shown in Table 7.

Table 6. Material constants for Orthotropic equation of state.

	Value	Units	Parameters	Value	Units
Reference density	0.98	g/cm^3	Shear modulus 12	2.0×10^6	kPa
Young's modulus 11	3.62×10^6	kPa	Shear modulus 23	1.92×10^5	kPa
Young's modulus 22	5.11×10^7	kPa	Shear modulus 31	2.0×10^6	kPa
Young's modulus 33	5.11×10^7	kPa	Volumetric response: shock Grüneisen coefficient	1.64	-
Poisson's ratio 12	0.013	-	Parameter C1	3.57×10^3	m/s
Poisson's ratio 31	0.5	-	Parameter S1	1.3	-
Reference temperature	293	K	Specific heat	1.85×10^3	J/kgK

Table 7. Material constants for Orthotropic yield strength.

Parameters	Value	Units	Parameters	Value	Units
Plasticity constant 11	0.016	-	Eff. plastic strain #1	0	-
Plasticity constant 22	6×10^{-4}	-	Eff. plastic strain #2	0.01	-
Plasticity constant 33	6×10^{-4}	-	Eff. plastic strain #3	0.1	-
Plasticity constant 12	0	-	Eff. plastic strain #4	0.15	-
Plasticity constant 13	0	-	Eff. plastic strain #5	0.175	-
Plasticity constant 23	0	-	Eff. plastic strain #6	0.19	-
Plasticity constant 44	1	-	Eff. plastic strain #7	0.2	-
Plasticity constant 55	1.7	-	Eff. plastic strain #8	0.205	-
Plasticity constant 66	1.7	-	Eff. plastic strain #9	0.21	-
/	/		Eff. plastic strain #10	0.215	-
Eff. stress #1	1.48×10^3	kPa	Eff. stress #6	6.0×10^4	kPa
Eff. stress #2	7.0×10^3	kPa	Eff. stress #7	8.0×10^4	kPa
Eff. stress #3	2.7×10^4	kPa	Eff. stress #8	9.8×10^4	kPa
Eff. stress #4	4.0×10^4	kPa	Eff. stress #9	2.0×10^5	kPa
Eff. stress #5	5.0×10^4	kPa	Eff. stress #10	1.0×10^6	kPa

4.2. Numerical Results and Analysis

Table 8 presents the numerical simulation results of the A–P core penetrating the composite armor. v_i and v_r are the impacts and residual velocities of the ogive–nose penetrator. p is the depth of penetration. Due to the experimental results, the impact velocity is set from 430 m/s to 700 m/s. With the increased impact velocity, the penetration depth gradually increased. When the impact velocity reached 500 m/s, the ogive–nose penetrator could just perforate the composite armor.

Table 8. Numerical simulation results of perforation.

v_i (m/s)	State of Perforation and Deformation	p (mm)	v_i (m/s)	State of Perforation and Deformation	v_r (m/s)
430	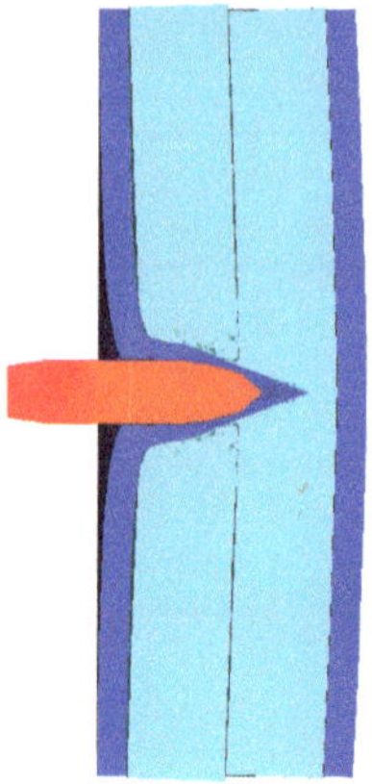	41.55	530	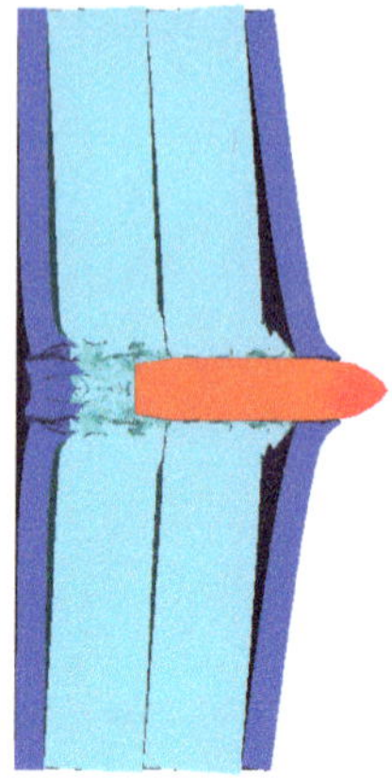	174
450	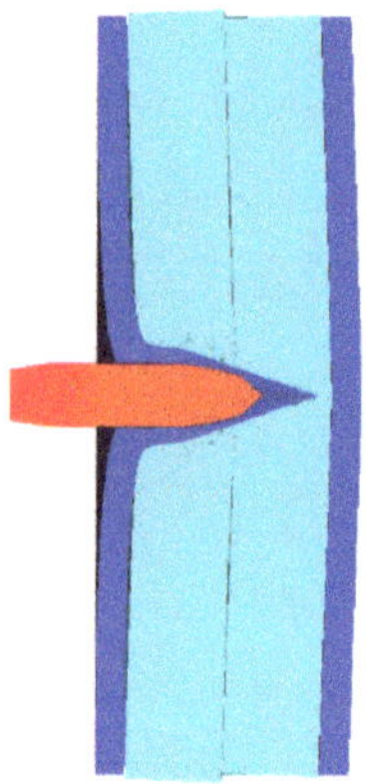	43.15	600	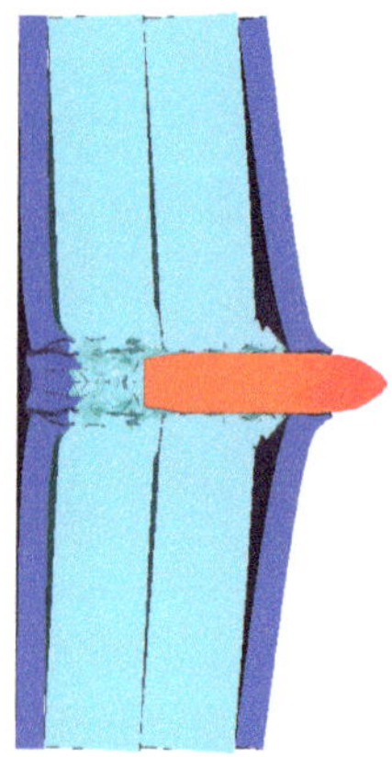	329
500	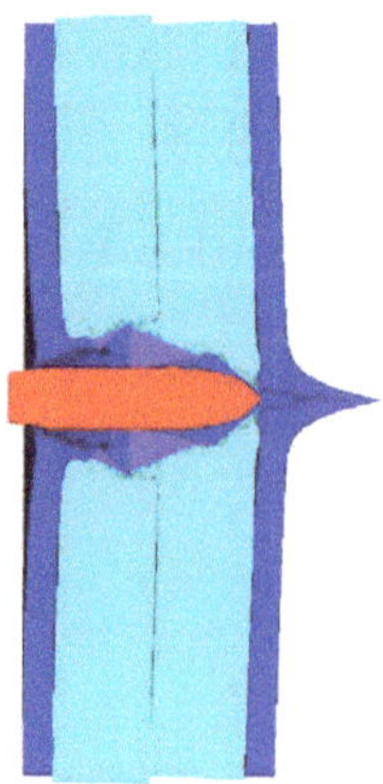	53	700	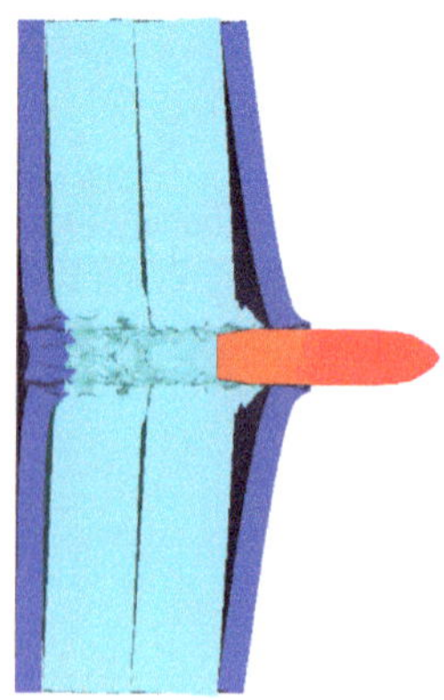	490

The contour of Von–Mises stress at the impact velocity of 500 m/s is shown in Figure 11. It can be inferred that the maximum stress exceeds the yield stress of the steel plate, and

Q235 back plate is pierced. Therefore, the velocity of 500 m/s can be considered as the ballistic limit of the composite armor, which is much higher than the 467.2 m/s obtained from the experimental results. The numerical simulation results are acceptable, with a relative error of 7.02%.

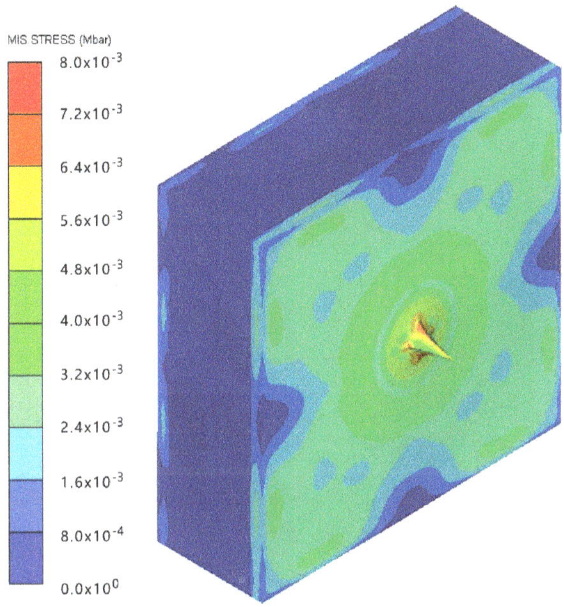

Figure 11. The contour of Von–Mises stress at the impact velocity of 500 m/s.

4.3. *Perforation Models and Analysis*

(1) Principle of energy conservation

The energy balance for the perforation is given by

$$\frac{1}{2}mv_i^2 = \frac{1}{2}mv_r^2 + W \tag{10}$$

$$W = W_{Q235} + W_{PE} \tag{11}$$

where m is the mass of the projectile, v_i is the impact velocity, v_r is the residual velocity, and W is the work performed during perforation. The mass of the A–P core was set at 40.4 g, then the work conducted during the perforation of the composite armor could be calculated, as listed in Table 9. The value of W stayed stable from 5.05 kJ to 5.09 kJ, which means that dissipated energy in the petaling stays stable at around 5 kJ. At the ballistic limit from the numerical results, 500 m/s, the dissipated energy is the same as the work performed at a higher velocity after perforation. So, the principle of energy conservation can be applied here.

Table 9. Results of calculated work W in the perforation.

v_i (m/s)	500	530	600	700	800	900	1000
v_r (m/s)	0	174	329	490	623	748	866
W (kJ)	5.05	5.06	5.09	5.05	5.09	5.06	5.05

(2) Lambert–Jonas model

The Lambert–Jonas model [26,30–32] can provide a reasonable fit to predict the residual velocity of the penetrator after perforation. The model can be expressed as

$$v_r = \begin{cases} 0, 0 \leq v_i \leq v_{bl} \\ \alpha\left(v_i^p - v_{bl}^p\right)^{1/p}, v_i \geq v_{bl} \end{cases} \tag{12}$$

where v_i, v_r, and v_{bl} are the impact, residual, and ballistic limit velocity in normal impact. α and p are the coefficients, where $0 \leq \alpha \leq 1$ and $p > 1$. Based on the numerical simulation results, the Lambert–Jonas model can be established to predict the residual velocity of the A–P core after perforating the PE composite armor.

When the model with $p = 2$, the coefficient α can be set as 1, and the model can be justified based on the energy conservation law [33]. This model can be written as

$$v_r = \begin{cases} 0, 0 \leq v_i \leq v_{bl} \\ \left(v_i^2 - v_{bl}^2\right)^{1/2}, v_i \geq v_{bl} \end{cases} \tag{13}$$

the predicted $v_r - v_s$ curve and the simulation results are presented below. As shown in Figure 12, the Lambert–Jonas model can be an effective method in predicting the residual velocity of the A–P core after perforation. In addition, the perforation process can be regarded as a rigid body penetration.

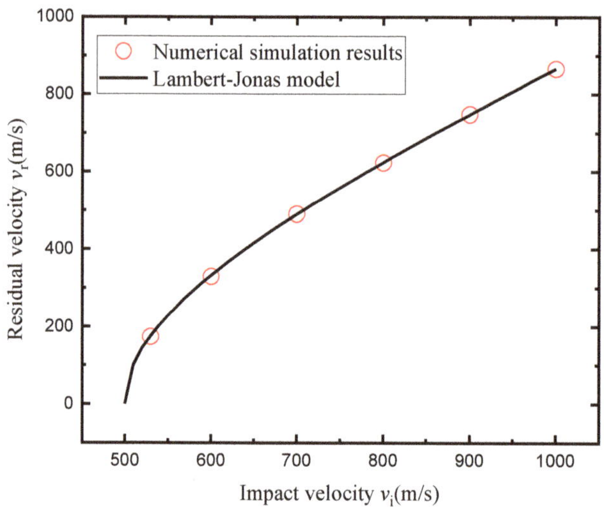

Figure 12. Comparison between the Lambert–Jonas model and the numerical simulations.

(3) Cavity–Expansion Model

As the A–P core has a diameter of 12.48 mm and a length of 53.4 mm, the composite armor with a thickness of 53 mm can be considered an intermediate target. The square armor has a width of 300 mm, which is about 24 times the diameter of the A–P core. Thus, the cylindrical cavity expansion can be used to predict the ballistic limit of the A–P core. Figure 13 shows the dimensions of the A–P core. The caliber–radius–head (CRH) is 3.05, which is also denoted as ψ.

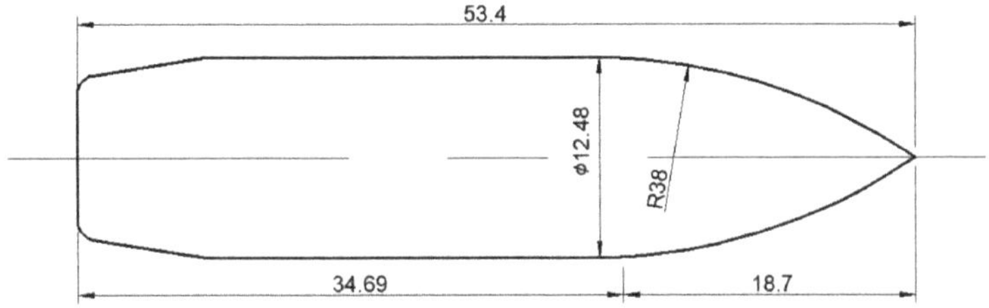

Figure 13. Geometry and dimensions (in mm).

Coefficient k_1 is expressed as

$$k_1 = \left(4\psi^2 - 4\psi/3 + 1/3\right) - \frac{4\psi^2(2\psi - 1)}{\sqrt{4\psi - 1}} \sin^{-1}\left[\frac{\sqrt{4\psi - 1}}{2\psi}\right] \tag{14}$$

The radial stress σ_r at the cavity surface versus cavity expansion velocity V is given by [34]

$$\sigma_r = \sigma_s + \rho_t B V^2 \tag{15}$$

where σ_s is the quasi–static radial stress required to open the cylindrical cavity, ρ_t is the density of the target, and B is a dimensionless constant. σ_s, b, and B are obtained from [23]

$$\sigma_s = \frac{Y}{\sqrt{3}}\left\{1 + \left[\frac{E}{\sqrt{3}Y}\right]^n \int_0^b \frac{(-\ln x)^n}{1 - x} dx\right\} \tag{16}$$

$$b = 1 - \gamma^2 \tag{17}$$

$$B = \frac{1}{2}\left\{\frac{1}{(1 - \nu)\sqrt{1 - \alpha^2}} \ln\left[\frac{1 + \sqrt{1 - \alpha^2}}{\alpha}\right] + \gamma^2 - 2\ln[\gamma] - 1\right\} \tag{18}$$

where Y is, the yield stress and ν is Poisson's ratio of the target. α and γ are given by

$$\alpha^2 = \frac{\sqrt{3}(1 - 2\nu)}{2(1 - \nu)}\left(\frac{\rho_t V^2}{Y}\right) \tag{19}$$

$$\gamma^2 = \frac{2(1 + \nu)Y}{\sqrt{3}E} \tag{20}$$

Furthermore, a rigid ogive–nosed projectile, with the impact velocity of v_i, the ballistic limit of v_{bl} and the residual velocity v_r, is given by

$$v_{bl} = \left(\frac{2\sigma_s}{\rho_p}\frac{h}{(L + k_1 l)}\right)^{1/2}\left[1 + C + \frac{2}{3}C^2\right]^{1/2} \tag{21}$$

$$v_r = v_{bl}\left[\left(\frac{v_i}{v_{bl}}\right)^2 - 1\right]^{1/2}\left[1 - C + \frac{1}{2}C^2\right]^{1/2} \tag{22}$$

where C is a small parameter related to the target inertia. When target inertia is neglected, the ballistic limit of v_{bl} and the residual velocity v_r can be simplified as [23,25,35] as

$$v_{bl} = \left(\frac{2\sigma_s}{\rho_p}\frac{h}{(L + k_1 l)}\right)^{1/2} \tag{23}$$

$$v_r = v_{bl} \left[\left(\frac{v_i}{v_{bl}} \right)^2 - 1 \right]^{1/2} \tag{24}$$

where the residual velocity v_r is the same as the Lambert–Jones model in Equation (13).

Based on the constitutive models of the target materials, the quasi–static radial stress σ_s can be expressed as [36]

$$\sigma_s = \frac{Y}{\sqrt{3}} \left[1 + \ln\left(\frac{E}{\sqrt{3}Y} \right) \right] + \frac{\pi^2 H}{18} \tag{25}$$

where E and H are Young's modulus and the constant tangent modulus in the plastic region if the stress versus strain curve of the target can be expressed as

$$\sigma = \begin{cases} E\varepsilon\,, \sigma < Y \\ Y + H\varepsilon,\ \sigma \geq Y \end{cases} \tag{26}$$

Thus, the value of σ_s for the Q235 face sheets can be calculated. For UHMWPE laminates, there may not be a mature model to predict the quasi–static radial stress required to open the cylindrical cavity, but the range of the σ_s can be estimated from the empirical formula [37,38] below,

$$\sigma_s = (1.33 \sim 2)Y_t \tag{27}$$

When the coefficient is set as the minimum value of 1.33, the value at a relatively low level can be obtained, as listed in Table 10.

Table 10. The predicted value of quasi–static radial stress σ_s.

Materials	*E* (GPa)	Ultimate Tensile/ Compressive Strength (MPa)	*H* (GPa)	σ_s Predicted from Equation (25) (GPa)	σ_s Predicted from Equation (27) (GPa)
Q235	200	305	281	2.76	/
UHMWPE	113~124	2700~3200	/	/	3.59~4.26

For the composite armor composed of Q235 face sheets and UHMWPE laminates, the effective value of σ_s can range from 2.76 GPa to 4.26 GPa. When the value of effective σ_s is set as 3.08 GPa, the ballistic limit of the composite armor calculated from Equation (23) is 467 m/s, which is consistent with the value obtained from the numerical simulation results.

In conclusion, the principle of energy conservation and the Lambert–Jonas model can be applied to calculate the work performed during the perforation and the residual velocities of the A–P core after perforation. In addition, the quasi–static radial stress σ_s required to open the cylindrical cavity can be estimated from the cavity–expansion model. With the value of 3.08 GPa, the predicted ballistic limit is consistent with the numerical simulation results.

5. Conclusions

A UHMWPE composite armor made up of two pieces of UHMWPE laminates in the middle and Q235 steel face sheets is proposed, and a study of the ballistic limit of the composite armor under the impact of a typical ogive–nose penetrator was carried out. (1) According to the experimental results, the average velocity of 501.1 m/s and 433.3 m/s can be taken as the ballistic limit of UHMWPE composite armor under the impact of the ogive–nose projectile, which is 467.2 m/s. In comparison, the ballistic limit obtained from the numerical simulation results was 500 m/s, which is acceptable with a relative error of 7.02%. (2) Petaling, as the main dominant failure mode for both face sheets, could be observed within the impact velocity range of 501 m/s to 1026 m/s. Delamination and shear failure dominated the penetration process of UHMWPE laminates. In addition, the charring layer on the front steel plate could be observed, and more severe ablation could be noticed at the impact velocity of around 1000 m/s. (3) Through theoretical models, the perforation mechanism of composite armor under the impact of A–P cores was analyzed. The principle of energy conservation and the Lambert–Jonas model was applied to calculate the work performed during the perforation and the residual velocities. In addition, the quasi–static radial stress σ_s required to open the cylindrical cavity were estimated from the cavity–expansion model. With the value of 3.08 GPa, the predicted ballistic limit was consistent with the numerical simulation results.

The ballistic limit of the UHMWPE composite armor under the impact of the ogive–nose projectile was considered to be 467.2 m/s, which indicates that the composite armor may not have a strong ability to resist the penetration of sharp head penetrators. In order to enhance the resistance against bullets such as A–P projectiles, UHMWPE should be strengthened, and the structure should be further optimized in future studies.

Author Contributions: Methodology, validation, formal analysis, writing—original draft preparation, funding acquisition, L.D.; supervision, visualization; writing—review, validation, X.G.; writing—review and editing, supervision, project administration, P.S.; supervision, data curation, resources, X.K. All authors have read and agreed to the published version of the manuscript.

Funding: The authors thank the National Natural Science Foundation of China (Grant No. 11802142) and the project of the State Key Laboratory of Explosion Science and Technology (Grant No. KFJJ20-08M).

Institutional Review Board Statement: The study did not require ethical approval. This statement can be excluded.

Data Availability Statement: The raw and processed data generated during this study will be made available upon reasonable request.

Conflicts of Interest: The authors declare no conflict of interest.

References

1. Ramadhan, A.; Talib, A.A.; Rafie, A.M.; Zahari, R. High velocity impact response of Kevlar-29/epoxy and 6061-T6 aluminum laminated panels. *Mater. Des.* **2013**, *43*, 307–321. [CrossRef]
2. Supian, A.; Sapuan, S.; Jawaid, M.; Zuhri, M.; Ilyas, R.; Syamsir, A. Crashworthiness response of filament wound kenaf/glass fibre-reinforced epoxy composite tubes with influence of stacking sequence under intermediate-velocity impact load. *Fibers Polym.* **2022**, *23*, 222–233. [CrossRef]
3. Aisyah, H.; Paridah, M.; Sapuan, S.; Ilyas, R.; Khalina, A.; Nurazzi, N.; Lee, S.; Lee, C. A comprehensive review on advanced sustainable woven natural fibre polymer composites. *Polymers* **2021**, *13*, 471. [CrossRef]
4. Suriani, M.; Rapi, H.Z.; Ilyas, R.; Petrů, M.; Sapuan, S. Delamination and manufacturing defects in natural fiber-reinforced hybrid composite: A review. *Polymers* **2021**, *13*, 1323. [CrossRef] [PubMed]
5. Nurazzi, N.; Asyraf, M.; Khalina, A.; Abdullah, N.; Aisyah, H.; Rafiqah, S.A.; Sabaruddin, F.; Kamarudin, S.; Norrrahim, M.; Ilyas, R. A review on natural fiber reinforced polymer composite for bullet proof and ballistic applications. *Polymers* **2021**, *13*, 646. [CrossRef] [PubMed]
6. Pereira, A.C.; de Assis, F.S.; da Costa Garcia Filho, F.; Oliveira, M.S.; da Cruz Demosthenes, L.C.; Lopera, H.A.C.; Monteiro, S.N. Ballistic performance of multilayered armor with intermediate polyester composite reinforced with fique natural fabric and fibers. *J. Mater. Res. Technol.* **2019**, *8*, 4221–4226. [CrossRef]
7. Phoenix, S.L.; Porwal, P.K. A new membrane model for the ballistic impact response and V50 performance of multi-ply fibrous systems. *Int. J. Solids Struct.* **2003**, *40*, 6723–6765. [CrossRef]
8. Mei, Z.; Zhu, X.; Zhang, L. Ballistic protective mechanism of FRC laminates. *Acta Mater. Compos. Sin.* **2006**, *23*, 143–149.
9. Chen, C.; Zhu, X.; Wang, J.; Hou, H.; Tang, T. Mechanism of high-velocity blunt-nosed projectiles penetrating moderately thick UHMWPE fiber reinforced plastic laminate. *Acta Mater. Compos. Sin.* **2013**, *30*, 10.
10. Laessig, T.; Long, N.; May, M.; Riedel, W.; Heisserer, U.; Werff, H.; Hiermaier, S. A non-linear orthotropic hydrocode model for ultra-high molecular weight polyethylene in impact simulations. *Int. J. Impact Eng.* **2015**, *75*, 110–122. [CrossRef]
11. Long, H.N.; Ryan, S.; Cimpoeru, S.J.; Mouritz, A.P.; Orifici, A.C. The effect of target thickness on the ballistic performance of ultra high molecular weight polyethylene composite. *Int. J. Impact Eng.* **2015**, *75*, 174–183.
12. Nguyen, L.H.; Lassig, T.R.; Ryan, S.; Riedel, W.; Mouritz, A.P.; Orifici, A.C. Numerical Modelling of Ultra-High Molecular Weight Polyethylene Composite under Impact Loading. *Procedia Eng.* **2015**, *103*, 436–443. [CrossRef]
13. Nguyen, L.H.; Ryan, S.; Cimpoeru, S.J.; Mouritz, A.P.; Orifici, A.C. The Efficiency of Ultra-High Molecular Weight Polyethylene Composite Against Fragment Impact. *Exp. Mech.* **2016**, *56*, 595–605. [CrossRef]
14. Nguyen Long, H.; Laessig Torsten, R.; Shannon, R.; Werner, R.; Mouritz Adrian, P. A methodology for hydrocode analysis of ultra-high molecular weight polyethylene composite under ballistic impact. *Compos. Part A* **2016**, *84*, 224–235. [CrossRef]

15. Chen, Z.; Xu, Y.; Li, M.; Li, B.; Song, W.; Xiao, L.; Cheng, Y.; Jia, S. Investigation on Residual Strength and Failure Mechanism of the Ceramic/UHMWPE Armors after Ballistic Tests. *Materials* **2022**, *15*, 901. [CrossRef]
16. Yuansheng, C.; Jieke, X.; Zhe, L.; Jun, L.; Pan, Z. Damage Response Characteristics of UHMWPE/aluminum Foam Composite Sandwich Panel Subjected to CombinedBlast and Fragment Loadings. *Acta Armamentarii* **2021**, *42*, 1753.
17. Hu, P.; Cheng, Y.; Zhang, P.; Liu, J.; Yang, H.; Chen, J. A metal/UHMWPE/SiC multi-layered composite armor against ballistic impact of flat-nosed projectile. *Ceram. Int.* **2021**, *47*, 22497–22513. [CrossRef]
18. Yu, Z.; Li, G.; Fan, Q.; Wang, Y.; Xuan, X. Study on protection mechanism of 30CrMnMo-UHMWPE composite armor. *Materials* **2017**, *10*, 405.
19. Ding, L.; Gu, X.; Shen, P.; Kong, X.; Zhou, Y. Dynamic Response of UHMW-PE Composite Armors under Ballistic Impact of Blunt Projectiles. *Materials* **2022**, *15*, 5594. [CrossRef]
20. Zochowski, P.; Bajkowski, M.; Grygoruk, R.; Magier, M.; Burian, W.; Pyka, D.; Bocian, M.; Jamroziak, K. Comparison of Numerical Simulation Techniques of Ballistic Ceramics under Projectile Impact Conditions. *Materials* **2022**, *15*, 18. [CrossRef]
21. Zochowski, P.; Bajkowski, M.; Grygoruk, R.; Magier, M.; Burian, W.; Pyka, D.; Bocian, M.; Jamroziak, K. Ballistic Impact Resistance of Bulletproof Vest Inserts Containing Printed Titanium Structures. *Metals* **2021**, *11*, 225. [CrossRef]
22. Ammarullah, M.I.; Santoso, G.; Sugiharto, S.; Supriyono, T.; Wibowo, D.B.; Kurdi, O.; Tauviqirrahman, M.; Jamari, J. Minimizing Risk of Failure from Ceramic-on-Ceramic Total Hip Prosthesis by Selecting Ceramic Materials Based on Tresca Stress. *Sustainability* **2022**, *14*, 13413. [CrossRef]
23. Børvik, T.; Forrestal, M.J.; Warren, T.L. Perforation of 5083-H116 Aluminum Armor Plates with Ogive-Nose Rods and 7.62 mm APM2 Bullets. *Exp. Mech.* **2010**, *50*, 969–978. [CrossRef]
24. Forrestal, M.J.; Børvik, T.; Warren, T.L. Perforation of 7075-T651 Aluminum Armor Plates with 7.62 mm APM2 Bullets. *Exp. Mech.* **2010**, *50*, 1245–1251. [CrossRef]
25. Forrestal, M.J.; Warren, T.L.; Børvik, T.; Chen, W. *Perforation of 6082-T651 Aluminum Plates with 7.62 mm APM2 Bullets at Normal and Oblique Impacts*; Springer International Publishing: Berlin/Heidelberg, Germany, 2015.
26. Carlucci, D.E.; Jacobson, S.S. *Ballistics: Theory and Design of Guns and Ammunition*; CRC Press: Boca Raton, FL, USA, 2007.
27. Andre, M.M. *Dynamic Behavior of Materials*; John Wiley & Sons: Hoboken, NJ, USA, 1994.
28. Johnson, G.R.; Cook, W.H. A constitutive model and data for metals subjected to large strains, high strain-rates and high temperatures. In Proceedings of the Seventh International Symposium on Ballistics, Hague, The Netherlands, 19–21 April 1983; p. 541.
29. Johnson, G.R.; Cook, W.H. Fracture characteristics of three metals subjected to various strains, strain rates, temperatures, and pressres. *Eng. Fract. Mech.* **1985**, *21*, 31–48. [CrossRef]
30. Lambert, J.P.; Jonas, G.H. *Towards Standardization in Terminal Ballistics Testing: Velocity Representation*; Defence Technical Information Center: Fort Belvoir, VA, USA, 1976.
31. Zukas, J.A. *High Velocity Impact Dynamics*; Wiley-Interscience: Hoboken, NJ, USA, 1990.
32. Hazell, P.J. *Armour: Materials, Theory, and Design*; CRC Press: Boca Raton, FL, USA, 2015.
33. Recht, R.F.; Ipson, T.W. Ballistic Perforation Dynamics. *J. Appl. Mech.* **1963**, *30*, 384. [CrossRef]
34. Forrestal, M.J.; Warren, T.L. Perforation equations for conical and ogival nose rigid projectiles into aluminum target plates. *Int. J. Impact Eng.* **2009**, *36*, 220–225. [CrossRef]
35. Piekutowski, A.; Forrestal, M.; Poormon, K.; Warren, T. Perforation of aluminum plates with ogive-nose steel rods at normal and oblique impacts. *Int. J. Impact Eng.* **1996**, *18*, 877–887. [CrossRef]
36. Forrestal, M.; Warren, T. Brief Technical Note Perforation of Steel Plates with 7.62 mm APM2 Bullets. *J. Dyn. Behav. Mater.* **2022**, *8*, 418–421. [CrossRef]
37. Taylor, G. The formation and enlargement of a circular hole in a thin plastic sheet. *Q. J. Mech. Appl. Math.* **1948**, *1*, 103–124. [CrossRef]
38. Rosenberg, Z.; Dekel, E. *Terminal Ballistics*; Springer: Berlin/Heidelberg, Germany, 2012.

polymers

Article

Chemical and Enzymatic Fiber Modification to Enhance the Mechanical Properties of CMC Composite Films

Xiaobao Li [1,†], Zhengjie Tang [1,†], Zhenbing Sun [1], John Simonsen [2], Zhinan Luo [1], Xiaoping Li [1,3,*] and Jeffery J. Morrell [4,*]

1 Yunnan Key Laboratory of Wood Adhesives and Glue Products, Southwest Forestry University, Kunming 650224, China
2 Department of Wood Science and Engineering, Oregon State University, Corvallis, OR 97331, USA
3 International Joint Research Center for Biomass Materials, Southwest Forestry University, Kunming 650224, China
4 National Centre for Timber Durability and Design Life, University of the Sunshine Coast, Brisbane, QLD 4102, Australia
* Correspondence: lxp810525@163.com (X.L.); jmorrell@usc.edu.au (J.J.M.)
† These authors contributed equally to this work.

Abstract: Carboxymethyl cellulose (CMC) is a cellulose derivative that can be obtained from wood, bamboo, rattan, straw, and other cellulosic materials. CMC can be used to produce biofilms for many purposes, but the properties of these resulting films make them unsuitable for some applications. The effects of three kinds of plant fiber addition on CMC film properties was investigated using CMC derived from eucalyptus bark cellulose. Tensile strength (TS) and elongation at break (EB) of CMC/sodium alginate/glycerol composite films were 26.2 MPa and 7.35%, respectively. Tensile strength of CMC composite films substantially increased, reaching an optimum at 0.50 g of fiber. The enhancement due to industrial hemp hurd fiber on CMC composite films was more obvious. Pretreatment with hydrogen peroxide (H_2O_2) and glacial acetic acid (CH_3COOH) produced films with a TS of 35.9 MPa and an EB of 1.61%. TS values with pectinase pretreated fiber films was 41.3 MPa and EB was 1.76%. TS of films pretreated with pectinase and hemicellulase was 45.2 MPa and EB was 4.18%. Chemical and enzymatic treatment both improved fiber crystallinity, but film tensile strength was improved to a greater extent by enzymatic treatment. Surface roughness and pyrolysis residue of the film increased after fiber addition, but Fourier transform infrared spectroscopy (FTIR), opacity, and water vapor transmission coefficients were largely unchanged. Adding fiber improved tensile strength of CMC/sodium alginate/glycerol composite films and broadened the application range of CMC composite films without adversely affecting film performance.

Keywords: eucalyptus bark; Yunnan pine wood; bamboo culms; industrial hemp hurd; FTIR; XRD; TG; mechanical properties

Citation: Li, X.; Tang, Z.; Sun, Z.; Simonsen, J.; Luo, Z.; Li, X.; Morrell, J.J. Chemical and Enzymatic Fiber Modification to Enhance the Mechanical Properties of CMC Composite Films. *Polymers* **2022**, *14*, 4127. https://doi.org/10.3390/polym14194127

Academic Editors: R.A. Ilyas, S.M. Sapuan, Emin Bayraktar, Shukur Abu Hassan, Nabil Hayeemasae and Khubab Shaker

Received: 17 August 2022
Accepted: 26 September 2022
Published: 2 October 2022

Publisher's Note: MDPI stays neutral with regard to jurisdictional claims in published maps and institutional affiliations.

1. Introduction

Petroleum-based composite films are widely used in the food, pharmaceutical, and chemical industries due to their good properties and low cost. However, there is increasing interest in moving away from fossil fuel-based materials to renewable natural polymers such as cellulose. Cellulose can be modified to produce carboxymethyl cellulose (CMC), an odorless, tasteless, non-toxic, neutral or slightly alkaline, white or yellowish powder. CMC is hygroscopic, relatively light and heat stable, and transparent in aqueous solutions [1]. CMC is widely used in oil drilling, food packaging, concrete modification, and soil improvement [2–7]. The degree of substitution (DS, average number of hydroxyl groups substituted with carboxymethyl groups per anhydroglucose unit (AGU)) has a major influence on the properties, and therefore the potential uses, of CMC. The theoretical maximum DS of CMC is three; the degree of substitution of CMC directly affects the solubility, emulsification, thickening, stability, acid resistance, and salt resistance of CMC [8]. CMC with a super high degree of substitution (DS = 1.7~3.0) is often used in the textile, printing, and dyeing industry. CMC with a high degree of substitution (DS = 1.0~1.2) is often used as a food additive. CMC (DS = 0.6~0.9) with a low degree of substitution is commonly used in industrial drilling, ceramics, detergents, building materials, etc. [9]. In past research reports, some rich and underutilized plant cellulose sources have been used as raw materials for CMC production to replace cellulose materials obtained from cotton linter or wood bleached pulp. However, there is little information about CMC from eucalyptus bark cellulose. Eucalyptus bark comes from the eucalyptus tree, fallen off every year and is rich in cellulose. Eucalyptus wood is widely used in wood-based panel manufacturing, pulp and paper, etc., but the bark is a by-product of eucalyptus that typically has no use [10].

CMC also has great potential to create degradable films. However, single CMC composites are not suitable for all applications due to their poor mechanical properties and water permeability. Thickeners such as starch, sodium alginate, or gelatin can improve the mechanical properties and reduce moisture absorption of CMC films. When the ratio of CMC to corn starch is 4:6, the tensile strength of single starch film can be significantly improved, from 3.8 to 17.0 MPa [11], whereas addition of 1.5% sodium alginate to a CMC/chitosan mixture produced antibacterial food packaging films with tensile strengths and elongation at break of 65.32 MPa and 17.85%, respectively [12]. Adding 3.2% gelatin to a 0.8% CMC solution, the tensile strength of CMC composite film became 7.84 $\pm$ 0.30 MPa [13].Therefore, sodium alginate is often used as a thickener in CMC composite films. Plasticizers such as sorbitol, polyethylene glycol, or glycerin help improve ductility and tensile strength of composite samples. After 1.2% sorbitol was blended with CMC–gelatin–chitosan as plasticizer, the elongation at break of the composite film became 9.23% [14]. Adding 25% (*w*/*w*) glycerol to starch–alginate–CMC increased elongation at break by 58.6% [15]; 5 wt% polyethylene glycol blended with clay minerals–CMC, the maximum elongation at break was only 8.0% [16]. This comparison shows that the plasticizing effect of glycerin is the best.

CMC composite film properties can also be enhanced via addition of fibers derived from a variety of natural sources. For example, tensile strength increased 1.93 times with the addition of 8% sugar cane fiber to a polyvinyl alcohol (PVA) composite film [17]. The addition of wheat bran fiber to a corn starch composite film was associated with an increase in tensile strength from 2 to 5.07 MPa and a decrease in elongation at break from 60 to 28% [18]. The addition of cassava bagasse fiber to a cassava starch/glycerol composite film significantly increased maximum tensile strength (from 1.23 $\pm$ 0.15 to 7.78 $\pm$ 0.83 MPa) [19].The ability of small amounts of fiber to enhance film properties have seen these products used in construction, automotive, packaging, sports, and biomedicine. These applications highlight the potential for adding fibers to improve the properties of CMC films. However, there is little information about the effects of fiber-separation methods on the properties of CMC film. Fiber can be separated by mechanical or chemical methods. The mechanical processing method often results in ripped or torn fibers with a high elastic modulus and elongation at break, but poor tensile strength [20]. Chemical methods include nitric acid + potassium chlorate (HNO_3 + $KClO_3$), sodium hypochlorite (NaClO), hydrogen peroxide + glacial acetic acid (H_2O_2 + HAc), and sodium hydroxide (NaOH); sulfate can represent a gentler separation method [21]. For example, fibers prepared by H_2O_2 + HAc were not hollow, and resulted in separated whole fibers with high tensile strength. There have been few reports on enzymatic separation of plant fibers; however, enzymes have catalytic efficiencies that are 10^7–10^{13} times higher than non-enzymatic catalysts [22,23]. Enzymatic catalytic reactions are substrate-specific substrates, do not affect other raw materials, and cause less damage to raw materials. As enzymes are proteins, they are biodegradable, more environmentally friendly, and are a better choice for fiber separation [20,24,25].

In this study, chemical and enzymatic methods were used to separate plant fibers from different raw materials. The fibers were characterized and the properties of fiber-amended

CMC composites were studied. We also suggest a method for making CMC films of high quality and improving the value of eucalyptus bark.

2. Materials and Methods

2.1. Materials

Eucalyptus bark (Eucalyptus globulus Labill.), Yunnan pine wood (Pinus yunnanensis), bamboo culms (Neosinocalamus affinis), and industrial hemp hurd (Cannabis sativa) were all obtained locally (Kunming, Yunnan Province, China). They were ground to pass through a 40–60 mesh screen (250–420 μm) before cellulose extraction and chemical analysis. The contents of benzene-alcohol extract, holo-cellulose, cellulose, hemicellulose, and lignin were determined according to Chinese Standards GB/T 2677.6-1994 (Determination of organic solvent extract in paper raw materials), GB/T 2677.10-1995 (Determination of holo-cellulose in paper raw materials), GB/T 744-1989 (Determination of α-cellulose in pulp), and GB/T 2677.8-94 (Determination of acid-insoluble lignin in paper raw materials).

2.2. Plant Fiber Separation

Fibers of Yunnan pine wood, industrial hemp hurd, and bamboo culms were cut into pieces 3–5 by 3–5 by 5–10 mm (width by thickness by length) prior to chemical and enzymatic fiber extraction.

For chemical fiber separation: five grams of Yunnan pine wood, Bamboo culms, or industrial hemp hurd were placed in beakers and immersed in 100 mL of a 50:50 mixture of hydrogen peroxide (H_2O_2): glacial acetic acid (CH_3COOH) at 70 °C until the sample turned white. The samples were washed with distilled water until the pH was 7, then the samples were shaken slightly to separate the fibers. These procedures were performed in triplicate.

Enzymatic fiber separation: three enzymatic methods were used to obtain fibers. In the first method, five grams of a given material was treated with 100 mL of 5.00% lipase solution at 50 °C, stirred (900–1000 r/min) for 3 h, then the solution was filtered off and the residual materials were treated with GB2677.10-1995 (Determination of the content of holo-cellulose from paper raw materials) to remove most of the lignin. The residual material was immersed in 40 mL of 50 °C distilled water and stirred (900–1000 r/min) for more than 3 h until most of the fibers were separated, then washed with distilled water to obtain residues. The second method used the same procedure, but then immersed the fibers in 40 mL of 5% pectinase solution instead of distilled water. The final method used the same procedure, but then immersed the materials in 40 mL of 5% a 20:20 mixture of 5% pectinase solution and 5% hemicellulase solution instead of distilled water. Lipase (CAS:9001-62-1), pectinase (CAS:9032-75-1), and hemicellulose (CAS:9025-56-3) were obtained from Aladdin Biotechnology Co., Ltd., (Shanghai, China) and had enzyme activities of 100,000; 30,000; and 5000 U/g, respectively. Each fiber material/enzymatic treatment combination was prepared in triplicate.

2.3. Preparation of Cellulose, CMC, and CMC Composite Films

Four grams of cellulose extracted from eucalyptus bark according to Chinese Standard GB/T744-1989 (Determination of α-cellulose from pulp) was mixed with 80 mL of 100% ethanol(CH_3CH_2OH) and 20 mL of 30% NaOH solution and then stirred(900–1000 r/min) for 60 min at 30 °C. Five grams of sodium chloroacetate ($C_2H_2ClNaO_2$) were added and heated at 65 °C for 3 h. The sample was washed with 90% glacial acetic acid to a pH of 7, then washed with 80% ethanol 3 times and 95% ethanol once, before being oven-dried at 65 °C for 3 h to obtain CMC.

One gram of CMC was mixed with 0.40 g sodium alginate ($C_6H_7NaO_6$), 0.15 g glycerol ($C_3H_8O_3$), and 49.00 g distilled water and stirred (900–1000 r/min) at 70 °C for 15 min. The mixture was treated in an ultrasonic bath for 10 min and vacuumed for 45 min to remove air bubbles. The solution was poured into a polytetrafluoroethylene (PTFE) mold and cured at 30 °C for 48 h to obtain CMC composite films. The fiber-modified CMC composite film process was similar except that 0.1, 0.3, and 0.5 g of a given plant fiber was added during mixing (shown in Figure 1).

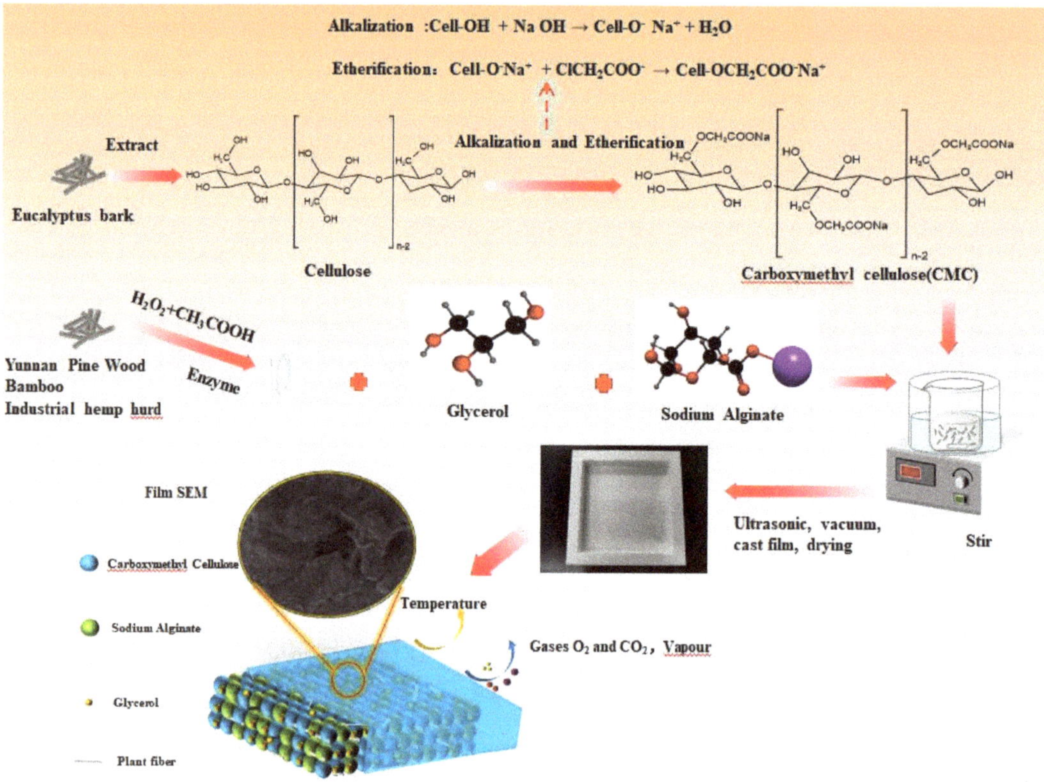

Figure 1. Preparation of CMC composite films.

2.4. Fiber and Composite Film Characterization

Fiber dimensions: The length and width of 100 fibers, as well as the cell wall thickness and cell cavity width of each of the chemically or enzymatically prepared fibers were measured under a light microscope by ImageJ software.

Sample microstructure: The microstructure of eucalyptus bark powder, cellulose, CMC, and CMC composite films were observed by placing a sample on an aluminium stub and coating with gold/palladium before observation with the Czech TESCAN MIRA LMS field emission scanning electron microscope at an accelerating voltage of 200 eV to 30 KeV. A minimum of five fields were examined per material.

Fourier-transform infrared spectroscopy (FTIR): Eucalyptus bark powder, cellulose, and CMC were mixed with KBr and formed into a pellet while the CMC composite films were directly analyzed on a Nicolet i50 FTIR analyzer (Thermo Nicolet Corporation, Madison, WI, USA) with a scanning range of 500 to 4000 cm^{-1} and 64 scans. Baseline correction was performed to analyze the spectral differences between plant fibers obtained by different treatments.

X-ray diffractometer analysis (XRD): The crystal structures of eucalyptus bark, cellulose, and the CMC films were studied by X-ray diffraction (XRD) on an Ultima IV X-ray diffractometer (Rigaku Corporation, Tokyo, Japan) using a scanning angle from 5 to 60°, a step size of 0.026° (accelerating current = 30 mA and voltage = 40 kV), and Cu-Kα radiation of λ = 0.154 nm. The degree of crystallinity (DOC, %) was calculated according to the formula:

$$\text{DOC}\% = \frac{I_{Max} - I_{Am}}{I_{Max}} \times 100 \quad (1)$$

I_{Max} is the maximum intensity of the main peak (about 22°), and I_{Am} is the diffraction intensity of amorphous cellulose (about 15°).

Thermogravimetric (TG) analysis: Approximately 5 to 6 mg of sample powder ground to pass through an 80 to 120 mesh screen and placed into sample holders for analysis on a TGA92 thermo gravimetric analyzer (KEP Technologies EMEA, Caluire, France). N_2 was used as the shielding gas and Al_2O_3 as the reference compound. The temperature was increased from 35 to 800 °C at a rate of 20 °C/min to generate a thermogravimetric curve.

Degree substitution (DS) of CMC: The degree of substitution of hydroxyl groups has an important influence on resulting CMC properties. The degree of substitution was determined by the acidimeter method by weighing 0.2 g (accuracy 0.1 mg) of the sample, dissolving it in 80 mL of water, stirring for 10 min, and adjusting the pH to 8.0. The sample was titrated using sulphuric acid (H_2SO_4) with continuous stirring to pH 3.74. The volume (mL) of sulphuric acid titration solution used was recorded (to the nearest 0.05 mL). The degree of substitution (*DS*) was then calculated using the amount required to reach the end point according to Equations (1) and (2), as follows (Table 1).

$$B = \frac{2cV}{m} \quad (2)$$

$$DS = \frac{0.162B}{1 - 0.08B} \quad (3)$$

where B = amount of carboxymethyl substance contained in the sample, mmol/g;
m = quality of the sample, g;
c = concentration of sulphuric acid standard titration solution, mol/L;
V= volume value of standard titration solution of sulphuric acid, mL.

Table 1. Chemical composition of eucalyptus bark, Yunnan pine wood, bamboo culms, and industrial hemp hurd [a].

Chemical Compound	Eucalyptus Bark	Yunnan Pine Wood	Bamboo Culms	Industrial Hemp Hurd
Phenyl alcohol extract (%)	2.63 (0.31)	4.79 (0.47)	2.09 (0.53)	3.83 (0.29)
Holo-cellulose (%)	71.5 (0.16)	62.5 (0.15)	76.0 (0.05)	75.5 (0.15)
Celluloses (%)	44.9 (0.20)	42.9 (0.06)	55.9 (0.16)	54.6 (0.19)
Hemicellulose (%)	26.6 (0.12)	19.5 (0.08)	20.1 (0.06)	20.9 (0.25)
Lignin (%)	27.2 (0.05)	32.4 (0.09)	25.2 (0.31)	21.1 (0.06)

[a] Values represent means of 3 replicates. The numbers in parentheses are one standard deviations.

Physical Properties: Tensile strength (MPa) and elongation at break (%) were measured on ten 0.089 to 0.098 mm by 150 mm dog-bone samples of each material on a universal testing machine according to procedures described in GB/T 1040.1-2006 (Plastics—Determination of tensile properties). The load was applied to failure at a rate of 1 mm/min.

Film Opacity and Water Vapor Transmission: Opacity of the CMC composite films was tested by cutting 10 by 40 mm samples and placing them on the inner surface on one side of a cuvette and measuring absorbance at 600 nm on an XP Spectrum 752# ultraviolet spectrophotometer (XP-Spectrum Company, Shanghai, China). Five measurements were made for each sample.

Water vapor transmission rate was assessed under controlled temperature and relative humidity conditions using unit time, unit water vapor pressure difference, and thickness through the unit, and expressed as the unit area of the water vapor volume of the specimen.

The water vapor transmission coefficient of the specimen was calculated according to Equation (3).

$$P = \frac{\Delta m \times d}{A \times t \times \Delta p} \quad (4)$$

where P is the water vapor transmission coefficient of the sample in grams/square centimeter per second Pascal [g cm/(cm^2·s·Pa)].

Δm is the mass change of the sample in grams (g) during the period t.

A is the sample area through the water vapor in square meters (m^2).

t is the difference in time between two intervals after the mass change has stabilized in hours (h).

d is the thickness of the specimen in centimeters (cm).

Δp is the difference in water vapor pressure between the two sides of the specimen in Pascals (Pa).

2.5. Statistical Analysis

Equality of variance was confirmed using Fisher's test for raw material chemical analyses, fiber size measurements, and physical properties measurements of CMC composite films. Student's t test was carried out to compare the samples in pairs at $p < 0.05$. Data were analyzed using SPSS 25.0 statistical package (IBM, Armonk, NY, USA).

3. Results and Discussion

3.1. Chemical Composition

Cellulose content of eucalyptus bark was 44.9% higher than that of Yunnan pine wood and lower than that of bamboo culms or industrial hemp hurd (Table 1). Hemicellulose content of eucalyptus bark was 26.6% and was higher than that of other three materials. Lignin content of the bark was 27.2% lower than that of Yunnan pine wood and higher than either bamboo culms or industrial hemp hurd. These results were consistent with previous research and indicate that eucalyptus bark-derived cellulose is a suitable alternative [26]. Using this material as a substitute would reduce chemical consumption and allow utilization of a waste product.

3.2. Effects of Different Pretreatment Methods on Fiber Yield and Dimensions

Chemical treatment resulted in 100% fiber yield after 10 h of treatment at 70 °C (Table 2). Enzyme treatments produced lower yields. Distilled water treatment resulted in less than 5% fiber yields from bamboo culms and industrial hemp hurd and only 10–20% yield from Yunnan pine wood, despite the 12 h total treatment time. The subsequent use of pectinase alone or in combination with hemicellulase resulted in 90 to 95% fiber yield. Pectin plays important roles in cell wall interactions, especially in primary cell wall formation, and its disruption may facilitate fiber separation. Treatment times for the pectinase treatments were only 6 to 8 h whereas they were 5.5 to 8.5 h for the pectinase and hemicellulase treatments. These results illustrate the potential for producing high fiber yields using enzymatic treatments [27].

The distilled water treatment resulted in little fiber recovery and will not be further discussed. The other treatments had varying effects on the properties of the resulting fibers (Table 3). Pectinase treatment resulted in the shortest Yunnan pine wood fibers whereas the chemical treatment resulted in fibers that were nearly 50% longer. Similarly, pectinase and hemicellulase treated bamboo culms and industrial hemp hurd fibers were only half as long as those from the chemical treatment. Fiber length tended to be greater in all of the chemical treatments compared with the enzymatic treatments, although the pectinase treatment was sometimes similar to the chemical treatment (Table 3). The largest difference in fiber length was found between chemically and enzymatically treated industrial hemp hurd fibers. Decreased fiber lengths may reflect a tendency for enzymatically treated fibers to break more easily as they are separated, which would be detrimental to increasing the tensile strength of any composite. Fiber widths and lumen diameters tended to be similar for the same material regardless of whether the samples were chemically or enzymatically treated. The treatments are less likely to affect fiber width or lumen size, given that their primary effects would be on the cell walls themselves. Cell walls were slightly thicker in pectinase-treated Yunnan pine wood fibers than chemically treated fibers, whereas cell wall thicknesses of pectinase and hemicellulase and chemically treated Yunnan pine wood fibers were similar. Pectin is an important component in primary wood cell wall formation but becomes less important with subsequent lignification. However, the specificity of pectinase for pectin could lead to more efficient separation with reduced breakage.

Table 2. Effect of chemical treatment alone or coupled with three sequential enzyme treatments on fiber yield and total treatment time of Yunnan pine wood, bamboo culms, and industrial hemp hurd fibers.

Material	Degree of Fiber Separation (%) and Total Treatment Time (h) [a]							
	Chemical		Distilled Water		Pectinase		Pectinase and Hemicellulase	
	Fiber Yield (%)	Total Time (h)	Fiber Yield (%)	Total Time (h)	Fiber Yield (%)	Total Time (h)	Fiber Yield (%)	Total Time (h)
Yunnan pine wood	100	10.0	10–20	12.0	90–95	8.0	90–95	8.5
Bamboo culms	100	10.0	<5	12.0	90–95	6.0	90–95	5.5
Industrial hemp hurd	100	10.0	<5	12.0	90–95	8.0	90–95	8.5

[a] Values represent results from three replicates per material per treatment.

Table 3. Effect of chemical or enzymatic treatment on characteristics of fibers derived from Yunnan pine wood, bamboo culms, and industrial hemp hurd [a].

Material	Pretreatment Method	Fiber Length (µm)	Fiber Width (µm)	Lumen Width (µm)	Cell Wall Thickness (µm)	Length:Width	Wall Thickness:Lumen Width
Yunnan pine wood	Chemical	938.6 (463.6–2118.4)	35.4 (20.4–58.5)	20.6 (9.80–48.0)	8.38 (3.61–13.2)	27.4 (13.1–60.4)	0.46 (0.10–0.79)
	Pectinase	667.6 (312.7–1537.3)	31.7 (16.7–57.6)	23.6 (9.25–35.4)	12.3 (6.08–23.7)	22.0 (9.94–48.1)	0.57 (0.19–1.06)
	Pectinase + Hemicellulase	741.9 (398.5–1507.1)	35.8 (21.1–53.0)	19.6 (9.10–32.0)	9.64 (4.77–17.4)	21.3 (10.3–45.8)	0.53 (0.18–1.15)
Bamboo culms	Chemical	1096.2 (357.7–2104.1)	18.4 (5.76–40.2)	10.9 (2.77–29.5)	4.13 (1.17–9.11)	69.4 (13.1–195.8)	0.44 (0.09–0.97)
	Pectinase	984.6 (344.7–2090.7)	18.5 (9.22–34.4)	9.79 (2.88–22.6)	5.26 (2.17–12.6)	60.0 (19.3–177.4)	0.63 (0.16–1.43)
	Pectinase + Hemicellulase	590.9 (293.9–1657.2)	18.8 (7.07–49.1)	10.1 (2.53–37.0)	4.96 (2.03–13.3)	36.7 (9.63–81.9)	0.62 (0.20–1.26)
Industrial hemp hurd	Chemical	1139.1 (537.3–1963.4)	30.4 (14.2–47.3)	21.6 (5.45–40.4)	3.99 (0.98–11.0)	38.0 (15.9–62.8)	0.19 (0.05–0.56)
	Pectinase	514.7 (243.4–797.1)	36.0 (16.6–54.1)	24.9 (7.97–50.9)	4.71 (0.31–11.9)	15.2 (6.85–22.4)	0.21 (0.02–0.53)
	Pectinase + Hemicellulase	513.1 (257.7–968.6)	35.5 (13.6–60.5)	21.7 (7.61–45.4)	4.64 (0.17–13.0)	15.2 (7.68–28.4)	0.28 (0.08–0.62)

[a] Values represent means of 100 fibers per material per treatment while figures in parentheses are the range.

Fiber length to width ratios can be a useful indicator of potential effects of fiber addition on tensile properties. The addition of fibers with higher length to width ratios may have a greater effect on tensile strength. Length to width ratios tended to be smaller in enzymatically treated fibers than in chemically treated fibers of the same species. As noted earlier, this may reflect a tendency for enzymatically treated fibers to be more brittle and produce shorter fibers, which would reduce tensile properties.

The relative crystallinity of untreated Yunnan pine wood, bamboo culms, and industrial hemp hurd were 38.8, 49.1, and 47.4%, respectively (Table 4). Almost all of the lignin was removed from materials treated chemically or enzymatically and crystallinity was increased. The degree of crystallinity was greatest in Yunnan pine wood but crystallinity also increased in bamboo culms and industrial hemp hurd, although the differences were not significant. Increased crystallinity indicates that chemical and enzymatic treatments removed some amorphous cellulose, resulting in an increase in the proportion of crystalline cellulose. Increased crystallinity may result in stronger reinforcing fibers.

Table 4. Effect of chemical or enzymatic treatment on relative crystallinity of Yunnan pine wood, bamboo culms, and industrial hemp hurd fibers [a].

Materials	Relative Crystallinity (%)		
	Yunnan Pine Wood	Bamboo Culms	Industrial Hemp Hurd
Chemical treatment	55.1	53.5	55.8
Distilled water treatment	45.1	53.5	51.7
Pectinase treatment	47.1	56.0	54.5
Pectinase + hemicellulase treatment	53.9	50.6	51.7
Raw material	38.8	49.1	47.4

[a] Values represent means of 3 replicates.

3.3. CMC Characterization

Degree of Substitution: The substitution degree (DS) on CMC from eucalyptus bark cellulose was 0.89, which is similar to the values obtained for corn stover, straw, and reed CMCs, which ranged from 0.6 to 1.0 [28,29].

SEM: SEM examination of eucalyptus bark revealed that it consisted of many substances tightly aggregated together in small granular form (Figure 2a-1). The cellulose recovered from this material was in the form of polymerized fibrous bundles (Figure 2a-2), which became more discrete when the materials were reacted to form CMC with differing degrees of fiber breakage (Figure 2a-3). These changes reflect the effects of alkaline treatment and subsequent esterification, coupled with water penetration into the cellulose bundles, with resulting chain separation.

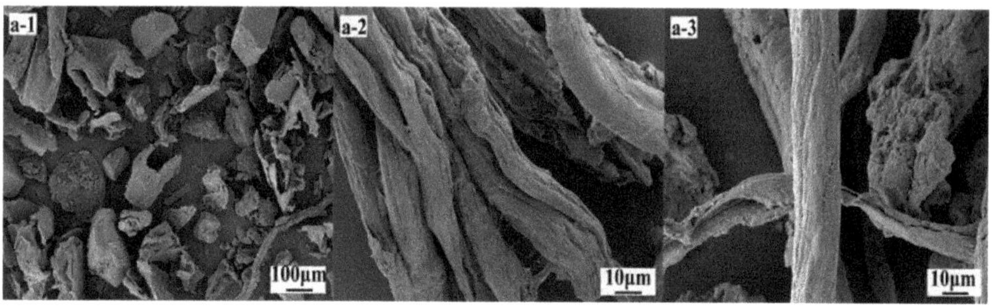

Figure 2. (**a-1**) SEM images showing microstructure differences for unprocessed eucalyptus bark (**a-2**), cellulose microfibrils after chemical treatment, and (**a-3**) CMC produced from the eucalyptus bark cellulose.

FTIR Analysis: FTIR spectra of commercially available CMC and eucalyptus bark CMC both contained the stretching vibrations of the CMC carboxylate anion COO^- at about 1630 cm^{-1}, with two characteristic absorption peaks at 1410 cm^{-1} corresponding to the asymmetric (C=O) and symmetric (C-O) stretching vibrations caused by the carboxylic acid group, and an absorption peak at 1030 cm^{-1} corresponding to the stretching vibration of the cellulose C-O-C group. The absorption peak at 898 cm^{-1} is characteristic of the β-glycosidic bond in cellulose [30–37]. No absorption peaks were observed at 1518 and 1320 cm^{-1}, which are attributed to the aromatic vibration of the lignin ring and the C-O stretching vibration of the syringyl group, respectively [38], nor was there a peak at 1730 cm^{-1}, which is attributed to hemicellulose. These results indicate that cellulose extraction was nearly complete with little evidence of residual lignin or hemicelluloses, and the cellulose was successfully transformed into CMC (Figure 3).

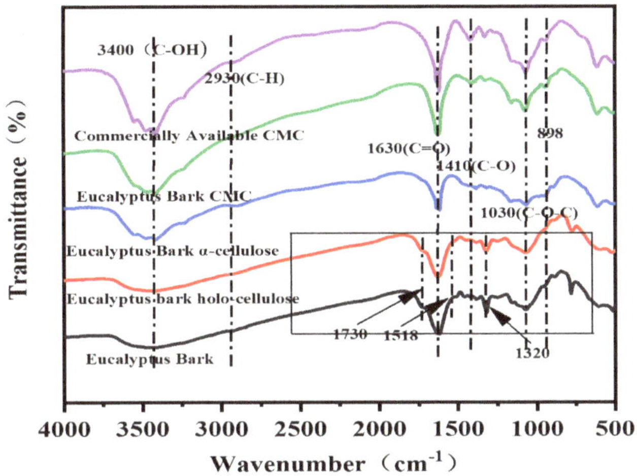

Figure 3. FTIR spectra of bark, holo−cellulose, cellulose, and CMC prepared from eucalyptus bark, as well as commercially available CMC.

X-ray Diffraction: Eucalyptus bark and holo-cellulose showed diffraction peaks at $2\theta = 15.65$, 16.45, 22.15, and 34.35°, whereas the cellulose diffraction peaks were at 14.85, 15.4, 21.75, and 34.35°, respectively (Figure 4). All three of the latter peaks correspond to the crystal planes of (101), $(10\bar{1})$, (200), and (004), which are typical reflections of cellulose Type I [39–43]. The relative crystallinity of eucalyptus bark and cellulose were 46.39 and 55.07%, respectively, again indicating that the cellulose was successfully extracted. The diffraction peaks of bark-derived CMC were at $2\theta = 19.95$ and 31.8°, which were similar to the diffraction peaks obtained from commercially available CMC, indicating successful transformation of cellulose to CMC (Figure 4) [4,43].

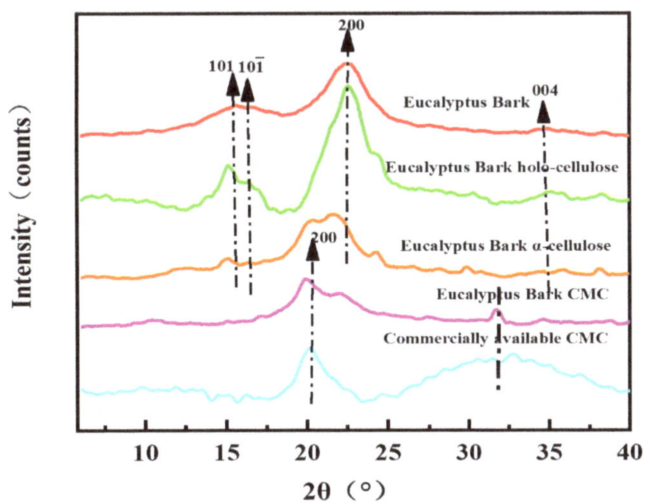

Figure 4. XRD diffractogram showing bark, holo-cellulose, cellulose, and CMC from eucalyptus bark, as well as commercially available CMC.

3.4. Effects of Fiber Treatment Method and Addition of Composite Film Properties

The addition of chemically treated Yunnan pine wood or bamboo culm fibers to the CMC/sodium alginate/glycerol film was associated with decreased tensile strength with increasing fiber content, although the difference in bamboo culm was small. Conversely, tensile strength of the CMC composite film increased with increased industrial hemp hurd fiber. The addition of enzyme-treated Yunnan pine wood, bamboo culms, or industrial hemp hurd fibers enhanced tensile strength of CMC composite films, although this enhancement was small at the lowest addition level (0.1% wt/wt). The results showed that enzymatically treated plant fibers were more likely to produce better CMC composite films (Table 5). Industrial hemp hurd fiber has a smaller wall lumen ratio than Yunnan pine and bamboo fibers. Small compounds such as CMC, sodium alginate, or glycerol are more likely to enter the fiber lumen, especially on enzymatically modified shorter fibers. This should improve tensile strength of the resulting composite.

Table 5. Effect of addition of chemically or enzymatically derived Yunnan pine wood, bamboo culms, and industrial hemp hurd fibers to CMC film on the tensile strength, elongation at break, opacity, or vapor transmission [a].

Material	Pretreatment Method	Plant Fiber (g)	Tensile Strength (MPa)	Elongation at Break (%)	Opacity (A/mm)	Water Vapor Transmission Coefficient (Pv) [g·cm/(cm^2·s·Pa)]
Control		0	**26.2 (0.56) $_{B}$**	**7.35 (0.62) $_{A}$**	**7.55 (0.56) $_{A}$**	**0.20 (0.02) $_{AB}$**
Yunnan pinewood	Chemical	0.1	18.2 (3.68) $_{BC}$	2.87 (1.23) $_{BCD}$	8.06 (1.14) $_{A}$	0.23 (0.03) $_{AB}$
		0.3	15.0 (5.89) $_{BC}$	1.29 (0.27) $_{CD}$	8.17 (0.31) $_{A}$	-
		0.5	11.4 (3.13) $_{BC}$	1.32 (0.22) $_{CD}$	7.63 (0.28) $_{AB}$	-
	Pectinase	0.1	14.4 (3.13) $_{BC}$	0.77 (0.34) $_{CD}$	6.51 (0.46) $_{AB}$	-
		0.3	36.6 (8.51) $_{AB}$	1.11 (0.15) $_{CD}$	7.14 (0.49) $_{AB}$	-
		0.5	38.9 (6.53) $_{AB}$	0.96 (0.10) $_{CD}$	7.18 (0.79) $_{AB}$	0.24 (0.01) $_{B}$
	Pectinase + Hemicellulase	0.1	25.0 (3.71) $_{BC}$	1.06 (0.29) $_{CD}$	7.30 (0.44) $_{AB}$	-
		0.3	25.5 (0.47) $_{BC}$	1.62 (0.03) $_{CD}$	7.51 (0.76) $_{AB}$	-
		0.5	37.4 (4.09) $_{A}$	1.30 (0.30) $_{CD}$	6.75 (0.35) $_{AB}$	0.20 (0.03) $_{BC}$
Bamboo culms	Chemical	0.1	26.8 (3.89) $_{B}$	4.32 (0.41) $_{B}$	8.02 (1.02) $_{AB}$	-
		0.3	28.1 (2.50) $_{AB}$	2.22 (0.38) $_{CD}$	6.82 (0.97) $_{AB}$	0.16 (0.03) $_{AB}$
		0.5	23.6 (4.62) $_{BC}$	2.23 (0.55) $_{CD}$	7.80 (1.51) $_{AB}$	-
	Pectinase	0.1	23.3 (5.08) $_{BC}$	1.13 (0.48) $_{CD}$	6.78 (0.36) $_{AB}$	-
		0.3	36.6 (5.14) $_{AB}$	1.64 (0.37) $_{CD}$	7.50 (0.41) $_{AB}$	-
		0.5	45.8 (0.41) $_{A}$	1.34 (0.02) $_{CD}$	7.47 (0.22) $_{AB}$	0.31 (0.02) $_{A}$
	Pectinase + Hemicellulase	0.1	25.3 (4.99) $_{B}$	2.01 (0.64) $_{CD}$	8.11 (0.65) $_{A}$	-
		0.3	29.1 (6.39) $_{AB}$	3.68 (0.53) $_{BC}$	7.97 (1.07) $_{A}$	-
		0.5	38.0 (4.66) $_{AB}$	2.95 (0.61) $_{BC}$	6.63 (0.34) $_{AB}$	0.20 (0.02) $_{C}$
Industrial hemp hurd	Chemical	0.1	26.4 (4.30) $_{B}$	3.18 (0.95) $_{BC}$	7.64 (0.03) $_{AB}$	-
		0.3	32.9 (4.93) $_{AB}$	1.09 (0.23) $_{CD}$	7.60 (0.67) $_{AB}$	-
		0.5	35.9 (7.18) $_{AB}$	1.61 (0.28) $_{CD}$	6.09 (2.10) $_{AB}$	0.25 (0.03) $_{B}$
	Pectinase	0.1	21.0 (2.40) $_{BC}$	1.71 (0.30) $_{CD}$	8.55 (0.65) $_{A}$	-
		0.3	27.2 (8.86) $_{AB}$	1.58 (0.34) $_{CD}$	7.67 (0.21) $_{AB}$	-
		0.5	41.3 (3.25) $_{A}$	1.76 (0.28) $_{CD}$	7.73 (0.72) $_{AB}$	0.24 (0.01) $_{BC}$
	Pectinase + Hemicellulase	0.1	23.6 (0.69) $_{BC}$	3.84 (1.48) $_{BC}$	7.57 (0.49) $_{AB}$	-
		0.3	38.9 (4.69) $_{AB}$	4.27 (0.35) $_{B}$	7.24 (0.88) $_{AB}$	-
		0.5	45.2 (3.66) $_{A}$	4.18 (0.43) $_{B}$	6.30 (0.57) $_{AB}$	0.24 (0.04) $_{B}$

[a] Values represent means of replicates per treatment and figures in parentheses represent one standard deviation. Values followed by the same letter(s) do not differ-significantly from one another by Tukey's pairwise comparisons ($\alpha = 0.05$).

The addition of chemically or enzymatically treated fibers produced more variable results with elongation at break. Values dropped sharply from the non-modified control but there were no consistent trends associated with fiber additive level or chemical vs. enzymatic pre-treatment. Although fiber addition clearly altered elongation at break, there were no consistent trends with regard to pre-treatment method or additive level (Table 5). Plant fibers usually have high rigidity and should improve tensile strength, but elongation at break could also decrease sharply, as observed in other materials [44].

Opacity is a useful measure for assessing the suitability of a film for commercial purposes. Although opacity varied widely with fiber pre-treatment and additive level, there were no consistent trends with regard to pre-treatment method or concentration. These results suggest that addition of low levels of chemically or enzymatically treated fibers had no consistent effect on opacity, likely because the overall fiber levels remained low.

The background water vapor coefficient for non-modified CMC was 0.20 $g \cdot cm/(cm^2 \cdot s \cdot Pa)$, which is in line with previous reports for CMC films [45]. Addition of 0.1, 0.3, or 0.5 g of chemically or enzymatically recovered fibers to the CMC film had no significant effect on water vapor transmission regardless of plant source (Table 5). These results suggest that these fibers have the potential to improve tensile strength without negatively impacting the ability of the film to function as a water barrier. These attributes would make the films more suitable for food storage.

The data were subjected to an analysis of variance and means were examined using a Tukey's pairwise comparison test ($\alpha = 0.05$).

Pyrolytic Properties: The addition of Yunnan pine wood, bamboo culms, or industrial hemp hurd obtained by pectinase and hemicellulase treatment to the CMC was associated with a higher first decomposition peak compared with the non-modified CMC (Table 6). Mass loss at that time was lower for the Yunnan pine wood and bamboo culms but similar for the industrial hemp hurd. The second decomposition peak was slightly lower with the addition of either Yunnan pine wood or bamboo culms fiber but nearly the same as non-amended CMC with addition of industrial hemp hurd. The mass losses at the second decomposition peak were all slightly higher with addition of Yunnan pine wood and industrial hemp hurd losing the most mass. The addition of fibers to the CMC altered both peak temperatures, whereas it decreased mass losses for the first peak and increased them for the second. The final residual weights of the films were 28, 32, 35, and 22 wt%, respectively. The addition of plant fibers resulted in an increase in the required decomposition temperature and an increase in the final weight residue (Figure 5) [4,46].

Table 6. Effect of addition of Yunnan pine wood, bamboo culms, and industrial hemp hurd fibers obtained using pectinase + hemicellulase treatment on decomposition peaks and mass losses of CMC film.

Sample	Decomposition Peak (°C)	Mass Loss (%)	Decomposition Peak (°C)	Mass Loss (%)
Control	231	16	296	52
Yunnan pine wood	236	11	300	57
Bamboo culms	236	11	303	54
Industrial hemp hurd	235	15	295	57

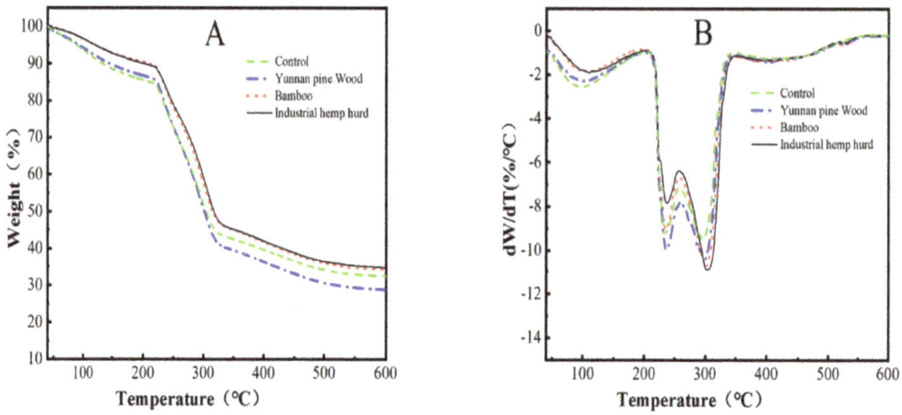

Figure 5. Effect of addition of Yunnan pine wood, bamboo culms, or industrial hemp hurd fibers obtained by pectinase + hemicellulase treatment on (**A**) TG and (**B**) DTG curves of eucalyptus bark CMC composite films.

4. Conclusions

Fibers prepared by chemical and enzymatic methods differed in length, width, cell wall thickness, and lumen width, which improved the crystallinity of the fibers. The length:width ratio of the enzymatically prepared fibers was smaller than that of chemically prepared fibers. Wall thickness:lumen width ratios of industrial hemp hurd fibers were the smallest among the three materials. Tensile strength (TS) and elongation at break (EB) of CMC composite films without plant fiber were 26.2 MPa and 7.35%, respectively. TS of CMC composite films was greatly improved by addition of industrial hemp hurd fiber, especially after enzyme treatment. TS of CMC composite films increased to 35.9, 41.3, and 45.2 MPa for chemical, pectinase, and pectinase + hemicellulase treatments, respectively, after adding 0.5 g industrial hemp hurd fiber. EB changed to 1.61, 1.76, and 4.18%, respectively. Addition of modified fibers did not affect opacity or water vapor permeability, indicating that adding low levels of fiber to CMC significantly improved film characteristics making them potentially suitable for food storage.

Author Contributions: Conceptualization, X.L. (Xiaoping Li) and J.J.M.; methodology, X.L. (Xiaoping Li) and J.J.M.; validation, X.L. (Xiaobao Li). Z.T., Z.S., Z.L. and J.S.; formal analysis, X.L. (Xiaoping Li) and J.J.M.; investigation, X.L. (Xiaobao Li), Z.T., Z.S., Z.L. and J.S.; resources, X.L. (Xiaoping Li); data curation, Z.T.; writing—original draft preparation, X.L. (Xiaobao Li); writing—review and editing, J.J.M., X.L. (Xiaoping Li); supervision, X.L. (Xiaoping Li); project administration, X.L. (Xiaoping Li) and J.J.M.; funding acquisition, X.L. (Xiaoping Li). All authors have read and agreed to the published version of the manuscript.

Funding: This study was supported by the National Nature Science Foundation (31870551), Top Young Talents in Yunnan Province (YNWR-QNBJ-2018-120) and 111 Project (D21027).

Institutional Review Board Statement: Not applicable.

Data Availability Statement: The data presented in this study are available from the listed authors.

Conflicts of Interest: The authors declare no conflict of interest.

References

1. Li, W.; Zhao, X.; Ji, Y.; Jia, J.; Zhao, W. Progresses in preparation and production technology for carboxymethylcellulose. *China Pet. Process. Petrochem. Technol.* **2013**, *42*, 693–702.
2. Zhang, Y.; Huang, R.; Yu, J.; Jiang, X. Facile fabrication of sodium carboxymethyl cellulose/reduced graphene oxide composite hydrogel and its application for Pb(II) removal. *Micro Nano Lett.* **2021**, *16*, 30–38. [CrossRef]

3. Kawasaki, T.; Nakaji-Hirabayashi, T.; Masuyama, K. Complex film of chitosan and carboxymethyl cellulose nanofibers. *Colloids Surf. B* **2016**, *139*, 95–99. [CrossRef] [PubMed]
4. Yao, Y.; Sun, Z.; Li, X.; Tang, Z.; Li, X. Effects of Raw Material Source on the Properties of CMC Composite Films. *Polymers* **2021**, *14*, 32. [CrossRef]
5. Zhao, X.; Clifford, A.; Poon, R. Carboxymethyl cellulose and composite films prepared by electrophoretic deposition and liquid-liquid particle extraction. *Colloid Polym. Sci.* **2018**, *296*, 927–934. [CrossRef]
6. Yao, Y.; Sun, Z.; Li, X.; Tang, Z.; Li, X. Research status of carboxymethyl cellulose composite film. *Packag. Eng. (Chongqing, China) Pao Chuang Kung Ch'eng* **2021**, *9*, 1–10.
7. Nguyen-Sy, T.; Tran-Le, A.D.; Nguyen-Thoi, T.; Langlet, T. A multi-scale homogenization approach for the effective thermal conductivity of dry lime–hemp concrete. *J. Build. Perform. Simul.* **2018**, *11*, 179–189. [CrossRef]
8. Casaburi, A.; Rojo, Ú.M.; Cerrutti, P. Carboxymethyl cellulose with tailored degree of substitution obtained from bacterial cellulose. *Food Hydrocoll.* **2018**, *75*, 147–156. [CrossRef]
9. Rahman, M.; Hasan, M.; Nitai, A.S. Recent developments of carboxymethyl cellulose. *Polymers* **2021**, *13*, 1345. [CrossRef]
10. Yadav, K.R.; Sharma, R.K.; Kothari, R.M. Bioconversion of eucalyptus bark waste into soil conditioner. *Bioresour. Technol.* **2002**, *81*, 163–165. [CrossRef]
11. Tavares, K.M.; de Campos, A.; Luchesi, B.R. Effect of carboxymethyl cellulose concentration on mechanical and water vapor barrier properties of corn starch films. *Carbohydr. Polym.* **2020**, *246*, 116521. [CrossRef]
12. Lan, W.; He, L.; Liu, Y. Preparation and properties of sodium carboxymethyl cellulose/sodium alginate/chitosan composite film. *Coatings* **2018**, *8*, 291. [CrossRef]
13. Hazirah, M.N.; Isa, M.I.N.; Sarbon, N.M. Effect of xanthan gum on the physical and mechanical properties of gelatin-carboxymethyl cellulose film blends. *Food Packag. Shelf Life* **2016**, *9*, 55–63. [CrossRef]
14. Bakry, N.F.; Isa, M.I.N.; Sarbon, N.M. Effect of sorbitol at different concentrations on the functional properties of gelatin/carboxymethyl cellulose (CMC)/chitosan composite films. *Int. Food Res. J.* **2017**, *24*, 1753–1762.
15. Tong, Q.; Xiao, Q.; Lim, L.T. Effects of glycerol, sorbitol, xylitol and fructose plasticisers on mechanical and moisture barrier properties of pullulan–alginate–carboxymethylcellulose blend films. *Int. J. Food Sci. Technol.* **2013**, *48*, 870–878. [CrossRef]
16. de Melo Fiori, A.P.S.; Camani, P.H.; dos Santos Rosa, D. Combined effects of clay minerals and polyethylene glycol in the mechanical and water barrier properties of carboxymethylcellulose films. *Ind. Crops Prod.* **2019**, *140*, 111644. [CrossRef]
17. Guo, G.; Xiang, A.; Tian, H. Thermal and mechanical properties of eco-friendly poly (vinyl alcohol) films with surface treated bagasse fibers. *J. Polym. Environ.* **2018**, *26*, 3949–3956. [CrossRef]
18. Zong, Q.; Hong, J.; Min, W.; Zhi, G.; Miao, Z. Effect of wheat bran fiber on the behaviors of maize starch based films. *Starch-Stärke* **2020**, *72*, 1900319.
19. Carvalho, G.R.; Marques, G.S.; Matos Jorge, L.M. Cassava bagasse as a reinforcement agent in the polymeric blend of biodegradable films. *J. Appl. Polym. Sci.* **2019**, *136*, 47224. [CrossRef]
20. Kumar, K.P.; Sekaran, A.S.J. Some natural fibers used in polymer composites and their extraction processes: A review. *J. Reinf. Plast. Compos.* **2014**, *33*, 1879–1892. [CrossRef]
21. Hong, C.; Ge, W.; Hai-Tao, C. Properties of single bamboo fibers isolated by different chemical methods. *Wood Fiber Sci.* **2011**, *43*, 111–120.
22. Tang, Z.; Yang, M.; Qiang, M. Preparation of cellulose nanoparticles from foliage by bio-enzyme methods. *Materials* **2021**, *14*, 4557. [CrossRef] [PubMed]
23. Li, X.; Qiang, M.; Yang, M. Combining Fiber Enzymatic Pretreatments and Coupling Agents to Improve Physical and Mechanical Properties of Hemp Hurd/Wood/Polypropylene Composite. *Materials* **2021**, *14*, 6384. [CrossRef]
24. Lynd, L.R.; Weimer, P.J.; Van, W.H. Microbial cellu-lose utilization: Fundamentals and biotechnology. *Microbiol. Mol. Biol. Rev.* **2002**, *66*, 506–577. [CrossRef] [PubMed]
25. Ibarra, D.; Monte, M.C.; Blanco, A. Enzymatic deinking of secondary fibers: Cellulases/hemicellulases versus lac-case-mediator system. *J. Ind. Microbiol. Biotechnol.* **2012**, *39*, 1–9. [CrossRef]
26. Periera, H.; Miranda, I. The chemical composition of wood and bark of fast-grown *Eucalyptus globulus* L. trees during the first 3 years. In Proceedings of the 6th International Conference on Biomass for Energy, Industry and Environment, Athens, Greece, 22–26 April 1991.
27. Fu, J.; Nyanhongo, G.; Silva, C. Bamboo fibre processing: Insights into hemicellulase and cellulase substrate accessibility. *Biocatal. Biotransform.* **2012**, *30*, 27–37. [CrossRef]
28. Wang, W. Preparation of carboxymethyl cellulose from straw of crop. *Tianjin Huagong* **2004**, *18*, 10–12.
29. Dai, X.; Xue, Y.; Peng, W.; Liu, Z.; Ren, H. Synthesis and characterization of carboxymethylated cellulose from straw and reed. *Guangzhou Chem. Ind. Technol.* **2014**, *42*, 36–39.
30. Alemdar, A.; Sain, M. Isolation and characterization of nanofibers from agricultural residues-wheat straw and soy hulls. *Bioresour. Technol.* **2008**, *199*, 1664–1671. [CrossRef]
31. Lu, Q.; Lin, W.; Tang, L.; Wang, S.; Chen, X. A mechanochemical approach to manufacturing bamboo cellulose nanocrystals. *J. Mater. Sci.* **2015**, *50*, 611–619. [CrossRef]
32. Park, S.; Lee, S.; Kim, S. The use of a nanocellulose reinforced polyacrylonitrile precursor for the production of carbon fibers. *J. Mater. Sci.* **2013**, *48*, 6952–6959. [CrossRef]

33. Troedec, M.; Sedan, D.; Peyratout, C.; Bonnet, J.; Smith, A. Influence of various chemical treatments on the composition and structure of hemp fibres. *Compos. Part A* **2008**, *39*, 514–522. [CrossRef]
34. Garside, P.; Wyeth, P. Identification of cellulosic fibres by FTIR spectroscopy: Thread and single fibre analysis by attenuated total reflectance. *Stud. Conserv.* **2003**, *48*, 269–275. [CrossRef]
35. Mihranyan, A.; Esmaeili, M.; Razaq, A.; Alexeichik, D. Influence of the nanocellulose raw material characteristics on the electrochemical and mechanical properties of conductive paper electrodes. *J. Mater. Sci.* **2011**, *47*, 4463–4472. [CrossRef]
36. Kuo, C.H.; Lee, C. Enhancement of enzymatic saccharification of cellulose by cellulose dissolution pretreatments. *Carbohydr. Polym.* **2009**, *77*, 41–46. [CrossRef]
37. Adsul, M.; Soni, S.; Bhargava, S. Facile approach for the dispersion of regenerated cellulose in aqueous system in the form of nanoparticles. *Biomacromolecules* **2012**, *13*, 2890–2895. [CrossRef]
38. Schooneveld-Bergmans, M.; Beldman, G.; Voragen, A. Structural features of (glucurono) arabinoxylans extracted from wheat bran by barium hydroxide. *J. Cereal Sci.* **1999**, *29*, 63–75. [CrossRef]
39. Besbes, I.; Vilar, M.; Boufi, S. Nanofibrillated cellulose from TEMPO-oxidized eucalyptus fibres: Effect of the carboxyl content. *Carbohydr. Polym.* **2011**, *84*, 975–983. [CrossRef]
40. Liu, H.; Liu, D.; Yao, F.; Wu, Q. Fabrication and properties of transparent polymethylmethacrylate/cellulose nanocrystals composites. *Bioresour. Technol.* **2010**, *101*, 5685–5692. [CrossRef]
41. Peter, Z. Order in cellulosics: Historical review of crystal structure research on cellulose. *Bioresour. Technol.* **2021**, *254*, 117417. [CrossRef]
42. Lu, Q.L.; Wu, J.; Li, Y.; Huang, B. Isolation of thermostable cellulose II nanocrystals and their molecular bridging for electroresponsive and pH-sensitive bio-nanocomposite. *Ind. Crops Prod.* **2021**, *173*, 114127. [CrossRef]
43. Jawad, Y.M.; Al-Kadhemy, M.F.H.; Salman, J.A.S. Synthesis, Structural and Optical Properties of CMC/MgO Nanocomposites. *Mater. Sci. Forum.* **2021**, *1039*, 104–114. [CrossRef]
44. Bourmaud, A.; Beaugrand, J.; Shah, D.U.; Placet, V.; Baley, C. Towards the design of high-performance plant fibre composites. *Prog. Mater. Sci.* **2018**, *97*, 347–408. [CrossRef]
45. Oun, A.; Rhim, J. Preparation and characterization of sodium carboxymethyl cellulose/cotton linter cellulose nanofibril composite films. *Carbohydr. Polym.* **2015**, *127*, 101–109. [CrossRef]
46. Badry, R.; Ezzat, H.A.; El-Khodary, S.; Morsy, M.; Elhaes, H.; Nada, N.; Ibrahim, M. Spectroscopic and thermal analyses for the effect of acetic acid on the plasticized sodium carboxymethyl cellulose. *J. Mol. Struct.* **2021**, *1224*, 129013. [CrossRef]

polymers

MDPI

Article

Effects of Polypropylene Fibers on the Frost Resistance of Natural Sand Concrete and Machine-Made Sand Concrete

Yan Tan [1,*], Junyu Long [1], Wei Xiong [2], Xingxiang Chen [1] and Ben Zhao [1]

1 Department of Architecture and Environment, Hubei University of Technology, Wuhan 430068, China
2 Department of Architecture and Engineering, Wuhan City Polytechnic, Wuhan 430064, China
* Correspondence: tanyan@hbut.edu.cn

Abstract: In order to study the effect of polypropylene fibers on the frost resistance of natural sand and machine-made sand concrete, polypropylene fibers (PPF) of different volumes and lengths were mixed into natural sand and machine-made sand concrete, respectively. The freeze–thaw cycle test was carried out on polypropylene-fiber-impregnated natural sand concrete (PFNSC) and polypropylene-fiber-impregnated manufactured sand concrete (PFMSC), respectively, and the apparent structural changes before and after freezing and thawing were observed. Its strength damage was analyzed. A freeze–thaw damage model and a response surface model (RSM) were established used to analyze the antifreeze performance of PFMSC, and the effects of the fiber content, fiber length, and freeze–thaw times on the antifreeze performance of PFMSC were studied. The results show that with the increase in the number of freeze–thaw cycles, the apparent structures of the PFMSC gradually deteriorated, the strength decreased, and the degree of freeze–thaw damage increased. According to the strength damage model, the optimum volume of PPF for the PFNSC specimens is 1.2%, and the optimum volume of PPF for the PFMSC specimens is 1.0%. According to the prediction of RSM, PFNSC can maintain good antifreeze performance within 105 freeze–thaw cycles, and when the PPF length is 11.8 mm, the antifreeze performance of PFNSC reaches the maximum, its maximum compressive strength value is 33.8 MPa, and the split tensile strength value is 3.1 MPa; PFMSC can maintain a good antifreeze performance within 96 freeze–thaw cycles. When the length of PPF is 9.1 mm, the antifreeze performance of PFMSC reaches the maximum, its maximum compressive strength value is 45.8 MPa, and its split tensile strength value is 3.2 MPa. The predicted values are in good agreement with the measured values, and the model has high reliability.

Keywords: polypropylene fiber mechanism sand concrete; frost resistance; freeze–thaw damage model; RSM strength model

Citation: Tan, Y.; Long, J.; Xiong, W.; Chen, X.; Zhao, B. Effects of Polypropylene Fibers on the Frost Resistance of Natural Sand Concrete and Machine-Made Sand Concrete. *Polymers* **2022**, *14*, 4054. https://doi.org/10.3390/polym14194054

Academic Editors: R.A. Ilyas, S.M. Sapuan, Emin Bayraktar, Shukur Abu Hassan, Nabil Hayeemasae and Khubab Shaker

Received: 18 August 2022
Accepted: 15 September 2022
Published: 27 September 2022

1. Introduction

Durability is an important indicator for measuring the ability of a material to resist the long-term damage of both itself and the natural environment [1–4]. Freeze–thaw damage in cold regions has an important impact on the durability of concrete structures. Freeze–thaw damage to concrete in alpine regions of China is a common issue, and it poses a huge threat to the safety of building use and economic and environmental protection. Frost resistance is particularly important. The main reason for the freeze–thaw damage of concrete is that the water in the concrete pores, under the action of alternating dry and wet and freeze–thaw cycles, forms fatigue stress from the combined action of ice expansion pressure and osmotic pressure, which causes the concrete to produce denudation damage from the surface to the inside, thereby reducing the strength of the concrete [5–7]. Research on the antifreeze performance of concrete structures under low-temperature freeze–thaw cycle environment can not only reveal potential dangers and avoid major safety accidents but also provide a basis for durability design, testing, and reinforcement of concrete structures in low-temperature freeze–thaw environments. Improving the frost resistance of concrete has become a hot topic of research at home and abroad [8–10].

Compared with ordinary concrete, adding fiber is one of the methods that can effectively reduce the brittleness and improve the toughness of concrete [11–14]. Fiber materials have been widely studied by scholars at home and abroad. The results show that fiber structures have the characteristics of strong plasticity, high toughness, and strong adhesion to concrete, and adding fibers into concrete can offset the internal stress and play an effective role in frost resistance [15–18]. The different types of fibers have different effects on the frost resistance of concrete. Among them, not only is polypropylene fiber (PPF) small in diameter, low in quality, low in cost, and good at self-dispersion, but it also has the characteristics of inhibiting concrete plastic cracking, preventing crack propagation, limiting matrix damage, improving the long-term working performance of concrete structures, and improving structural durability [19–21]. In the process of freezing, the increase in the elastic modulus of PPF can effectively offset the ice expansion force, and the decrease in the elastic modulus of PPF during the thawing process helps it to release the accumulated expansion energy. Therefore, the change in the elastic modulus of PPF is enhanced as a whole. The strain capacity of concrete in the freeze–thaw environment can effectively inhibit the frost heave cracking of concrete. Different lengths and dosages of PPF have varying degrees of influence on the tensile properties, fatigue resistance, and wear resistance of concrete [22–24]. The performance of concrete has been improved in all aspects [25]. Cheng Hongqiang et al. [26] conducted a freeze–thaw damage test of polypropylene fiber concrete. Under the action of freeze–thaw cycles, the damage of polypropylene fiber concrete continued to accumulate, and the mass loss rate and split tensile strength continued to decrease. The splitting tensile strength has been continuously improved. Chen Liuzhuo et al. [27] studied the frost resistance of ordinary concrete by adding steel fibers and polypropylene fibers. The results show that controlling the amount and length of fibers can effectively improve tensile strength and flexural strength. Durability is enhanced, and polypropylene fibers are more damage-resistant than steel fibers. Salehi Parisa et al. [28] studied the effect of PPF on the mechanical properties and durability of reinforced lightweight concrete through the mixed design of three volume dosages (0%, 0.5%, and 1%) of PPF, and the results show that the incorporation of fibers can improve compressive, tensile, and flexural strength, and it also reduces the water absorption and permeability of concrete. Liu Bo et al. [29] analyzed and summarized the influence of polypropylene fibers on the mechanical properties and durability of concrete and proved that the addition of polypropylene fibers can make concrete mechanically resistant to compression, tension, bending, and impact. In terms of performance and durability, including frost resistance, impermeability, and carbonation resistance, it is superior to ordinary concrete.

Due to the over-exploitation of natural sand, natural sand resources are depleted, which seriously affects the ecological environment, and finding new alternative sources has become a hot topic. Machine-made sand is artificially mined industrial waste slag, tailings, etc., which are mechanically crushed and screened into rocks with a particle size of less than 4.75 mm [30–32]. The mud powder in natural sand has an adverse effect on the working performance, volume stability, and durability of concrete, while the fine-grained stone powder in machine-made sand can improve the gradation of machine-made sand and fill the pores of the particles. A large number of studies have shown that an appropriate amount of stone powder can improve the workability of concrete and improve the strength and durability of machine-made sand concrete [33–35]. Replacing natural sand with manufactured sand has different degrees of influence on the mechanical properties of concrete. Ding et al. [36] prepared manufactured sand concrete with a compressive strength of higher than 50 MPa, and some scholars have prepared ultra-high-pressure concrete through different treatment methods. The compressive strength of high-performance concrete exceeds 140 MPa [37–40]. Bonavetti and Irassar [41] studied the effect of machine-made sand and gravel powder on the properties of mortar. With an increase in age and in the replacement rate of machine-made sand, the compressive strength of concrete showed an increasing trend. Donza et al. [42] studied the effects of several different materials of manufactured sand and natural sand on the properties of high-strength concrete. The results showed that the mechanical properties of high-strength concrete were better, due to the mechanical occlusion of the manufactured sand. Zeghichil et al. [43] studied the effect of natural sand and multi-angled machine-made sand on the working performance and mechanical properties of self-compacting concrete. Multi-angled machine-made sand requires more water to achieve the same working performance, but the compressive strength of the concrete is lower than that of regular concrete. The flexural strength is higher than that of concrete prepared from natural sand.

Regarding the application of fiber materials in machine-made sand concrete, Kandasamy and Murugesan [44] studied the compressive strength and split tensile strength of machine-made sand concrete and concrete mixed with plastic fibers. The results show that the addition of plastic fibers can effectively improve the mechanical properties and durability of manufactured sand concrete. Some scholars have improved the shortcomings of easy cracking and poor frost resistance by admixing PPF and mineral admixtures in desert sand concrete [45]. In view of the excellent performance of polypropylene fibers in concrete compared with other fibers, there are relatively few studies on the freeze–thaw cycle of polypropylene fiber concrete in which continuous graded machine-made sand replaces 100% of natural fine aggregates.

In this paper, by incorporating polypropylene fibers of different lengths and volumes, natural sand concrete and manufactured sand concrete with a replacement rate of 100% were used as the objects to study the change law of the durability of PFNSC and PFMSC under freeze–thaw cycles. A fitting model corresponding to the fiber lifting rate and contribution rate [46] was established to quantitatively characterize the influence of PPF. At the same time, based on learning from the classical combination model of damage [47–51], the damage theory under freeze–thaw cycles was used [52,53]. A freeze–thaw damage model and a response surface model were established to analyze the frost resistance of PFNSC and PFMSC, and the effects of fiber content, fiber length, and freeze–thaw times on the frost resistances of PFNSC and PFMSC were studied. The PFNSC and PFMSC specimens were compared.

2. Materials and Methods

2.1. Raw Materials and Mix Ratio

The cementitious materials used in this test are cement and grade I fly ash, the cement type is standard P.O. 42.5 grade ordinary cement, and all the cement indexes meet the requirements of GB175-2007 "General Portland Cement". Among the materials, fly ash complies with the requirements of the standard GB/T 1596-2005 "fly ash used in cement-based concrete", its apparent density is 2500 kg/m^3, the specific surface area is 455.2 m^2/kg, and it is a 30 mm continuous graded natural aggregate; fine aggregates are natural river sand (NRS) and machine-made sand(MS) in zone II, for which the fineness modulus is 2.9, the apparent density of natural river sand is 2650 kg/m^3, and the apparent density of manufactured sand is 2610 kg/m^3. According to the standard GB/T14684-2011 "Sand for Construction", the MB value and stone powder content of machine-made sand meet the requirements of Class II machine-made sand. The superplasticizer is polycarboxylate, and the gradation curve is shown in Figure 1. The physical map and microscopic topography of the polypropylene fibers are shown in Figure 2a,b, and their performance indicators are shown in Table 1.

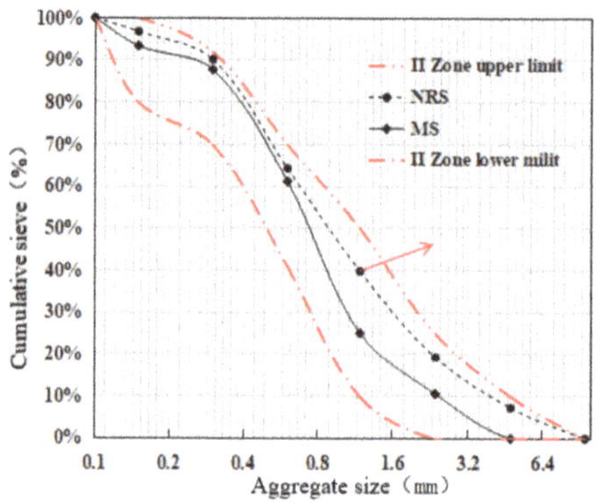

Figure 1. Cumulative sieve residue of fine sand.

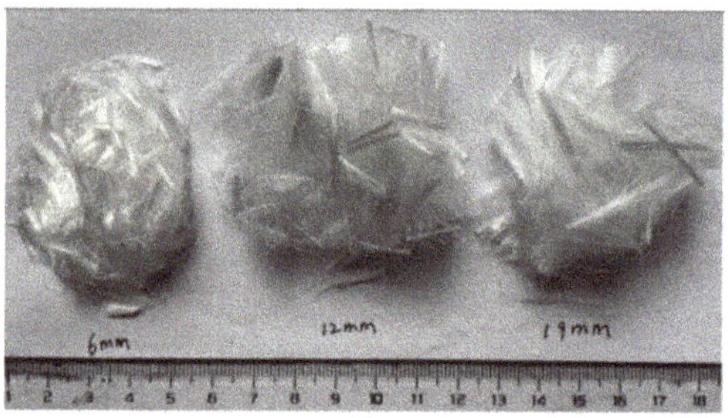

(a)

(b)

Figure 2. Physical and microscopic pictures of polypropylene fibers: (**a**) physical map, (**b**) micrograph.

Table 1. Physical indicators of polypropylene fiber.

Type	Density (g/cm^3)	Elastic Modulus (MPa)	Break Strength (MPa)	Elongation at Break (%)	Melting Point (°C)	Ignite (°C)	Acid and Al-kaliResistance
Fascicular	0.91	>4500	450	255±	165–175	590	Better

It can be seen from Figure 1 that the surface characteristics of the gradations of machine-made sand are that the occupancy of the middle section is higher than the occupancy of both ends, that is, "large in the middle and small at both ends", and the degree of gradation continuity is poorer than that of natural fine aggregate. In this test, the benchmark concrete strength grade is C30, and natural river sand and machine-made sand are used as fine aggregates. The mix ratio of fiber concrete is shown in Table 2. The percentage of polypropylene fibers should be controlled at about 1.0% and not more than 1.2% [54]. In this paper, the volume content of PPF is 0%, 1.0%, and 1.2%. PFNSC a–b and PFMSC a–b are 12 mm and 19 mm ("a" is the volume content of PPF and "b" is the length of PPF), respectively.

Table 2. Laboratory mix proportion of concrete.

NO.	Polypropylene Fiber		Material Dosage (kg/m^3)						
	Volume Dosage (%)	Length (mm)	Cement	Flash	Natural River Sand	Mechanism Sand	Coarse Aggregate	Water	Water Reducer
PFNSC0–0	0	0	398	80	763	0	920	200	4.71
PFNSC1.0–b	1.0	6/12/19	398	80	763	0	920	200	4.71
PFNSC1.2–b	1.2	6/12/19	398	80	763	0	920	200	4.71
PFMSC0–0	0	0	398	80	0	763	920	200	4.71
PFMSC1.0–b	1.0	6/12/19	398	80	0	763	920	200	4.71
PFMSC1.2–b	1.2	6/12/19	398	80	0	763	920	200	4.71

2.2. Test Equipment and Test Methods

A total of 14 sets of specimens were designed for the test, all of which conform to GB/T 50082-2009 "Standards for Test Methods for Long-term Performance and Durability of Ordinary Concrete".

The compressive test uses a DYE-2000A microcomputer servo pressure-testing machine (produced by Cangzhou Zhulong Engineering Instrument Co. Ltd., Cangzhou, China), as shown in Figure 3a, and a cube with a size of 100 mm × 100 mm × 100 mm to measure the compressive strength test. The test loading speed is 0.5 MPa/s, and the compressive strength correction coefficient is 0.95. After the test piece reaches the test age, it is taken out of the curing room. The surface of the test piece and the upper and lower bearing plates are wiped, and a check is made for whether the center of the test piece is aligned with the center of the lower pressure plate of the testing machine. After the inspection is completed and confirmed to be correct, the compressive strength test is carried out on the testing machine. The loading speed of the press is manually controlled. When the specimen is close to failure and begins to deform sharply, we stop adjusting the test throttle until failure and record the failure load. The arithmetic mean of the measured values of 3 test pieces is taken as the strength value of the test piece.

The split tensile test uses a DYE-2000A microcomputer servo pressure-testing machine (produced by Cangzhou Zhulong Engineering Instrument Co., Ltd., Cangzhou, China), as shown in Figure 3b. Cubes with dimensions of 100 mm × 100 mm × 100 mm were tested for split tensile strength. The test loading speed is 0.08 MPa/s. After the specimen is taken out, the dry and wet state of the specimen are kept unchanged, and measurements are taken to determine whether the size of the specimen meets the requirements. A line is drawn in the middle of the side of the specimen during forming to determine the position of the splitting surface, the lower cushion and gasket are placed in the center of the lower plate, and the specimen, upper gasket, and cushion are placed in sequence on the lower gasket. The contact busbar of the gasket is aligned with the load contact line on the specimen accurately, and the contact between the upper pressure plate and the cushion is adjusted. After checking the adjustment and correctness, turn on the split tensile testing machine. When the specimen is close to failure or deformation, we stop adjusting the test throttle until failure, and the value is recorded. Three test blocks are set for each group of mix ratios, and the value method is the same as that of the compressive strength value.The freeze–thaw test is performed using a TDR-III(produced by Hebei Huawei Co., Ltd., Cangzhou, China), automatic rapid freeze–thaw test machine, as shown in Figure 3c, with a cylindrical block with a size of 100 mm × 100 mm × 400 mm, and a freeze–thaw cycle test. This is carried out according to the "slow freezing method" in GB/T 50082-2009 "Standard for Long-term Performance and Durability Test of Ordinary Concrete". After the test piece reaches the test age, it is taken out from the curing room four days in advance and immersed in water with a temperature of 15~20 °C. After soaking, the test piece is taken out to dry the surface moisture, and its initial mass is weighed. A total of 150 freeze–thaw cycles were carried out in this test, and measurements after every 50 freeze–thaw cycles were used to collect relevant frost resistance durability parameters. The collected contents included: concrete specimen quality, compressive strength, and flexural strength.

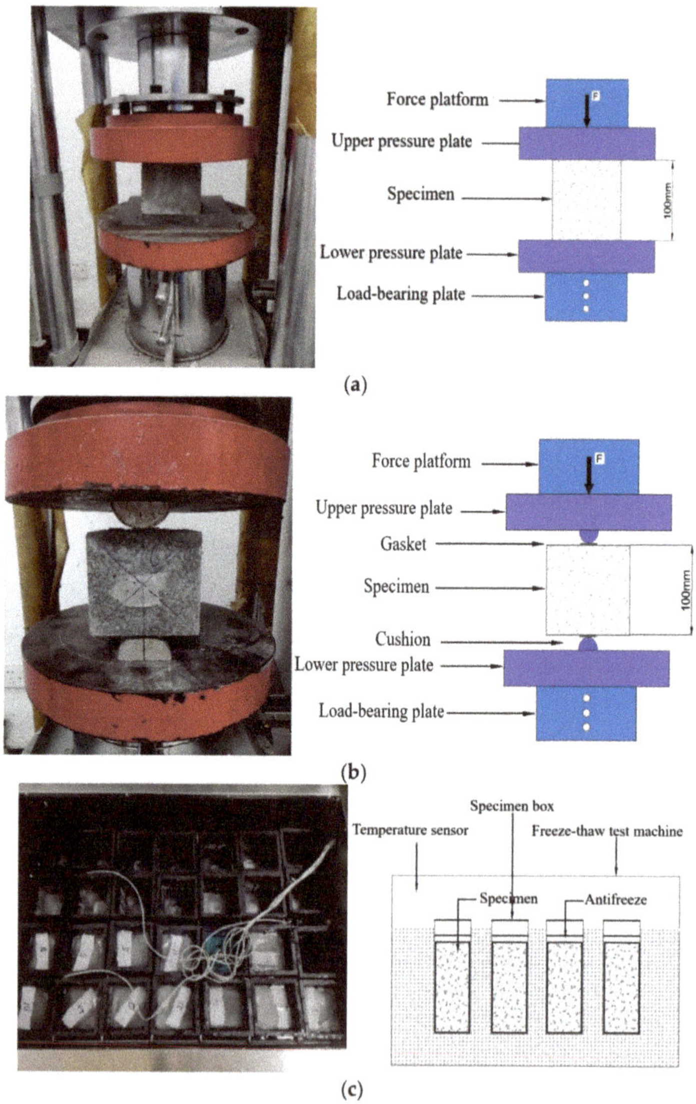

Figure 3. Testing device: (**a**) Compression test, (**b**) Flexural strength test, (**c**) Freeze–thaw cycle test.

3. Test Results and Discussion

3.1. Strength Analysis after Freeze–Thaw Cycle

The PFNSC and PFMSC specimens were set up in the experimental group and the control group with different fiber content, fiber length, and manufactured sand content. They were divided into 14 groups, and each group was set with three specimens, a total of 42 test blocks. The compressive strength, split tensile strength, and quality changes after 0, 50, 100, and 150 freeze–thaw cycles are shown in Table 3.

Table 3. Compressive strength, split tensile strength, and quality under different freeze–thaw cycles.

NO.	0 Freeze–Thaw Cycles			50 Freeze–Thaw Cycles			100 Freeze–Thaw Cycles			150 Freeze–Thaw Cycles		
	Strength (MPa)		Mass (kg)	Strength (MPa)		Mass (kg)	Strength (MPa)		Mass (kg)	Strength (MPa)		Mass (kg)
	Compression	Tensile		Compression	Tensile		Compression	Tensile		Compression	Tensile	
PFNSC0–0	40.7	2.8	2.405	39.7	2.6	2.402	35.6	2.1	2.398	27.5	1.8	2.381
PFNSC1.0–6	44.7	2.7	2.379	40.5	2.7	2.752	40.1	2.6	2.362	37.2	2.2	2.351
PFNSC1.0–12	47.1	3.8	2.382	43.6	3.5	2.375	42.4	2.6	2.360	41.8	2.4	2.347
PFNSC1.0–19	41	2.7	2.378	34.5	2.4	2.374	29.7	2.5	2.278	22.9	1.6	2.252
PFNSC1.2–6	47.7	3.5	2.382	44.2	3.4	2.381	42.3	3.3	2.326	33.2	2.1	2.261
PFNSC1.2–12	52	3.6	2.318	49.2	3.5	2.312	45.8	3.3	2.281	37.5	2.7	2.245
PFNSC1.2–19	44.9	3.2	2.408	41.1	3.1	2.389	38.3	2.9	2.342	34.8	2.2	2.294
PFMSC0–0	50.1	3.2	2.392	48.1	2.8	2.384	49.3	2.6	2.364	42.8	2.1	2.310
PFMSC1.0–6	46.5	3.3	2.333	45.8	3.4	2.323	43.6	3.1	2.317	40.8	2.6	2.247
PFMSC1.0–12	44.7	3.7	2.353	44.3	3.6	2.347	38.4	3.4	2.308	34.4	2.9	2.120
PFMSC1.0–19	48.1	3.4	2.391	42.7	3.9	2.383	32.9	3.7	2.293	30.8	2.8	2.105
PFMSC1.2–6	44.3	3.7	2.363	37.2	3.6	2.348	27.9	3.2	2.290	28.5	2.5	2.278
PFMSC1.2–12	46	3.3	2.397	40.5	3.1	2.394	35.9	2.8	2.381	31.4	2.9	2.294
PFMSC1.2–19	41.7	3.4	2.316	35.8	3.1	2.304	31.7	3.0	2.282	28.9	2.6	2.216
Variance	3.15	0.34	0.03	4.18	0.42	0.10	6.03	0.42	0.04	5.65	0.39	0.08

Using the results of the orthogonal test, the compressive strengths of the PFNSC and PFMSC specimens with different freezing and thawing times and fiber lengths under different polypropylene contents were generated using origin 2021 software (produced by OriginLab Co. Ltd., Northampton, MA, USA), as shown in Figure 4. It can be seen from Figure 4a,b that, with the increase in the number of freeze–thaw cycles, different PPF volume contents in PFMSC specimens had a greater impact on the compressive strengths of the specimens. With an increase in the number of freeze–thaw cycles, the strength of the specimen with a volume of 1.0% decreased rapidly. The strength of the concrete specimen with a volume of PPF of 1.2% was higher than that of the specimen with a 1.0% volume of PPF under different freeze–thaw cycles. The 1.0% concrete specimen shows that the 1.2% fiber volume content results in better frost resistance than the 1.0% fiber volume content. After 50–150 freeze–thaw cycles, the compressive strength of the specimen decreased first, then increased, and then decreased with the increase in fiber length, and the deceleration rate reached the maximum when the fiber length was 12 mm. The comparison of Figure 4a–d shows that with the increase in freeze–thaw times, the average reduction rate for the compressive strengths of the PFNSC and PFMSC specimens is close. The average reduction in the compressive strengths of PFNSC specimens is 26.4%; the average reduction in the compressive strengths of the PFMSC specimens is 26.2%. This shows that the frost resistance of the PFNSC and PFMSC specimens is equivalent.

Figure 5 shows the split tensile strength of PFNSC and PFMSC specimens with different freezing and thawing times and fiber lengths, respectively, for different polypropylene contents. It can be seen from Figure 5a,b that after 50–150 freeze–thaw cycles of the PFMSC specimens, the split tensile strength of the concrete is significantly lower than that of the control group without PPF. The split tensile strength decreases with the increase in fiber length, showing a trend of increasing first and then decreasing. For example, the split tensile strength of the PFMSC specimen with 50 freeze–thaw cycles and 1% PPF volume content is 1.47 times that of the plain concrete without a PPF resemblance. When the volume content of PPF is constant, the relative split tensile strengths of the concrete specimens gradually decrease with the increase in the number of freeze–thaw cycles, but after 100 freeze–thaw cycles, the split tensile strengths of PFMSC specimens decreases. Under the condition of constant fiber content, the relative split tensile strength decreased with the increase in fiber length, with the strength first increasing and then decreasing. By comparing Figure 5a–d, it is found that with the increase in freezing and thawing times, the average reduction rate of the split tensile strengths of PFNSC specimens is slightly higher than that of PFMSC specimens, indicating that the replacement of natural sand concrete by machine-made sand concrete can effectively enhance the durability of concrete. This is because appropriate stone powder in the PFMSC specimen can improve the grading of machine-made sand and fill the pores of particles, which improves the durability of concrete. After 50–150 freeze–thaw cycles, the average reduction in the splitting tensile strengths of PFNSC specimens was 32.5%; the average reduction in splitting tensile strengths of the PFMSC specimens was 23.2%. This shows that the antifreeze performance of PFMSC is better than that of the PFNSC specimens.

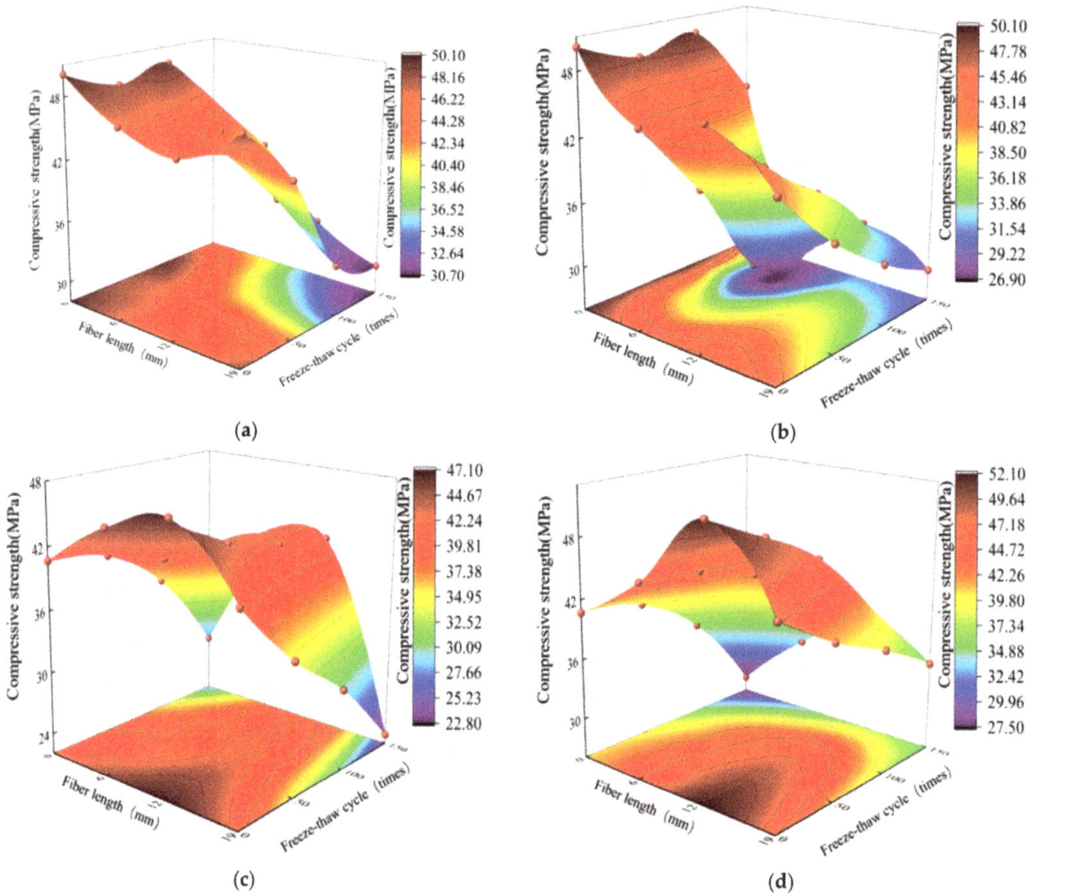

Figure 4. Compressive strength test value at different freeze–thaw cycles: (**a**) PFMSC compressive strength value when PPF volume is 1.0%, (**b**) PFMSC compressive strength value when PPF volume is 1.2%, (**c**) PFNSC compressive strength value when PPF volume is 1.0%, (**d**) PFNSC compressive strength value when PPF volume is 1.2%.

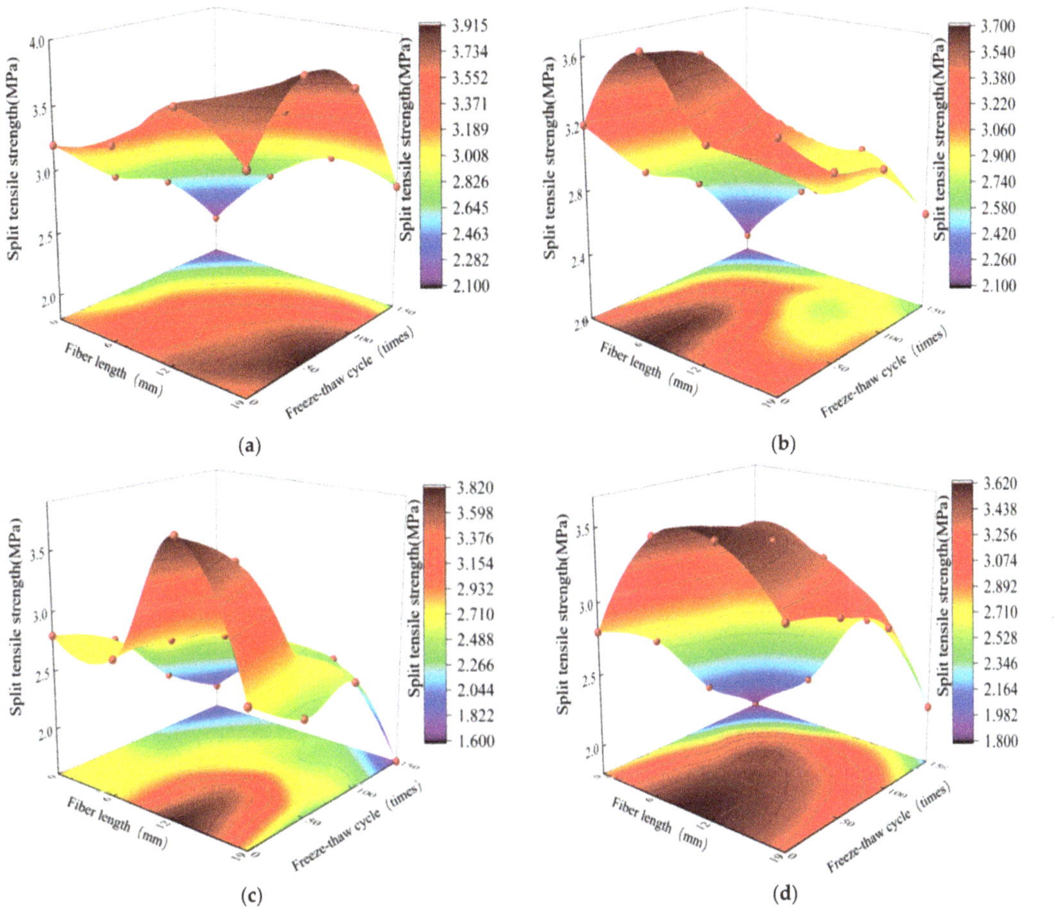

Figure 5. Split tensile strength at different freeze–thaw cycles: (**a**) PFMSC splitting tensile strength value when PPF volume is 1.0%, (**b**) PFMSC splitting tensile strength value when PPF volume is 1.2%, (**c**) PFNSC splitting tensile strength value when PPF volume is 1.0%, (**d**) PFNSC splitting tensile strength value when PPF volume is 1.2%.

The test results show that freeze–thaw cycles have a great influence on the compressive strengths and split tensile strengths of the PFNSC and PFMSC specimens. The addition improved the antifreeze properties of PFNSC and PFMSC. Different PPF volume contents and PPF lengths also have a certain degree of influence on the mechanical properties of concrete. It can be seen that, after 0–150 freeze–thaw cycles, when the volume of PPF is the same, the durability energies of PFMSC and PFNSC increase first and then decrease with the increase in the length of PPF; when the length of PPF is 6–12 mm, the mechanical properties of PFMSC and PFNSC increase with the increase in fiber volume content. When the PPF length is 19 mm, the mechanical properties of PFMSC and PFNSC decrease with the increase in fiber volume content.

Figure 6 shows the apparent state of the PFMSC specimen after 0, 50, 100, and 150 freeze–thaw cycles and the surface macro-state of the binarized concrete specimen under the same variables. It can be seen from Figure 6 that after 0 freeze–thaw cycles, there is a small area of cement material shedding on the surface of the specimen; that is, the black area marked in red is the area where a small part of the surface of the specimen has fallen off the cement material. With the increase in the number of freeze–thaw cycles, the quality of the PFMSC specimen continued to decrease, and the coarse aggregate inside the PFMSC specimen was gradually exposed. After 50 freeze–thaw cycles, a small area of the surface of the specimen had fallen off, and the mass loss was approximately 0.32%. After 100 freeze–thaw cycles, much of the surface of the test piece had fallen off, large coarse aggregate particles were exposed, and the mass loss was approximately 1.85%. When the freeze–thaw cycles reached 150, the surface of the specimen showed a covered black area, indicating that the surface of the concrete specimen was almost completely peeled off under 150 freeze–thaw cycles, the aggregate under the surface was completely exposed, and the mass loss was approximately 4.5%. With the increase in the number of freeze–thaw cycles, the surface peeling degree of the specimens gradually deepened, the strength damage of the PFMSC specimens was serious, and the mass loss was large.

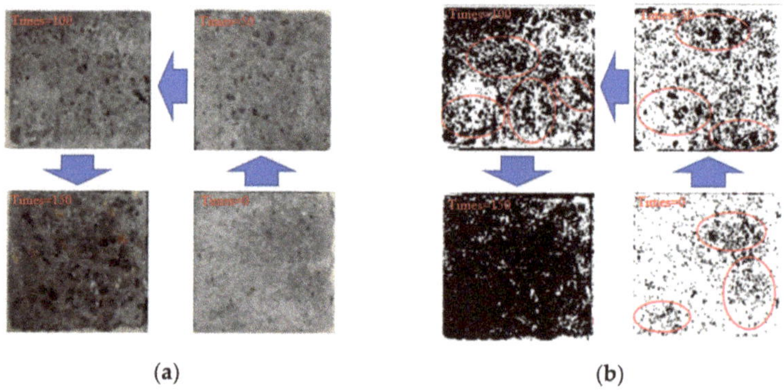

Figure 6. Binarization processing of specimen images with different freezing and thawing times. (**a**) Concrete apparent structure before and after freeze and thawing. (**b**) Binary treatment diagram before and after freeze and thawing.

3.2. Freeze–Thaw Injury of PFNSC and PFMSC under Strength Evaluation Index

In order to consider the change in freeze–thaw cycles and the decay law of concrete mechanical properties, the relationship between the compressive and split tensile strengths of the freeze–thaw cycle damage degree under different polypropylene fiber volume contents and lengths was analyzed. Based on the data changes in compressive strength and splitting tensile strength, according to "Concrete Damage Mechanics", the degree of freeze–thaw cycle damage of concrete is represented by D, and D_c and D_t are defined as the degree of freeze–thaw cycle damage under compression and splitting resistance, respectively, which correspond to strength index damage. They are calculated by Formulas (1) and (2).

$$D_{\mathrm{c,m}} = 1 - \frac{f_{c,n}}{f_{c,0}} \tag{1}$$

$$D_{\mathrm{t,m}} = 1 - \frac{f_{t,n}}{f_{t,0}} \tag{2}$$

In the formulas:

$D_{\mathrm{c,m}}$—Compressive strength damage variable (%); $D_{\mathrm{t,m}}$—Split tensile strength damage variable (%);

$f_{\mathrm{c,n}}$—Compressive strength (MPa) under corresponding freeze–thaw times; $f_{\mathrm{t,n}}$—Splitting strength (MPa) under corresponding freeze–thaw times.

In the formula, 'c' indicates that the specimen is a compression specimen, 't' indicates that the specimen is a split-pull specimen, 'n' is the number of freeze–thaw cycles, and 'm' is the content of manufactured sand. Through Formulas (1) and (2), the strength damage model of the polynomial function under the freeze–thaw cycles was established, and the strength damage of the PFNSC and PFMSC specimens was predicted using the model, as shown in Figure 7a–d under different freeze–thaw cycles.

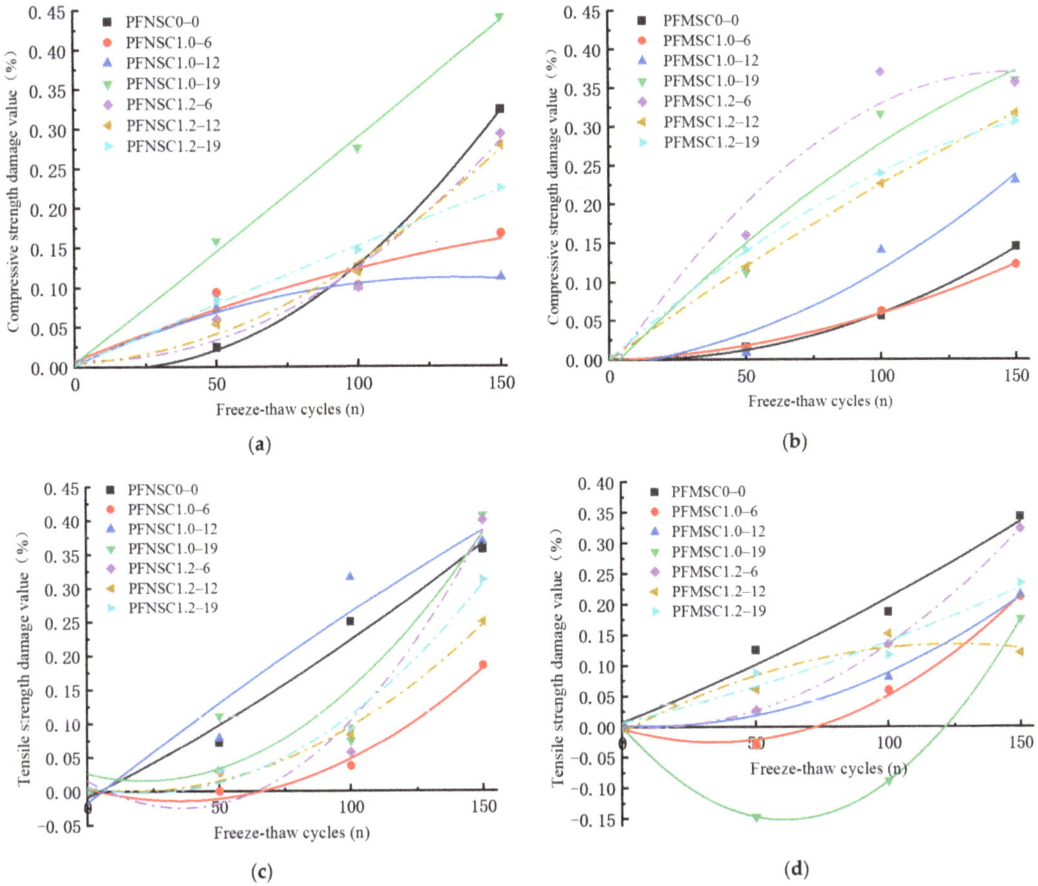

Figure 7. Strength damage under different freeze–thaw cycles: (**a**) PFNSC compressive strength damage rate, (**b**) PFMSC compressive strength damage rate, (**c**) PFNSC splitting tensile strength damage rate, (**d**) PFMSC splitting tensile strength damage rate.

It can be seen from Figure 7 that the extents of freeze–thaw damage of the compressive strengths and split tensile strengths of the PFNSC and PFMSC specimens increase with the increase in the number of freeze–thaw cycles. From Figure 7a,b on tensile strength damage, it can be seen that compared with PFNSC, the compressive strength of the PFMSC specimen is more damaged, because the shape of the machine-made sand is sharp and rough and has many edges and corners. After a certain number of freezing and thawing cycles, the machine-made sand will accelerate the material falling off. After PFNSC freeze–thaw cycles, the freeze–thaw damage rate without PPF is greater than that with PPF. This is because the polypropylene fibers are constrained in the concrete and bear part of the damage caused by the freeze–thaw cycles inside the specimen. With the effect of force, the freeze–thaw damage rate of PFMSC without PPF is lower than the freeze–thaw damage rate of the PPF-doped specimens, and generally, the longer the length of PPF, the greater the amount of freeze–thaw damage. This is due to the mechanism of the sand particles. The influence of shape characteristics on the compressive strengths of PFMSC specimens is greater than that of fibers, and the longer the fibers, the larger the contact area with machine-made sand will be. It can be seen from Figure 7c,d that the split tensile strength damage rate is significantly reduced by PPF. The freeze–thaw damage rate of the specimens without PPF is greater than the freeze–thaw damage rate of the specimens with PPF. Longitudinal pressure inside the specimen can improve the durability of concrete; PFNSC specimens have the same length of PPF, and the splitting tensile strength of the specimen when the volume content of PPF is 1.2% is less than that of the specimen when the volume content of PPF is 1%. The freeze–thaw damage rate of PFMSC is due to the bending of PPF when the inside of the specimen is greatly constrained, which increases the contact area and improves the anti-splitting ability; PFMSC specimens have the same length of PPF, and the volume content of PPF is 1.2%. The freeze–thaw damage rate of the specimen is greater than the freeze–thaw damage rate of the specimen with a volume content of 1% PPF. The mechanism is that the influence of the shape characteristics of the machine-made sand particles on the splitting tensile strength of PFMSC is greater than that of the fiber inside the specimen. The fitting function of the freeze–thaw damage rate is shown in Tables 4 and 5. It can be seen from Tables 4 and 5 that the fitted regression curve and the observed value of R^2 are both greater than 0.95, indicating that the fitted model is highly reliable.

Table 4. Fit function of compressive strength damage value.

PFNSC0–0	PFNSC1.0–6	PFNSC1.0–12	PFNSC1.0–19
$y = 2E - 5x^2 - 0.0005x + 0.0011$ $R^2 = 0.9996$	$y = 3E - 6x^2 + 0.0015x + 0.007$ $R^2 = 0.9307$	$y = 6E - 6x^2 + 0.0016x + 0.0018$ $R^2 = 0.9914$	$y = 7E - 7x^2 + 0.0028x + 0.0045$ $R^2 = 0.9961$
PFNSC1.2–6	**PFNSC1.2–12**	**PFNSC1.2–19**	
$y = E - 5x^2 + 0.0001x + 0.0092$ $R^2 = 0.9663$	$y = E - 5x^2 + 0.0002x + 0.0041$ $R^2 = 0.9996$	$y = 7E - 7x^2 + 0.0016x + 0.0019$ $R^2 = 0.9974$	
PFMSC0–0	**PFMSC1.0–6**	**PFMSC1.0–12**	**PFMSC1.0–19**
$y = 9E - 6x^2 - 0.0005x + 0.0056$ $R^2 = 0.9921$	$y=5E - 6x^2 + 0.0002x - 0.001$ $R^2 = 0.9979$	$y = 8E - 6x^2 + 0.0004x + 0.0083$ $R^2 = 0.9628$	$y = 7E - 6x^2 + 0.0036x + 0.0127$ $R^2 = 0.9634$
PFMSC1.2–6	**PFMSC1.2–12**	**PFMSC1.2–19**	
$y = 2E - 5x^2 - 0.0056x + 0.0163$ $R^2 = 0.9597$	$y = 6E - 6x^2 - 0.0028x + 0.0011$ $R^2 = 0.9999$	$y = 7E - 6x^2 - 0.0032x + 0.0006$ $R^2 = 0.9986$	

Table 5. Fit function of split-pull strength damage values.

PFNSC0–0	PFNSC1.0–6	PFNSC1.0–12	PFNSC1.0–19
$y = 4E - 6x^2 + 0.002x - 0.0089$ $R^2 = 0.9801$	$y = E5\,x^2 - 0.001x + 0.0037$ $R^2 = 0.9882$	$y = 3E - 6x^2 + 0.0031x - 0.0171$ $R^2 = 0.939$	$y = 2E - 5x^2 + 0.001x + 0.0259$ $R^2 = 0.8601$
PFNSC1.2–6	**PFNSC1.2–12**	**PFNSC1.2–19**	
$y = 3E - 5x^2 + 0.0023x + 0.0157$ $R^2 = 0.953$	$y = E - 5x^2 + 0.0005x + 0.0042$ $R^2 = 0.9908$	$y = 2E - 5x^2 + 0.008x + 0.0062$ $R^2 = 0.9869$	
PFMSC0–0	**PFMSC1.0–6**	**PFMSC1.0–12**	**PFMSC1.0–19**
$y = 3E - 6x^2 + 0.0017x + 0.0078$ $R^2 = 0.9801$	$y = 2E - 5x^2 - 0.0013x - 0.003$ R2 = 0.9952	$y = E - 5x^2 - 0.0002x + 0.0027$ $R^2 = 0.9947$	$y = 4E - 5x^2 - 0.005x$ $R^2 = 0.9999$
PFMSC1.2–6	**PFMSC1.2–12**	**PFMSC1.2–19**	
$y = 2E - 5x^2 - 0.0003x$ R2 = 0.9999	$y = -9E - 6x^2 + 0.0023x - 0.0076$ $R^2 = 0.9149$	$y = 3E - 6x^2 + 0.0031x - 0.0073$ R2 = 0.9619	

4. Model of the Composite Factor RSM Intensity

The RSM uses mathematical and statistical methods to model and analyze problems affected by multiple variables, with the ultimate goal of optimizing the response value [55,56]. Box and Wilson first proposed the response surface method. At that time, the research on the response surface method was limited to how to obtain an explicit function using statistical methods to approximate a complex implicit function. Fang et al. [57] used the

D-optimal design and a first-order response surface model to predict the dynamic response and damage identification of intact and damaged systems, and they also used numerical examples, reinforced concrete frame model tests, and I-40 real bridge test results to verify the effectiveness of the proposed method. Using the response surface method (RSM) to model the relationship between factors and levels can better analyze the accuracy, significance, and reliability of the experimental data. Zong Zhouhong et al. [58] used the center composite design (CCD) method and the response surface model to complete the finite element model correction of the Baishi Bridge, and they proved that the bridge finite element model correction based on the response surface method has a higher accuracy. Many scholars at home and abroad have established relational models through the response surface method and have obtained correspondingly optimal results [59–62].

The damage degree of PFNSC and PFMSC under different conditions was quantitatively analyzed via freeze–thaw damage. This showed that when the length of PPF is the same and the volume content of PPF is 1.2, the antifreeze performance of PFNSC is optimal; when the volume content of PPF is 1.0, the antifreeze performance of PFMSC is optimal. In order to better study the effect of PPF length and the number of freeze–thaw cycles on antifreeze performance, that is, taking the PPF length and the number of freeze–thaw cycles as two factors, a response surface strength model was established, and design-Expert software was used to conduct a multivariate analysis of the experimental data. Regression analysis was performed to establish the fitting model of Equations (3)–(6). Equations (3)–(6) were analyzed via variance analysis. The regression coefficient value R^2 was used to test the reliability of the model. When the regression coefficient value R^2 was close to 1, the model reliability was high.

$$f_{c0} = 63.91 - 2.63A + 0.13B + 2.95C + 1.59AB - 0.81AC - 2.36BC + 2.13A^2 - 8.27B^2 + 5.4C^2 \quad R^2 = 0.8851 \tag{3}$$

$$f_{c100} = 46.57 - 2.56A - 0.46B - 0.75C - 0.97AB - 1.24BC + 3.25A^2 - 2.8B^2 - 7.04C^2 \quad R^2 = 0.9050 \tag{4}$$

$$f_{t0} = 2.4 - 0.47A - 0.083B + 0.37C + 0.027AB - 0.1BC + 0.085A^2 - 0.17B^2 + 0.22C^2 \quad R^2 = 0.9368 \tag{5}$$

$$f_{t100} = 3.57 - 0.36A + 0.011B + 0.28C + 0.12AB + (2.85E - 003)BC - 0.27A^2 - 0.018B^2 - 0.59C^2 \quad R^2 = 0.9248 \tag{6}$$

In the formula:

f_{c0}—PFNSC compressive strength (MPa); f_{c100}—PFMSC compressive strength (MPa);

f_{t0} —PFNSC tensile strength (MPa); f_{t100}—PFMSC split tensile strength (MPa);

A—Freeze–thaw times (n); B—Polypropylene fiber length (mm); C—Polypropylene fiber volume content (%).

It can be seen from the fitting model that the values obtained by the fitting equation are all close to 1 and that the value of the coefficient of variation obtained by the model is very small, indicating that the correlation between the model and the test data is significant and the degree of fitting is high, which indicates the model can better analyze and predict freezing conditions. The effects of the PPF length and the number of freeze–thaw cycles on the antifreeze properties of PFNSC and PFMSC specimens within 150 thaw cycles are found. Figures 8 and 9 show the optimal three-dimensional response surfaces and contour maps of the compressive and split tensile strengths of the PFNSC specimen and the PFMSC specimen obtained, respectively, based on the RSM model within 150 freeze–thaw cycles.

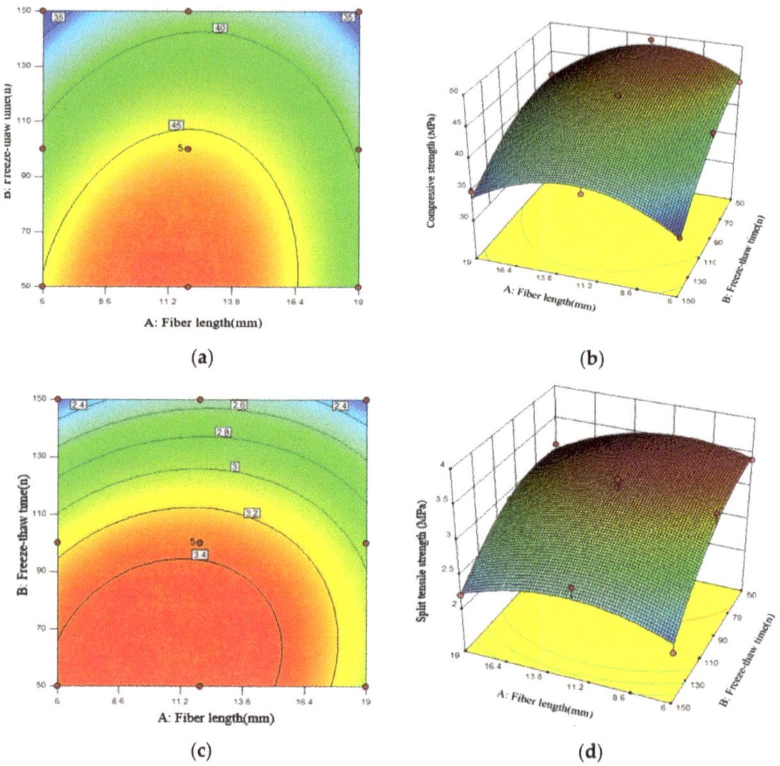

Figure 8. Effects of Factor A, B, and their interactions on R.(PFNSC): (**a**) Surface diagrams of compressive strength variation with A and B (2D), (**b**) Surface diagrams of compressive strength variation with A and B (3D), (**c**) Surface diagrams of split tensile strength variation with A and B (2D), (**d**) Surface diagrams of split tensile strength variation with A and B (3D).

When the volume content of PPF is 1.2%, the PFNSC contour map and 3D response surface are as shown in Figure 8. Through the analysis of Figure 8a,b, it can be seen from the contour map that the green on both sides gradually changes to orange in the center and that the center is the region with the highest responsivity, indicating that the material content corresponding to this region is the optimal doping value. In the three-dimensional response surface graph, the variation trend of the R value along factor B is greater than that of factor A, indicating that factor B (number of freeze–thaw cycles) has a greater impact on the R value than factor A (fiber length). When the number of freeze–thaw cycles increased from 50 to 150, the compressive strength and split tensile strength of PFNSC decreased by 27.5 MPa and 1.8 MPa, respectively; that is, the strength of the specimen decreased with the increase in the number of freeze–thaw cycles. When the fiber increased from 6 mm to 19 mm (1.8 MPa), the strength of the specimen showed a trend of first increase and then decrease, and when the fiber length was approximately 10 mm, the tensile strength reached the maximum. As can be seen from Figure 8c,d, the green on the top of the contour map gradually changes to the red on the bottom. The region below the center is the region with the highest response, and the corresponding polypropylene fiber length and the number of freeze–thaw cycles reach optimum values. The density of the contour lines on the ordinate in the split tensile strength contour diagram is greater than the density of the contour lines on the abscissa; that is, when the fiber volume content is 1.2%, the effect of fiber length on the split tensile strength is greater than that of freezing. The effect of the number of melting cycles is seen.

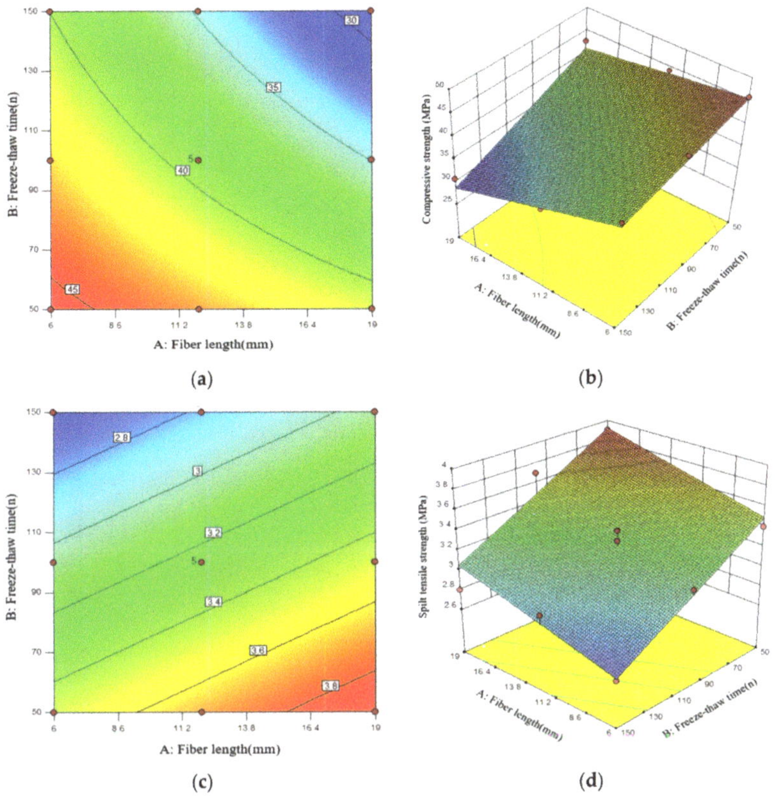

Figure 9. Effects of Factor A, B, and their interactions on R.(PFMSC): (**a**) Surface diagrams of compressive strength variation with A and B (2D), (**b**) Surface diagrams of compressive strength variation with A and B (3D), (**c**) Surface diagrams of split tensile strength variation with A and B (2D), (**d**) Surface diagrams of split tensile strength variation with A and B (3D).

As shown in Figure 10, when the volume content of the PFNSC fibers is 1.2%, the response surface model can predict that when the number of freeze–thaw cycles is 105 and the fiber length is 11.8 mm, the compressive strength and splitting tensile strength of PFNSC are both at maximum values of 33.8 MPa and 3.1 MPa, respectively, indicating that PFNSC can maintain a good antifreeze performance within 105 freeze–thaw cycles. The desirability is 0.975, indicating that the model has high prediction reliability.

As shown in Figure 11, when the volume content of the PFMSC fibers is 1.0%, the response surface model predicts that when the number of freeze–thaw times is 96 and the fiber length is 9.1 mm, the compressive strength and splitting tensile strength of PFMSC are both at maximum values of 41.21 MPa and 3.2 MPa, respectively, indicating that PFMSC can maintain a good antifreeze performance within 96 freeze–thaw cycles. The desirability is 0.826, indicating that the model has high prediction reliability.

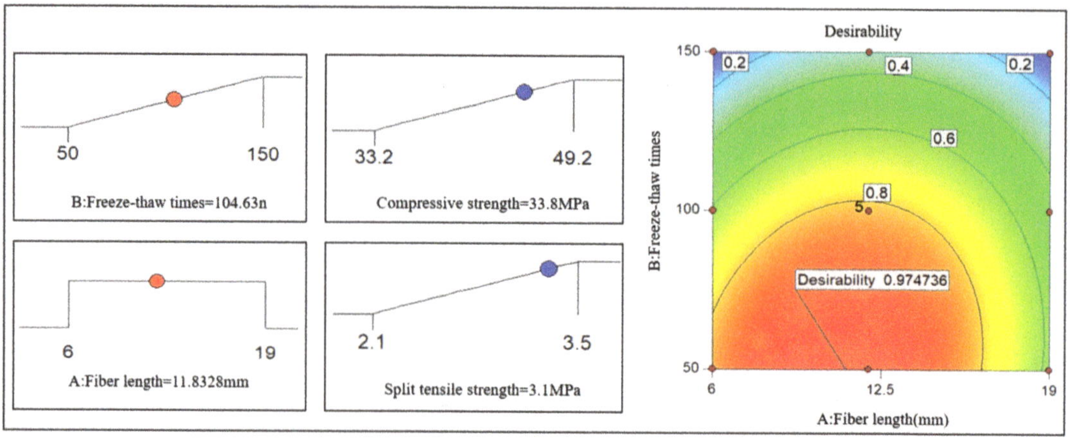

Figure 10. PFNSC and optimization results of strength maximization.

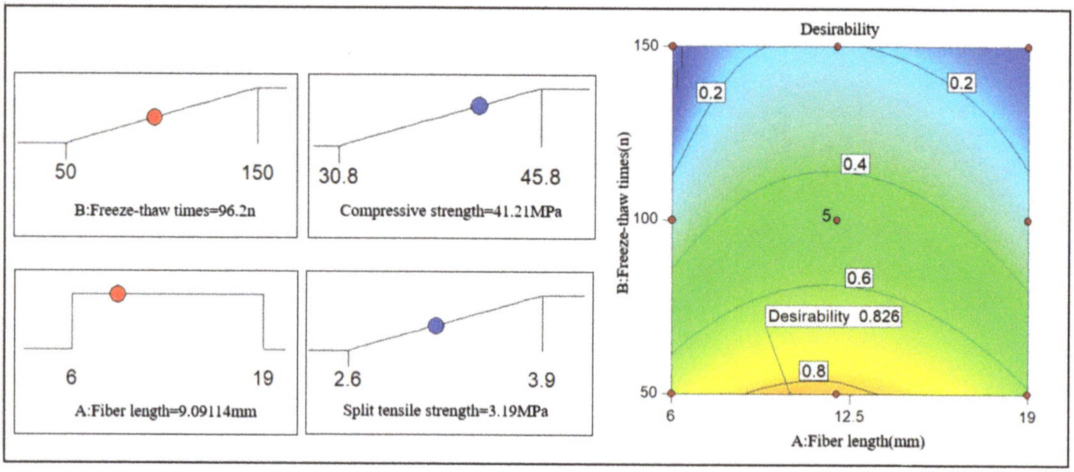

Figure 11. PFMSC and Optimization results of strength maximization 20.

5. Micromorphology Analysis

Scanning electron microscopy (SEM) using a Czech TESCAN MIRA LMS was used to observe the microscopic morphologies of manufactured sand concrete specimens without PPF and PFMSC specimens, before and after the freeze–thaw cycle test. The microscopic morphologies of the PFMSC0–0 specimen and the PFMSC specimen before and after freezing and thawing are shown in Figure 12. After freezing and thawing of PFMSC0–0, the matrix pores increase and increase, and deep cracks appear; the surface of the PFMSC specimen before the freezing and thawing test is relatively smooth and flat, with fewer cracks and pores. More cracks and pores appear on the surface of the matrix after the freeze–thaw test, but because PPF belongs to the class of bundled monofilament organic fibers, it is distributed in three-dimensional random directions in concrete, and a network-like reinforcement system is formed inside it. The cracking of concrete cracks and the expansion of cracks caused by drying and chemical shrinkage during the cement hydration process are inhibited. After the freeze–thaw cycle, the connection between the fibers and the matrix is still relatively tight, indicating that the addition of PPF can effectively improve the frost resistance of manufactured sand concrete.

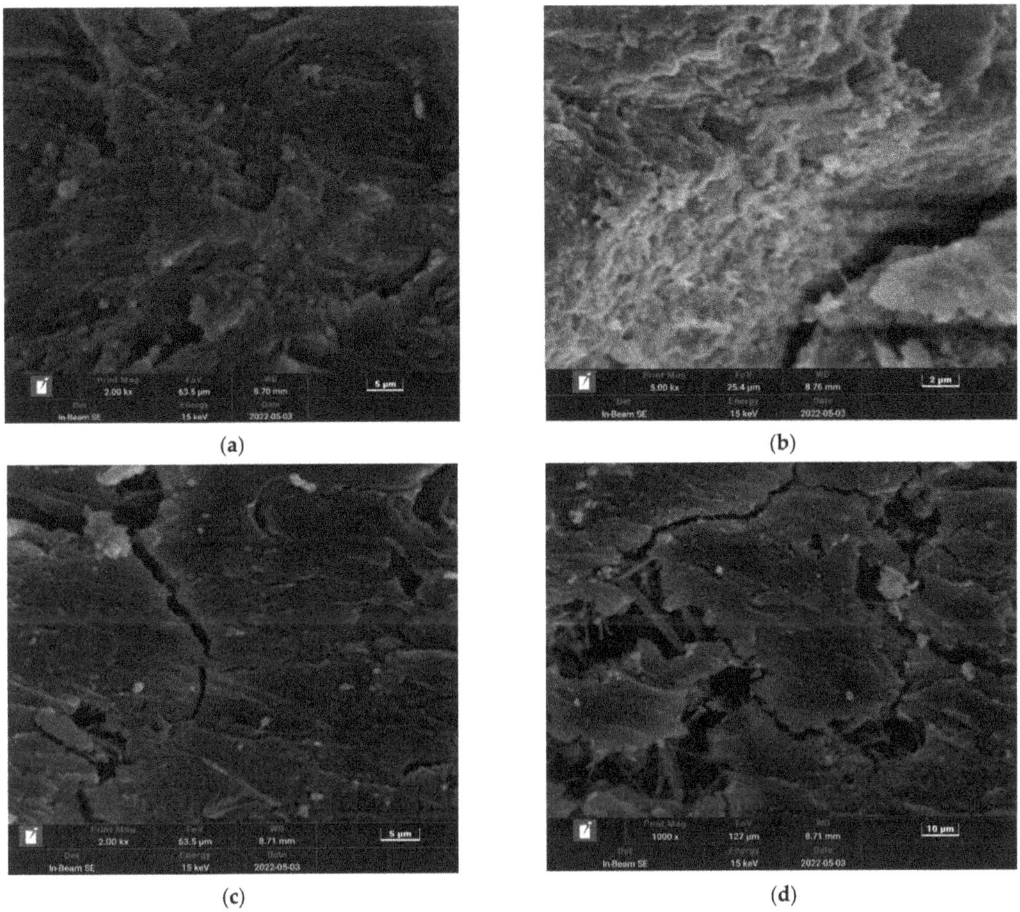

Figure 12. PFMSC microtopography of PFMSC before and after freeze–thawing: (**a**) PFMSC0–0: Before freezing and thawing, (**b**) PFMSC0–0: After freezing and thawing, (**c**) PFMSC1.0–9: Before freezing and thawing, (**d**) PFMSC1.0–9: After freezing and thawing.

6. Conclusions

1. Before the freeze–thaw cycle, the overall compressive strength of PFNSC increased with the increase in PPF volume content and length, and the overall PFMSC showed a decreasing trend with the increase in PPF volume content and length. The compressive strength is the best when the content of PPF is 1.2%, and the compressive strength of PFMSC is the best when the length of PPF is 9 mm and the volume content of PFMSC is 1.0%. The overall trend is an increase, and the splitting tensile strength is the best when the PPF length is 12 mm and the volume content is 1%.
2. With the increase in the number of freeze–thaw cycles, the exposed area of the internal aggregate of PFMSC gradually increased. After 50, 100, and 150 cycles, the average mass damage was 0.32%, 1.85%, and 4.5%, respectively. Mechanical properties and durability performance gradually deteriorated.
3. The mechanical properties and antifreeze properties of PFNSC and PFMSC were comprehensively evaluated with the strength value damage variable as the index, which could better reflect the evolution law of freeze–thaw damage. When the volume content of PPF is 1.0% and the length is 6 mm and 12 mm, the compressive strength

damage of the PFMSC specimen is lower; when the volume content of PPF is 1.0% and the length is 6 mm, the splitting tensile strength damage of the PFMSC specimen is higher. The strength damage of the PFMSC specimens is generally lower than that of the PFNSC specimens, and when the PPF length is the same, the volume content of 1% can better reduce the strength damage. According to the prediction results of PFNSC and PFMSC, the antifreeze performances of PFNSC and PFMSC are similar. This shows that polypropylene fibers have similar effects on PFNSC and PFMSC. It also shows that it is feasible to replace natural sand concrete with 100% artificial sand in practical engineering.

4. The optimal performances of PFNSC and PFMSC are predicted by the RSM strength composite model. PFNSC can maintain a good antifreeze performance within 105 cycles of freezing and thawing. When the volume content of PPF is 1.2% and the length is 11.82 mm, the freezing performance is optimal, the compressive strength value is 33.8 MPa, and the split tensile strength value is 3.1 MPa. PFMSC can maintain a good antifreeze performance within 96 freeze–thaw cycles. When the volume content of PPF is 1.2% and the length is 9.1 mm, the antifreeze performance of the specimen reaches its maximum, its maximum tensile strength value is 45.8 MPa, and the split tensile strength value is 3.2 MPa.

Author Contributions: Conceptualization, Y.T., J.L., X.C., W.X. and B.Z.; methodology and data curation, Y.T. and J.L.; validation and investigation, Y.T., J.L. and B.Z.; formal analysis, Y.T., J.L.; resources, B.Z.; writing—original draft preparation, Y.T.; writing—review and editing, J.L. and B.Z.; supervision, X.C.; and funding acquisition, X.C. All authors have read and agreed to the published version of the manuscript.

Funding: The central government guides local science and technology development projects (No. 2020CFA046); Demonstration Construction Project of Ecological Science and Technology Manor Base (No. 2018ZYYD037).

Data Availability Statement: The data used to support the findings of this study are available from the corresponding author upon request.

Acknowledgments: The authors acknowledge the Hubei University of Technology for funding this research.

Conflicts of Interest: The authors declare no conflict of interest.

References

1. Detwiler, R.J.; Dalgleish, B.J.; Williamson, R.B. Assessing the durability of concrete in freezing and thawing. *ACI Mater. J.* **1989**, *86*, 29–35. [CrossRef]
2. Alireza, F. Compressive behavior of concrete under environmental effects. In *Compressive Strength of Concrete*; IntechOpen: London, UK, 2020. [CrossRef]
3. Alireza, F. Temperature and humidity effects on behavior of grouts. *Adv. Concr. Constr.* **2017**, *5*, 659–669. [CrossRef]
4. Navid, C.; Alireza, F.; Nima, P. Nano silica and metakaolin effects on the behavior of concrete containing rubber crumbs. *Civil Eng.* **2020**, *1*, 17. [CrossRef]
5. Niu, D.T.; Yang, D.Z.; Luo, D. Durability assessment method of existing concrete structures. *Build. Struct.* **2021**, *51*, 115–121.
6. Soroushia, P.; Nagi, M.; Okwuegbu, A. Freeze-thaw durability of Lightweight carbon fiber reinforced cement composites. *ACI Mater. J.* **1992**, *89*, 491–494.
7. Mandal, S.; Chakarborty, S.; Gupta, A. Some studies on durability of recycled aggregate concrete. *Indian Concr. J.* **2002**, *76*, 385–388.
8. Metha, P.K.; Monteiro, P.J.M. *Concrete: Microstructure, Properties and Materials*, 3rd ed.; McGraw-Hill: New York, NY, USA, 2006; pp. 121–209.
9. Liu, D.; Wang, L.; Wang, Z.D.; Fang, J. Exploration on the durability test method of ultra–lightweight concrete. *Concrete* **2022**, *7*, 151–154. [CrossRef]
10. Wei, J.; Wu, X.H.; Zhao, X.L. A model for concrete durability degradation in freeze-thawing cycles. *Acta Mech. Solida Sin.* **2003**, *16*, 353–358.
11. Niu, D.T.; Jian, L.; Bai, M. Experimental analysis is on the frost-resistance of steel fiber reinforced concrete. *J. Civ. Archit. Environ. Eng.* **2012**, *34*, 80–84.

12. Yao, W. Property of fiber reinforced concrete at low temperature and its damage mechanism under freezing-thawing cycles. *J. Glaciol. Geocryol.* **2005**, *27*, 545–549.
13. Mansouri, I.; Shahheidari, F.S.; Hashemi, S.M.; Farzampour, A. Investigation of steel fiber effects on concrete abrasion resistance. *Adv. Concr. Constr.* **2020**, *9*, 367–374.
14. Fouad, T.; Shad, M.; Issam, K.; Husam, H. Material Properties of Synthetic Fiber–Reinforced Concrete under Freeze-Thaw Conditions. *J. Mater. Civ. Eng.* **2018**, *9*, 30.
15. Andreikiv, O.E.; Dolinska, I.Y.; Raiter, O.K. Computational Model for the Evaluation of the Service Life of Fiber-Reinforced Concrete Structures Under Long-Term Static Loading. *Mater. Sci.* **2020**, *56*, 291–300. [CrossRef]
16. Bhogone, M.V.; Subramaniam, K.V. Early-Age Tensile Constitutive Relationships for Steel and Polypropylene Fiber Reinforced Concrete. *Eng. Fract. Mech.* **2021**, *244*, 107556. [CrossRef]
17. Bindiganavilev, B. Polymer and steel fiber-rein-forced cementitious composites under impact loading—Part 1: Bondslip response. *ACI Mater. J.* **2001**, *98*, 10–16.
18. Seung, H.P.; Dong, J.K.; GumSung, R.; Kyung, T.K. Tensile behavior of Ultra High Performance Hybrid Fiber Reinforced Concrete. *Front. Mater.* **2012**, *34*, 172–184. [CrossRef]
19. de Sá, F.R.G.; de Silva, F.A.; Cardoso, D.C.T. Tensile and flexural performance of concrete members reinforced with polypropylene fibers and GFRP bars. *Compos. Struct.* **2020**, *253*, 112784. [CrossRef]
20. Luo, H.L.; Yang, D.Y.; Zhou, X.Y.; Shan, C.C.; Liu, X.; Zhao, F.Z. The mechanical properties of reinforced concrete with different length ratio polypropylene fiber. *Compos. Mater. J.* **2019**, *36*, 1935–1948.
21. Fu, Q.; Niu, D.; Li, D. Impact characterization and modelling of basalt-polypropylene fibre-reinforced concrete containing mineral admixtures. *Cem. Concr. Compos.* **2018**, *93*, 246–259. [CrossRef]
22. Shen, D.J.; Liu, X.Z.; Zeng, X. Effect of polypropylene plastic fibers length on cracking resistance of high performance concrete at early age. *Constr. Build. Mater.* **2020**, *244*, 117874. [CrossRef]
23. Zoran, J.G.; Gordana, A.T.C.; Nenad, S.R.; Iva, M.D. Abrasion resistance of concrete micro-reinforced with polypro-pylene fibers. *Constr. Build. Mater.* **2012**, *27*, 305–312. [CrossRef]
24. Rouhollah, R.; Mohammad, Z.; Khaled, S. An investigation into influence of physical and chemical surface modification of macro-polypropylene fibers on properties of cementitious composites. *Constr. Build. Mater.* **2020**, *244*, 118340.
25. Latifi, M.R.; Biricik, Ö.; Mardani, A.A. Effect of the addition of polypropylene fiber on concrete properties. *J. Adhes. Sci. Technol.* **2022**, *36*, 345–369. [CrossRef]
26. Cheng, H.Q.; Gao, D.Y. Experimental study on freeze-thaw damage of polypropylene fiber concrete. *J. Southeast Univ.* **2010**, *40*, 197–200.
27. Chen, L.Z.; Zhang, G.T.; Huang, W.M. The injury research of fiber rein forced concrete under freeze-thaw cycles. *Sci. Technol. Eng.* **2015**, *15*, 145–150.
28. Salehi, P.; Dabbagh, H.; Ashengroph, M. Effects of microbial strains on the mechanical and durability properties of lightweight concrete reinforced with polypropylene fiber. *Constr. Build. Mater.* **2022**, *322*, 126519. [CrossRef]
29. Liu, B.; Zhang, X.T.; Ying, R.J.; Gao, Y. Review of polypropylene fiber concrete research. *Cem. Concr. Prod.* **2021**, *1*, 5–6.
30. Mundra, S.; Sindhi, P.R.; Chandwani, V. Crushed rock sand an economical and ecological alternative to natural sand to optimize concrete mi. *Perspect. Sci.* **2016**, *8*, 345–347. [CrossRef]
31. Shen, W.; Liu, Y.; Cao, L.; Huo, X.; Yang, Y.Z.; Zhou, C.; He, P.; Lu, Z. Mixing design and microstructure of ultra high strength concrete with manufactured sand. *Constr. Build. Mater.* **2017**, *143*, 312–321. [CrossRef]
32. Qin, Y.; Luo, J.; Chen, Z.; Mei, G.; Yan, L.E. Measuring the albedo of limited-extent targets without the aid of known-albedo masks. *Sol. Energy* **2018**, *171*, 971–976. [CrossRef]
33. Bonavetti, V.; Donza, H.; Menendez, G. Limestone filler cement in low wle concrete: A rational use of energy. *Cem. Concr. Res.* **2003**, *33*, 865–871. [CrossRef]
34. Bonavetti, V.L.; Rahhal, V.F.; Irassar, E.F. Studies on the carboaluminate formation in limestone filler-blended cements. *Cem. Concr. Res.* **2001**, *31*, 853–859. [CrossRef]
35. Anne, M.P.; Geert, D.S. Cement hydration in the presence of high filler contents. *Cem. Concr. Res.* **2005**, *35*, 2290–2299.
36. Ding, X.X.; Li, C.Y.; Xu, Y.Y. Experimental study on long-term compressive strength of concrete with manufactured sand. *Constr. Build. Mater.* **2016**, *108*, 67–73. [CrossRef]
37. Wang, j.; Li, B.X.; Yang, J.B. Effects of stone powder content and steel fiber length on the properties of manufactured sand ultra-high performance concrete. *Acta Silicate* **2020**, *39*, 2120–2126. [CrossRef]
38. Zhang, H.R.; Ji, T.; Zeng, X.P. Mechanical behavior of ultra-high performance concrete (UHPC) using recycled fine aggregate cured under different conditions and the mechanism based on integrated microstructural parameters. *Constr. Build. Mater.* **2018**, *192*, 489–507. [CrossRef]
39. Pyo, S.; Kim, H.; Lee, B.Y. Effects of coarser fine aggregate on tensile properties of ultra high performance concrete. *Cem. Concr. Compos.* **2017**, *84*, 28–35. [CrossRef]
40. Zhang, J.F.; Li, D.; Wang, Y.H. Toward intelligent construction: Prediction of mechanical properties of manufactured-sand concrete using tree-based models. *J. Clean. Prod.* **2020**, *258*, 120665. [CrossRef]
41. Bonavetti, V.L.; Irassar, E.F. The Effect of Stone Dust Content in Sand. *Cem. Concr. Res.* **1994**, *24*, 580–590. [CrossRef]

42. Donza, H.; Carera, O.; Irassar, E.F. High-strength concrete with different fine aggregate. *Cem. Concr. Res.* **2002**, *32*, 1755–1761. [CrossRef]
43. Zeghichi, L.; Benghazi, Z.; Baali, L. The Effect of the Kind of Sands and Additions on the Mechanical Behaviour of SCC. *Phys. Procedia* **2014**, *55*, 485–492. [CrossRef]
44. Keisuke, M.; Koji, Y.; Kazushi, Y. Structure and fracture toughness of thin-wall polypropylene moulded at different injection speeds. *Thin-Walled Struct.* **2018**, *125*, 12–20. [CrossRef]
45. Thanasak, W.; Antoine, E. Naaman Unrestrained early age shrinkage of concrete with polypropylene, PVA, and carbon fibers. *Mater. Struct.* **2007**, *40*, 289–300. [CrossRef]
46. Zhao, Y.R.; Liu, F.F.; Wang, L.; Guo, Z.L. Study on compressive strength model of single-sided frozen concrete based on pore structure. *J. Build. Mater.* **2020**, *23*, 432. [CrossRef]
47. Ramalingam, K.; Murugesan, R. Study on fiber reinforced concrete using manufactured sand as fine aggregate and domestic waste plastics as fibres. *J. Struct. Eng.* **2014**, *40*, 521–527. [CrossRef]
48. Pu, C.Z.; Cao, P.; Zhang, C.Y.; Zhang, Q.C. Variable parameter nonlinear creep damage model considering aging damage degradation. *Eng. Mech.* **2017**, *34*, 17–27.
49. Dong, W.; Shen, X.D.; Xue, H.J. Research on the freeze-thaw cyclic test and damage model of Aeolian sand lightweight aggregate concrete. *Constr. Build. Mater.* **2016**, *123*, 792–799. [CrossRef]
50. ACI Committee 544. Measurement of Properties of Fiber Reinforced Concrete. *ACI Mater. J.* **1998**, *85*, 83–93. [CrossRef]
51. Raymond, H.M. Response Surface Methodology-Current Status and Future Directions. *J. Qual. Technol.* **1999**, *31*, 30–44. [CrossRef]
52. Li, Y.L.; Guo, H.Y.; Zhou, H.; Li, Y.; Chen, J.H. Damage characteristics and constitutive model of concrete under uniaxial compression after Freeze-Thaw damage. *Constr. Build. Mater.* **2022**, *345*, 128171. [CrossRef]
53. Kim, R.; Min, J.Y.; Ahn, E.; Choi, H. Assessment of degradation index in freeze-thaw damaged concrete using multi-channel contactless ultrasound. *Constr. Build. Mater.* **2022**, *349*, 128815. [CrossRef]
54. Ma, W.L.; Xu, Z.G.; Qin, Y.; Cao, C.; Wang, Y.X.; Zhou, H. Experimental investigation of nonlinear flow characteristics in cracked polypropylene fibre-reinforced concrete. *Mater. Struct.* **2021**, *54*, 215. [CrossRef]
55. Hill, W.J.; Hunter, W.G. A Review of Response Surface Methodology: A Literature Review. *Technometrics* **1966**, *8*, 571–590. [CrossRef]
56. Mead, R.; Pike, D.J. A Review of Response Surface Methodology from A Biometrics Viewpoint. *Biometrics* **1975**, *31*, 803–851. [CrossRef]
57. Fang, S.E.; Perera, R. Damage identification by response surface based model updating using D-optimal design. *Mech. Syst. Signal Process.* **2011**, *25*, 717–733. [CrossRef]
58. Zong, Z.H.; Sun, J.L.; Xu, L.Q. Design and study of structural health monitoring system of Xiabaishi bridge. *J. China Railw. Soc.* **2009**, *31*, 65–71.
59. Liu, Y.W.; Moses, F. A sequential response surface method and its application in the reliability analysis of aircraft structural systems. *Struct. Saf.* **1994**, *16*, 39–46. [CrossRef]
60. Strzelecki, P. Determination of fatigue life for low probability of failure for different stress levels using 3-parameter Weibull distribution. *Int. J. Fatigue* **2021**, *145*, 106080. [CrossRef]
61. Siamardi, K.; Shabani, S. Evaluation the effect of micro-synthetic fiber on mechanical and freeze-thaw behavior of non-air-entrained roller compacted concrete pavement using response surface methodology. *Constr. Build. Mater.* **2021**, *295*, 123628. [CrossRef]
62. Ai, Y.; Zhu, S.P.; Liao, D.; Correia, J.A.F.O.; Souto, C.; De Jesus, A.M.P.; Keshtegar, B. Probabilistic modeling of fatigue life distribution and size effect of components with random defects. *Int. J. Fatigue* **2019**, *126*, 165–173.

polymers

MDPI

Article

Reinforced Structure Effect on Thermo-Oxidative Stability of Polymer-Matrix Composites: 2-D Plain Woven Composites and 2.5-D Angle-Interlock Woven Composites

Xingzhong Gao [1,2], Tiancong Han [1,2], Bolin Tang [3], Jie Yi [3] and Miao Cao [3,*]

1 School of Textile Science and Engineering, Xi'an Polytechnic University, Xi'an 710048, China
2 Key Laboratory of Functional Textile Material and Product, Xi'an Polytechnic University, Ministry of Education, Xi'an 710048, China
3 Key Laboratory of Yarn Materials Forming and Composite Processing Technology of Zhejiang Province, College of Material and Textile Engineering, Jiaxing University, Jiaxing 314001, China
* Correspondence: 00007889@zjxu.edu.cn

Abstract: The thermo-oxidative stability of carbon fiber polymer matrix composites with different integral reinforced structures was investigated experimentally and numerically. Specimens of 2-D plain woven composites and 2.5-D angle-interlock woven composites were isothermally aged at 180 °C in hot air for various durations up to 32 days. The thermal oxidative ageing led to the degradation of the matrix and the fiber/matrix interface. The degradation mechanisms of the matrix were examined by ATR-FTIR and thermal analysis. The interface cracks caused by thermal oxidative ageing were sensitive to the reinforced structure. The thermo-oxidative stability of the two composites was numerically compared in terms of matrix shrinking and crack evolution and then experimentally validated by interlaminar shear tests.

Keywords: thermo-oxidative stability; woven composites; structure effect; finite element

Citation: Gao, X.; Han, T.; Tang, B.; Yi, J.; Cao, M. Reinforced Structure Effect on Thermo-Oxidative Stability of Polymer-Matrix Composites: 2-D Plain Woven Composites and 2.5-D Angle Interlock Woven Composites *Polymers* **2022**, *14*, 3454. https://doi.org/10.3390/polym14173454

Academic Editors: Rushdan Ahmad Ilyas, Salit Mohd Sapuan, Emin Bayraktar, Shukur Abu Hassan, Nabil Hayeemasae and Khubab Shaker

Received: 21 July 2022
Accepted: 18 August 2022
Published: 24 August 2022

1. Introduction

Advanced textile composites are gaining market share in various industries, including examples in the aerospace [1], maritime [2], automotive [3], civil infrastructure [4,5], and wearable electronics [6] industries, due to their exceptional electrical, mechanical, and thermal properties. Woven composites are becoming a research hotspot as they are one of the most advanced textile composites [7–9] and have shown great potential for aircraft applications, such as wings and engine blades, due to the high strength/weight ratio and impact resistance [10–12]. Aerospace applications require a long service life of materials while the thermal oxygen ambient will always be met during the service, which causes the reduction of composite properties and threatens aircraft safety [13].

The polymer matrix is susceptible to temperatures. The thermal properties of the polymer matrix have been widely reported by previous work [14,15] using dynamic mechanical analysis, thermogravimetric analysis, differential scanning calorimeter analysis, and so on. The obtained glass transition temperature or decomposition temperature could be regarded as an important indicator for thermal stability. The elevated temperature can also accelerate the oxidation rate, and the combined effect of thermolysis and oxidation promotes chain scission, accompanied by the departure of low-molecule volatiles, and finally leads to chemical shrinkage [16–18]. The matrix shrinkage has been quantitatively measured at a microscopic scale by many scholars [19–22]. While in the fiber reinforced polymer composites, most of the reinforcement, such as carbon fiber, is reasonably inert to thermal oxidation in service (<250 °C) [23]. Thus, the matrix shrinkage causes a mismatch of deformation between the fiber and matrix and then induces a tensile stress within the fiber–matrix interfacial phase [24]. The stress accumulates with ageing time and eventually leads to interface cracks, which create additional pathways for oxygen diffusion and compromise the structural integrity of composite materials [25,26].

The distribution of interface cracks shows the relationship of the architectural structure of fibers. For composite laminates, the ply orientation angles [27], fiber spacing [28,29] as well as stacking sequences [30] have a significant influence on the initiation and propagation of the cracks. The evolution of ageing cracks is much harder to predict in fabric reinforced composites due to the complex interlacing structures. Anisotropic distribution of ageing cracks has been reported in braided and woven composites due to different yarn spatial configurations [31–34]. The thermal oxidative stability of the composites could be affected by the reinforced structure accordingly. Wu [35] and Fan [36] compared the ageing properties of braided composites with unidirectional/laminated composites, respectively, and both results indicated that the braided composites could improve durability thanks to better structural integrity. Revealing the effect of reinforced structure on the thermo-oxidative stability of composites has good benefits for the durability design of the composite structure.

Most of the previous research has mainly focused on the ageing behaviors of composites with some specific reinforced structures. Only a few papers investigated the reinforced structure effect by comparing the degradation of overall mechanical properties. How the reinforced structure affects the distribution of matrix shrinkage and interface cracks has not yet been revealed. Herein, we present a comprehensive investigation into the thermal-oxidative degradation mechanism of 2-D plain woven composites (2-D PWC) and 2.5-D angle-interlock woven composites (2.5-D AWC). The thermal degradation mechanisms of the matrix were evaluated by chemical and thermal analysis. The reinforced structure effect on the thermal-oxidative stability of the two woven composites was explored by both numerical and experimental approaches. The structure effect was first compared by the distribution of shrinkage displacement as well as the evolution of interface cracks, and then validated by interlaminar shear tests.

2. Experimental

2.1. Material

In this paper, a diglycidyl ether of bisphenol-A (DGEBA) epoxy resin (JC-02A/JC-02B, Changshu Jaffa Chemical Inc., Suzhou, China) was selected as the resin matrix. The carbon fiber tows were provided by Toray Industries, Inc., Japan. The specifications of the epoxy matrix and carbon fiber were listed in Tables 1 and 2, respectively.

Table 1. Specifications of the epoxy matrix.

Name	Chemical Component	Viscosity at Room Temperature (MPa·s)	Epoxide Number (eq/100 g)	Blending Ratio
JC-02A	Bisphenol A epoxy resin	1000–3000	0.5–0.53	100:80
JC-02B	Modified anhydride	30–50	-	

Table 2. Specifications of the carbon fiber.

Name	Density (g/cm^3)	Diameter (um)	Carbon Content
Carbon fiber	1.8 ± 0.02	7	≥95%

2.2. Material Preparation

Figure 1 shows the structures of 2-D plain woven fabric and 2.5-D angle-interlock woven fabric. The specifications of the preforms are listed in Table 3. Epoxy resin was used to impregnate the fabric with the vacuum assisted resin transfer method (VARTM). The composite was consolidated with the curing process: 90 °C for 2 h, 110 °C for 1 h, and 130 °C for 4 h in sequence, and the vacuum was about 0.1 MPa. As the single plain woven fabric ply is very thin, the 2-D plain woven composite was obtained by stacking 20 plies of fabrics along the thickness direction (in a 0° direction). The fiber volume fraction is 44.6% and 45.2% for 2.5-D angle-interlock woven composite and 2-D plain woven composite, respectively, obtained by muffle furnace combustion.

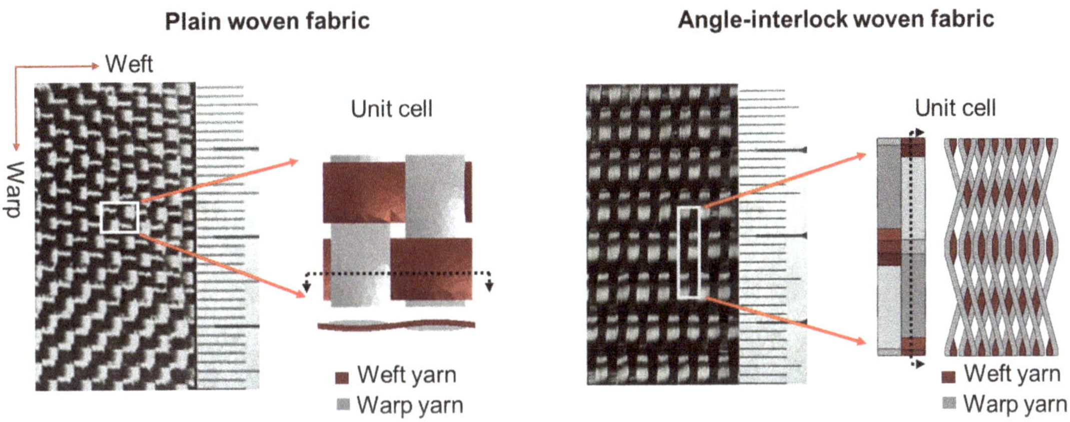

Figure 1. Structure of 2-D plain woven fabric and 2.5-D angle-interlock woven fabric.

Table 3. Manufacturer Specifications.

Preform	Parameter	Warp	Weft
2-D Plain woven	fiber type	T300-3K	T300-3K
	density/(ends·cm^{-1})	5.0	5.0
	layers	20 (in 0° direction)	
	thickness/mm	0.35 (single layer)	
2.5-D Angle-interlock woven	fiber type	T300-6K	T700-12K
	density/(ends·cm^{-1}) *	8.0	3.8
	layers	7	6/8
	thickness/mm	6.20	

* The density/(ends·cm^{-1}) means the number of the yarns per centimeter.

2.3. *Accelerated Ageing and Characterization*

2.3.1. Isothermal Ageing

The specimens were pretreated in an oven for 1 h at 80 °C and then divided into five groups. An unaged group (blank control group) and the other four groups were isothermally aged for 4, 8, 16, and 32 days at 180 °C in an air-circulating oven. Before ageing, all specimens were dried in the oven at 80 °C for 1 h. After ageing for a given time, the specimens were removed and cooled down to room temperature, then put into sealed bags to avoid moisture absorption.

2.3.2. Chemical Analysis

Attenuated total reflectance Fourier transform infrared spectroscopy (ATR-FTIR) analyses (Nicolet 6700 FTIR spectrometer, ThermoFisher, Waltham, MA, USA) were used to determine the functional characteristics of neat resin in the surface layer before and after ageing for 32 days.

2.3.3. Dynamic Mechanical Analysis (DMA)

DMA (Q800, TA Instruments, New Castle, DE, USA) was performed on resin casting in single cantilever mode with a frequency of 1 Hz over the temperature range 30 °C to 180 °C. The temperature ramping rate was 5 °C/min.

2.3.4. Thermogravimetric Analysis (TGA)

Thermogravimetric (TG) studies were carried out with a TA-Instrument (TGA 4000, PerkinElmer, Waltham, MA, USA). Powder samples of about 3 mg were heated under a nitrogen atmosphere. The samples were heated from 30 to 600 °C with a ramp rate of 20 °C/min.

2.3.5. Mechanical Test

The interlaminar shear tests of the specimens were conducted by the Instron universal testing machine (Instron 5967, Instron, Canton, MA, USA) at a test speed of 1 mm/min following ASTM D2344. The width-to-thickness ratio of the specimen is 2, while the span-to-thickness ratio is 4.

3. Numerical Analysis

3.1. Geometry Model

The mesoscale geometry models for both 2-D PWC (Figure 2a) and 2.5-D AWC (Figure 2b) were established based on measured structural parameters. The unit-cell of 2-D PWC consists of three fabric plies with random offset on account of the realistic staggered feature. The two models have a very close yarn volume fraction, which is 62.71% and 61.84% for PWC and AWC, respectively.

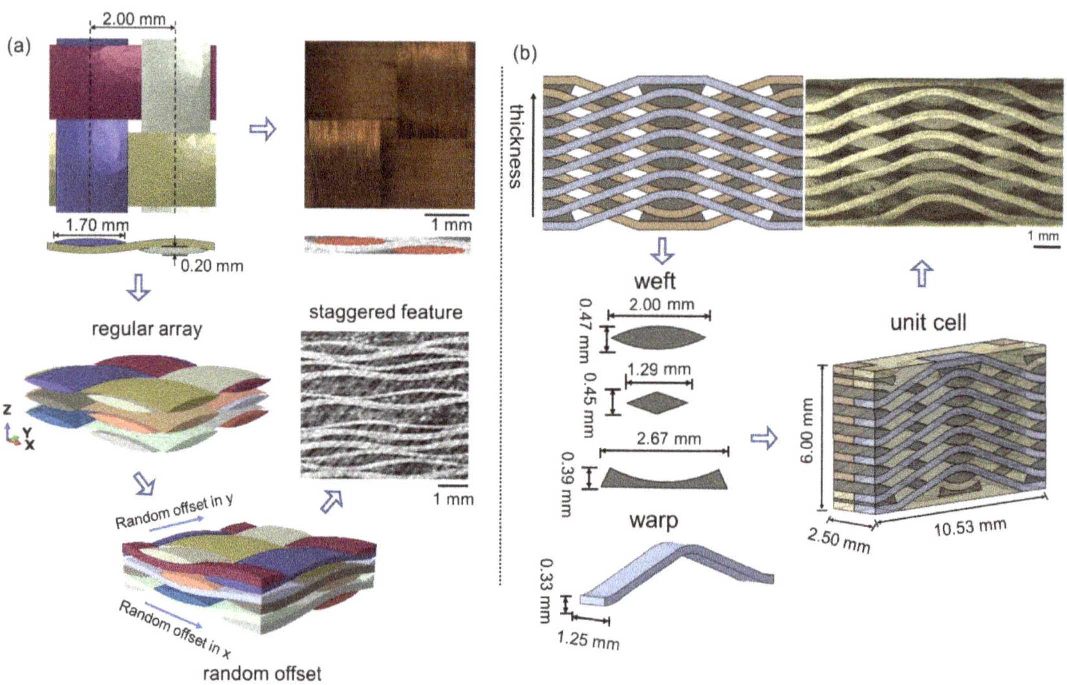

Figure 2. Mesoscale geometry models: (**a**) 2-D PWC; (**b**) 2.5-D AWC.

Linear tetrahedral elements (C3D4) were chosen to generate the mesh of the yarns and matrix for both PWC (Figure 3a) and AWC (Figure 3b) models. Zero-thickness cohesive elements (COH3D6) were generated to represent the interface between the yarns and the matrix. The average mesh size of the model was set to be 0.2 mm.

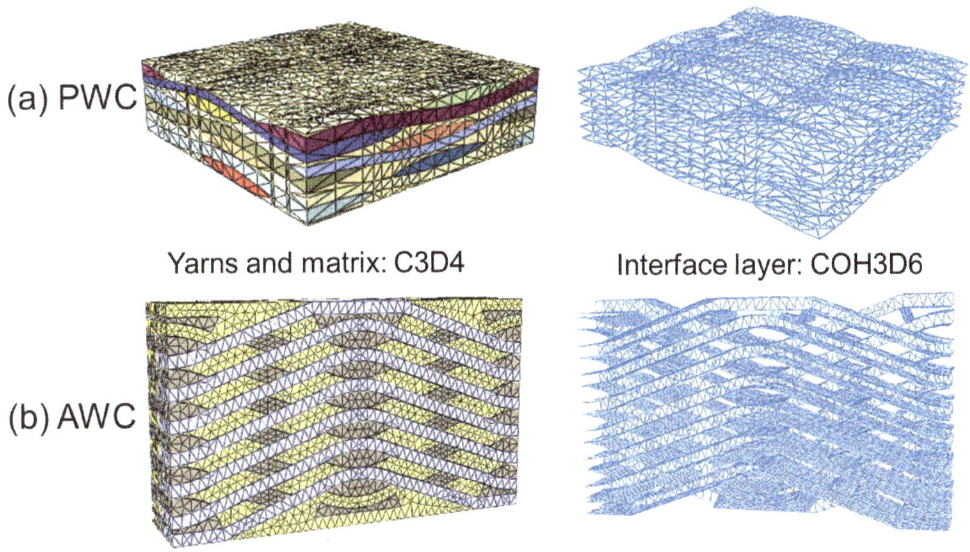

Figure 3. Finite element models: (**a**) 2-D PWC; (**b**) 2.5-D AWC.

3.2. *Constitutive Model*

3.2.1. Epoxy Resin

The epoxy resin is treated as an elastic-plastic solid obeying the J2-isotropic hardening plasticity theory with an associated flow rule and a von Mises yield criterion. The properties of the epoxy matrix were experimentally characterized by a previous study [37] and PWC, respectively.

3.2.2. Yarns

Woven yarns impregnated with the epoxy resin were regarded as transversely isotropic unidirectional composite lamina. The compliance matrix of the yarns was obtained by the bridging model [38].

$$[S] = \left(V_{fy}\left[S^{f}\right] + V_{my}[S^{m}][A]\right)\left(V_{fy}[I] + V_{my}[A]\right)^{-1} \tag{1}$$

where $\left[S^{f}\right]$ and $[S^{m}]$ refer to the compliance matrices of the fiber and resin. V_{fy} and V_{my} are the volume fractions of the fiber and resin matrix in yarns, respectively. $[A]$ is a bridging matrix and $[I]$ is a unit matrix. A fiber packing fraction V_{fy} of 72% was determined by dividing the realistic fiber volume fraction of the specimen by the yarn volume fraction. Table 4 lists the basic properties of constituents and calculated engineering constants of the yarns.

Table 4. Basic properties of constituents and yarns.

	Carbon Fiber	Epoxy Resin	Fiber Tows
E_{11} (GPa)	230	2.4	142.8
$E_{22} = E_{33}$ (GPa)	14		6.4
$G_{12} = G_{13}$ (GPa)	9	0.89	3.0
G_{23} (GPa)	5		2.3
$\nu_{12} = \nu_{13}$	0.25	0.35	0.11
ν_{23}	0.3		0.35

3.2.3. Interface

The yarn-matrix interfacial properties were described by a bilinear traction-separation constitutive model (Figure 4), relating the traction (t) and separation displacement (δ) between two adjacent faces:

$$t = k_p \tag{2}$$

where t is the nominal traction stress vector, consisting of three components: t_n, t_s, and t_t, which represent one normal and two in-plane shear tractions respectively. k_p is the penalty stiffness and δ is the vector of separations.

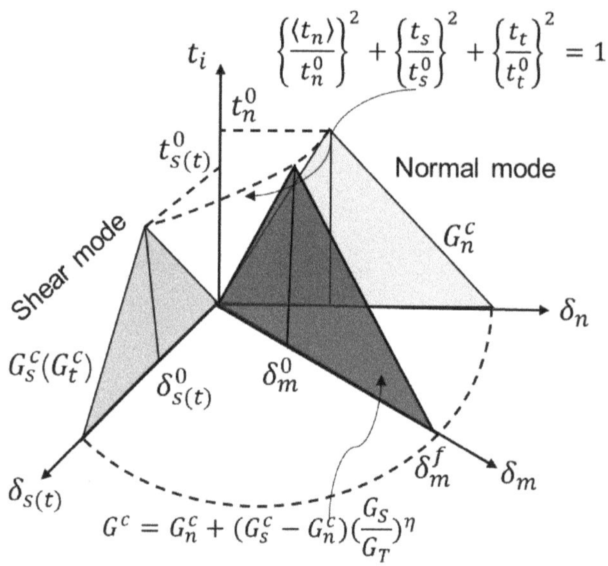

Figure 4. Bilinear traction–separation constitutive model.

Damage initiation is predicted by quadratic nominal stress criterion:

$$\left\{\frac{t_n}{t_n^0}\right\}^2 + \left\{\frac{t_s}{t_s^0}\right\}^2 + \left\{\frac{t_t}{t_t^0}\right\}^2 = 1 \tag{3}$$

The Benzeggagh–Kenane (BK) fracture criterion [39] is adopted to control the failure evolution process. The interfacial stiffness degrades accordingly based on the damage variable D varied between 0 (undamaged interface case) and 1 (complete decohesion case):

$$D = \frac{\delta_m^f\left(\delta_m^{max} - \delta_m^0\right)}{\delta_m^{max}\left(\delta_m^f - \delta_m^0\right)} \tag{4}$$

where δ_m^0, δ_m^f refers to effective separation at the initiation of damage and complete failure, respectively, and δ_m^{max} refers to the maximum value of the effective separation attained during the loading history. If unloading occurs before the complete decohesion, the penalty stiffness will become k_p', where $k_p' = (1-D)k_p$.

Table 5 listed the interfacial parameters for the model.

Table 5. Interfacial parameters for cohesive layers [32].

K_n (N/mm^3)	$K_t = K_s$ (N/mm^3)	t_n^0 (MPa)	$t_s^0 = t_t^0$ (MPa)	G_n^C (N/mm)	$G_s^C = G_t^C$ (N/mm)	β
4×10^6	1×10^6	120	150	0.25	1.0	1.0

3.2.3.1. Modeling of Matrix Shrinkage

Matrix shrinkage leads to interface cracks, which are regarded as one of the primary thermal ageing damage modes in composites. The distribution of the shrinkage displacement field and the residual stress are closely related to the architecture of the composites. In this paper, the architecture effect was compared with the initiation and propagation of interface cracks in 2-D PWC and 2.5-D AWC. As the oxidation reaction process in terms of molecular dynamics is hard to reproduce using the finite element software ABAQUS, we employed the "shrinkage equivalent temperature difference method" [21,32] to simplify the shrinkage process of the matrix. The shrinkage displacement was reproduced by the temperature difference, which could cause the same deformation. The equivalent temperature difference can be defined by:

$$\Delta T_{sh}(t) = \frac{\varepsilon_{sh}(t)}{\alpha_m} \tag{5}$$

where $\varepsilon_{sh}(t)$ refers to matrix shrinkage strain at different ageing days; α_m is the thermal expansion coefficient of the epoxy matrix; and $\Delta T_{sh}(t)$ is the equivalent temperature difference.

The final resin shrinkage displacement was obtained from previous work [32] using the same material system. The temperature field load was applied to the elements within the oxidized layer using ABAQUS/Standard analysis. The boundary conditions are shown in Figure 5.

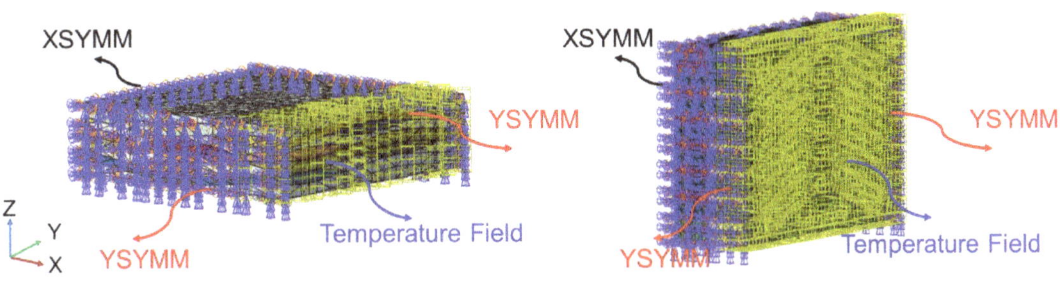

Figure 5. Boundary conditions for 2-D PWC and 2.5-D AWC.

4. Results and Discussion

4.1. Thermal Oxidative Degradation Mechanisms of Epoxy Resin

4.1.1. ATR-FTIR Analyses

The FTIR spectra of neat resin obtained before and after ageing for 32 days at 180 °C are shown in Figure 6. Compared with the unaged sample, several distinct changes take place in the spectrum of the unaged resin: The characteristic band of C-H near 2800~3000 cm^{-1} decreases in intensity. These phenomena demonstrate that C-H bonds were oxidized. Moreover, the characteristic absorption bands of the benzene ring near 1509 cm^{-1} (Stretching vibration of -C = C- skeleton in the benzene ring) and 828 cm^{-1} (in-plane deformation vibration of phenyl-H) decreased after 32 days of ageing, which shows that the benzene ring structure was partly destroyed. The decrease of the absorption band near 1181 cm^{-1}, which was derived from the bridge between the benzene rings [40], indicates that the backbone chains of epoxy resin can be cut by thermal aging. The bands at 1235 cm^{-1} attribute to aromatic ether increases after ageing. It can be inferred that aromatic ether is formed during the ageing process as an oxidation product [41].

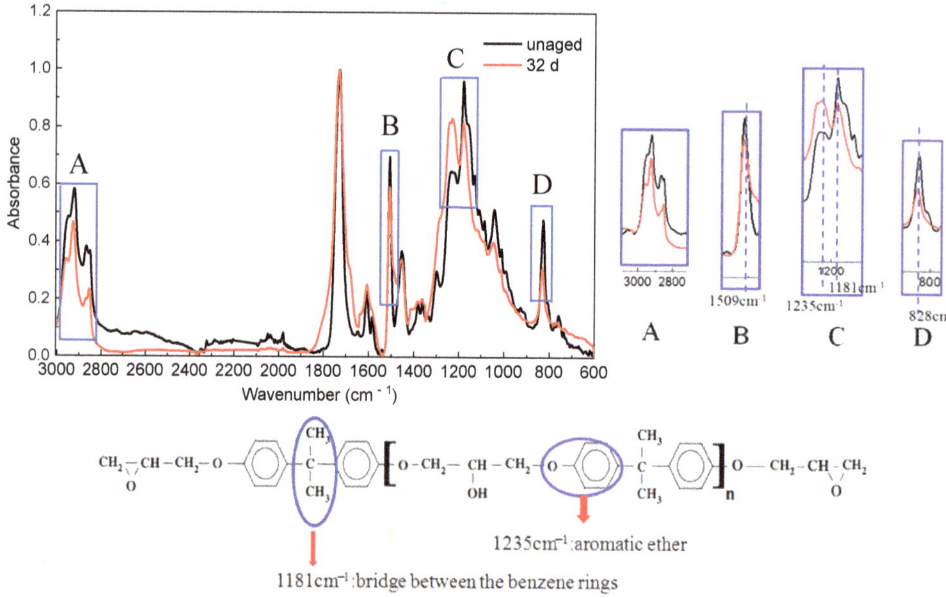

Figure 6. IR spectra of the un-aged resin and the surface of an epoxy cube after ageing at 180 °C for 32 days.

4.1.2. Dynamic Thermomechanical Behaviors

Figure 7a shows the storage modulus over temperature after ageing at 180 °C. The storage modulus decreases gradually with the increase of ageing time in the glassy state due to chain scission. While in the rubbery state, the situation is reversed. The formation of the oxidized layer may account for this result. As mentioned above, chain scission occurs in the bridge between the benzene rings in DGEBA. Some small liberated segments escape from the system and molecular rearrangements may occur among the remainders, forming new compounds with a higher concentration of benzene ring, which have a higher glass transition temperature [16].

Figure 7b shows the loss factor (tan) over temperature. In this paper, the temperature at the peak value of tan δ is treated as the T_g (glass transition temperature) of the sample. At the initial stage of ageing (0–8 d), the T_g increases from 135 °C to 140 °C due to structural changes, such as further crosslinking and the loss of dangling chains, occurring slowly during the stage prior to the onset of severe degradation [42]. The structural changes also decrease the damping ability of the material, as indicated by the drop in the maximum tan δ values [43]. After 8 days, the ageing is dominated by thermolysis and the Tg decreases accordingly.

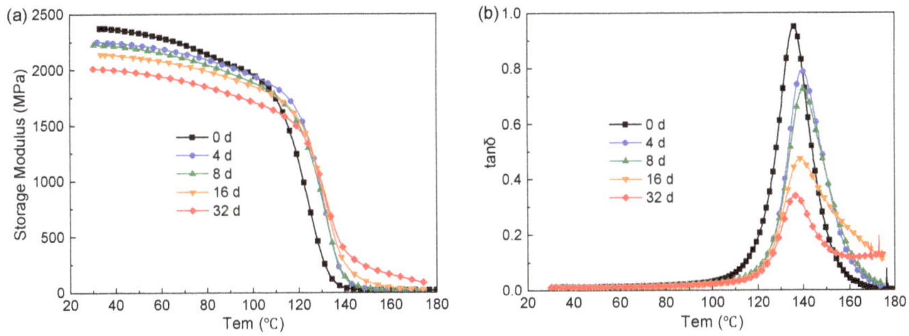

Figure 7. Dynamic thermomechanical behaviors. (**a**) storage modulus; (**b**) loss factor.

4.1.3. Thermogravimetric Analysis

As the "skin-core" structure forms after thermo-oxidative ageing [44], the samples cut from the "skin" (i.e., the oxidized layer) and the inner core were tested, respectively (Figure 8). The temperature with 10% weight loss, $T_{10\%}$, is regarded as the initial decomposition temperature. The initial decomposition temperatures for the unaged sample and the oxidized layer and inner core are 383 °C, 374 °C, and 388 °C, respectively. Compared with the unaged one, the oxidized layer has a lower initial decomposition temperature, while the inner core's is higher. During the ageing process, some chain scission occurs in the sample by thermolysis and then the liberated segments migrate towards the surface of the sample [17]. Some liberated segments have poorer thermal stability, which leads to the advance of the initial decomposition temperature of the oxide layer. Meanwhile, the migration of small molecules towards the surface leads to the backward decomposition of the initial temperature at the inner core. Additionally, the weight remaining in the oxidized layer was about 16% at a temperature of 600 °C, which is much higher than that of the unaged sample. This indicates the formation of a more stable compound in the oxidized layer.

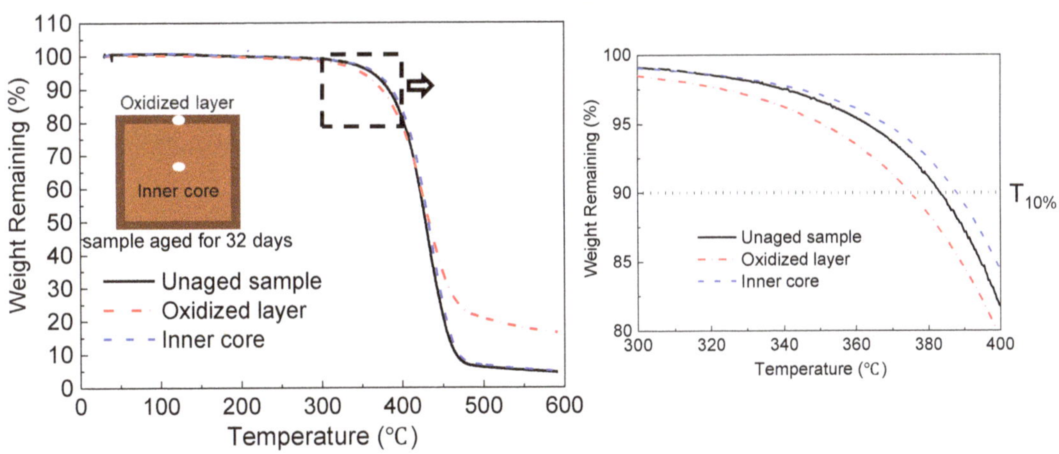

Figure 8. Thermogravimetric analysis.

4.2. *Reinforced Structure Effect*

The reinforcement architecture obviously affects the shrinkage displacement of the composite and further influences the distribution of ageing cracks [36]. In this section, the distributions of the matrix shrinkage and interface cracks in PWC and AWC were compared to reveal the effect of the reinforced structure on the thermal oxidative stability of the composites.

4.2.1. Matrix Shrinkage

Figure 9 compares the local shrinkage of the matrix in PWC and AWC after thermal oxidative ageing for 16 days. The maximum calculated shrinkage depth was around 25 μm, which occurred in the resin rich zone of the top and bottom sides in both PWC and AWC.

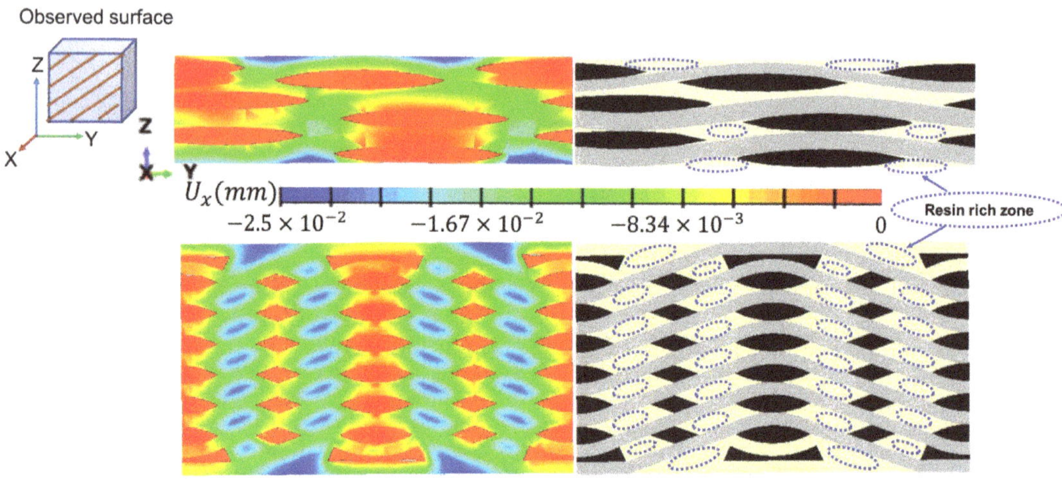

Figure 9. Local shrinkage displacement in PWC and AWC after thermal oxidative ageing for 16 days.

As the shrinkage occurred, the boundary of the matrix still stuck to the yarn and internal stress was induced accordingly. The shrinkage displacement and stress induced by shrinkage were compared in two local areas between AWC and PWC (Figure 10). The AWC sample showed larger shrinkage displacement as well as internal stress compared to that of the PWC, which could be attributed to the different arrangement of the yarns. Taking six cross-sections from the surface to the interior (Figure 11), the yarn volume fraction on each surface varied from 45.4% to 77.8% in the PWC, and the fraction was kept at the constant value of 62.2% in the AWC. Meanwhile, the AWC has a larger yarn to yarn space than that of PWC, and the maximum shrinkage depth increased as the yarn to yarn spacing increased (Figure 12), as the extent of the matrix rich zone increased [29].

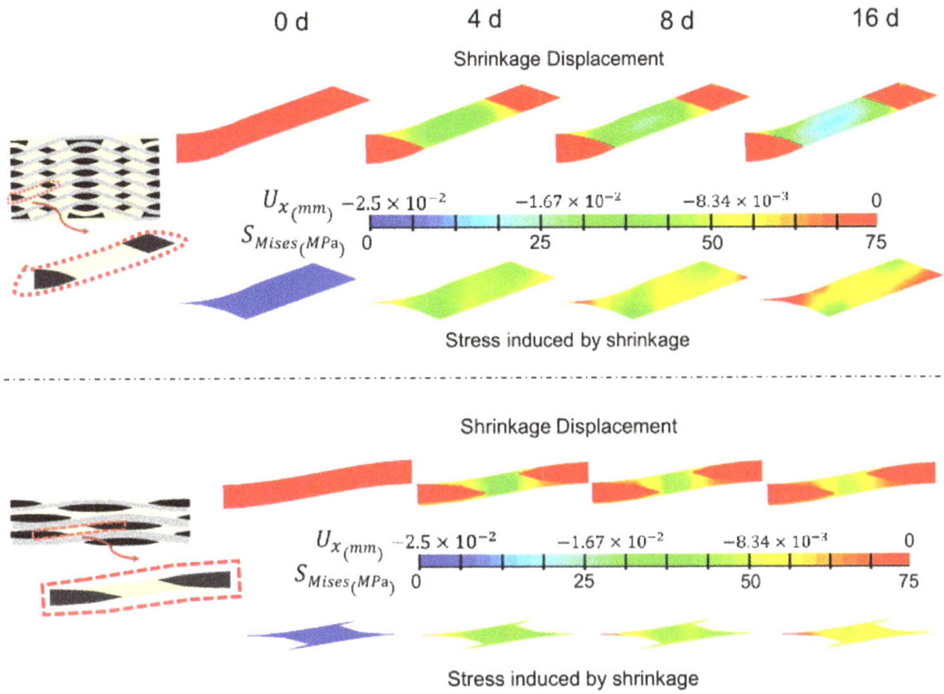

Figure 10. Local shrinkage and internal stress in PWC and AWC with different ageing time.

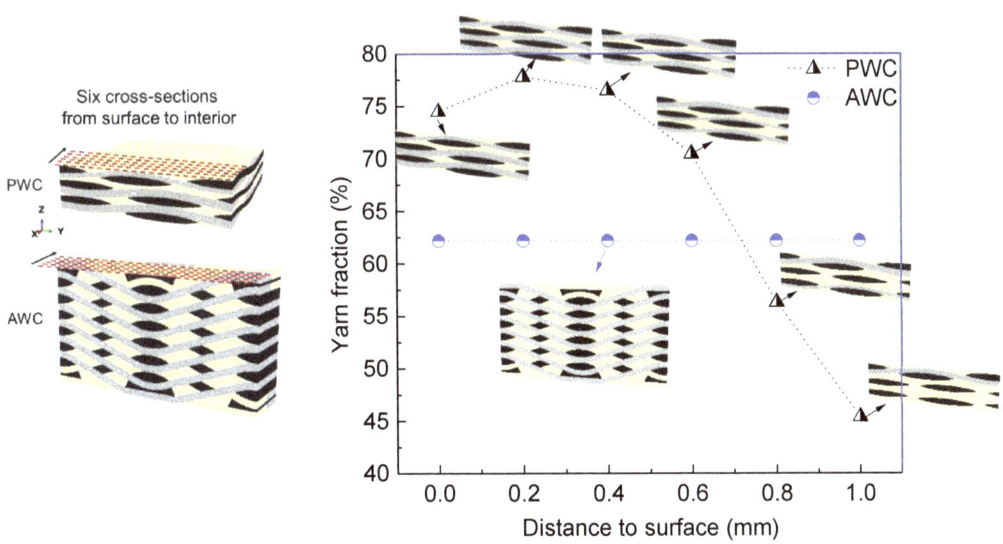

Figure 11. Variations of yarn volume fraction in six cross sections form the surface to interior.

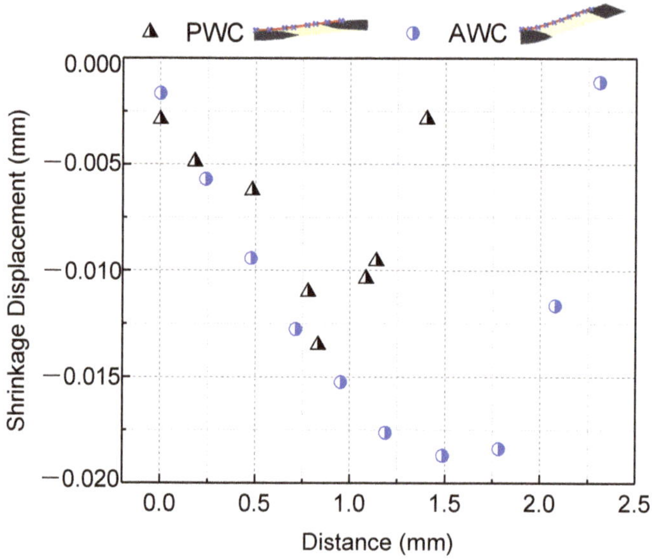

Figure 12. Comparison of shrinkage displacement with yarn distance in PWC and AWC.

4.2.2. Interface Crack Evolution

Figure 13 shows that the evolution of interface cracks as ageing time increased in both AWC and PWC. The cracks were initiated near the ends of weft yarns and propagated along the warp yarns as the shrinkage-induced internal stress accumulated. With the increase in ageing time, the interface cracks propagated and linked together in the PWC, forming continuous interlaminar cracks. The cracks are prone to extending under the external loading, causing delamination of the PWC specimen [45–47].

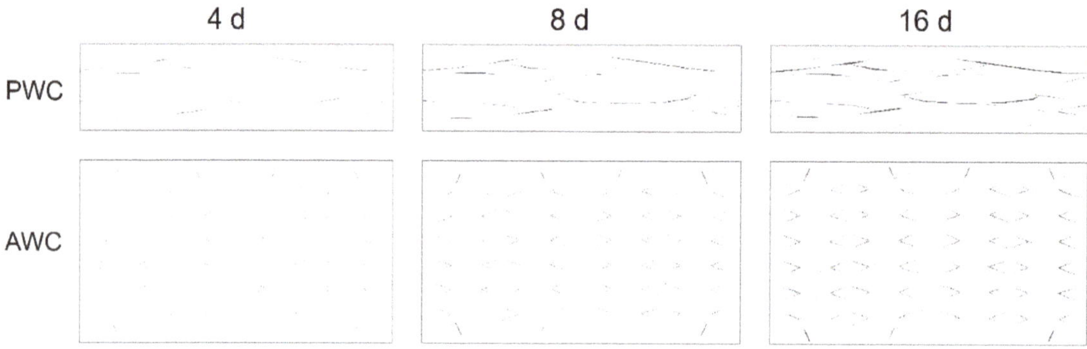

Figure 13. Evolution of interface cracks with the increase of ageing time.

The cracks in the AWC were relatively discrete due to the existence of reinforced yarns toward the thickness direction, which restrained the connection of the cracks (Figure 14). Thus, the interlocked yarns along the thickness have a positive effect in improving the interlaminar properties of the AWC after thermal oxidative ageing.

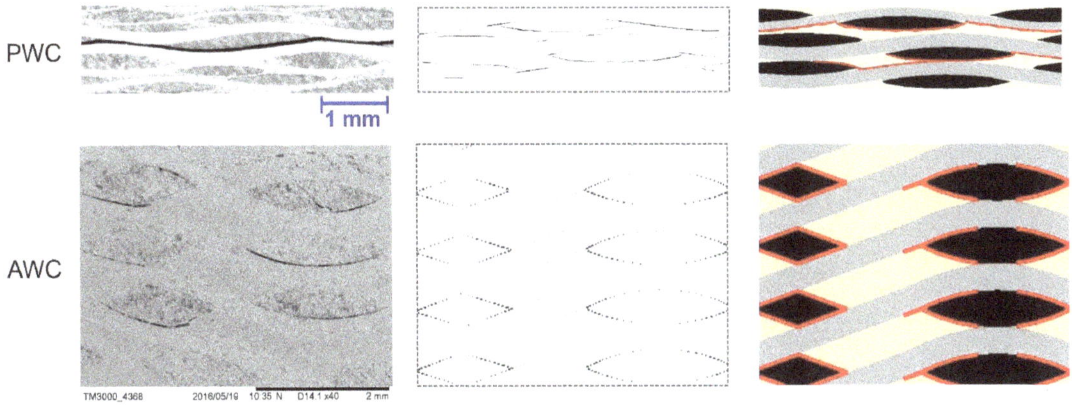

Figure 14. Distribution of interface cracks in AWC and PWC.

4.3. Experimental Validates-Interlaminar Performance

Figure 15a displays the typical load–displacement curves of AWC. The curves showed linear elastic in the initial stage before the first drop occurs after the peak load. After that, the load dropped in a "zigzag" manner due to the stress redistribution caused by local damage. In contrast to AWC, the curves of PWC (Figure 15b) showed a sudden drop in load after the peak load as failure took place catastrophically at the peak stress in a brittle manner. However, the brittleness decreased with the increase of the ageing time as indicated by the curves.

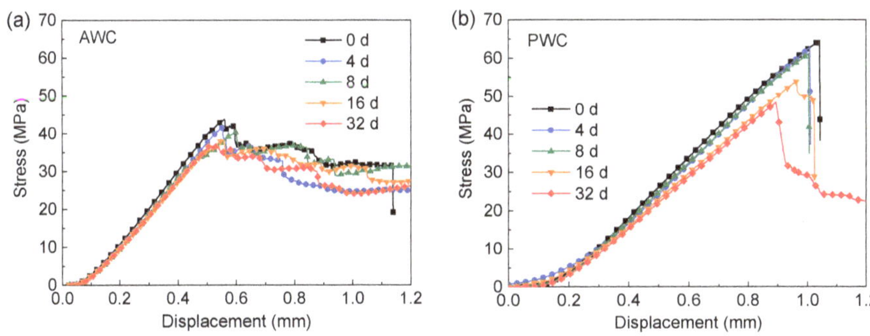

Figure 15. Load–displacement curves under the interlaminar shear tests: (**a**) AWC; (**b**) PWC.

Figure 16 compares the retention rate of the short beam strength along with ageing time. In the first eight days, the retention rate showed a slight declining tendency for both specimens, and the retention rate of the PWC was a bit higher than that of the AWC. While, with the increase of ageing time, the retention rate of PWC decreased significantly and became lower than that of the AWC after ageing for 16 days.

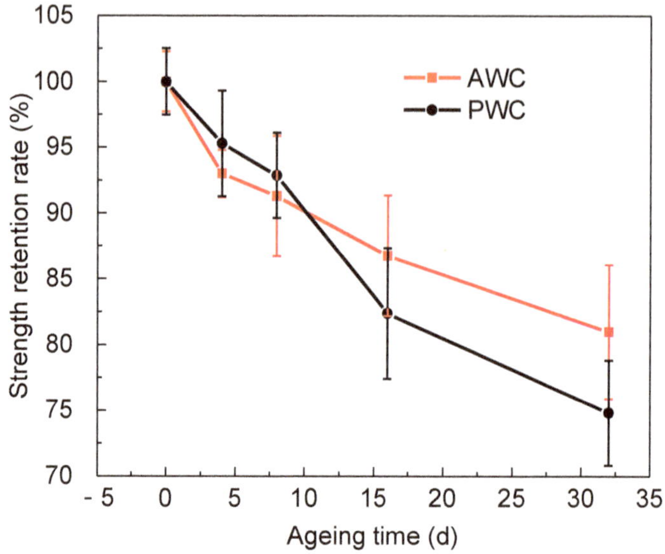

Figure 16. Retention rate of the short beam strength along with ageing time.

The obvious decline could be attributed to the change of failure modes after long-term exposure, as depicted in Figure 17.

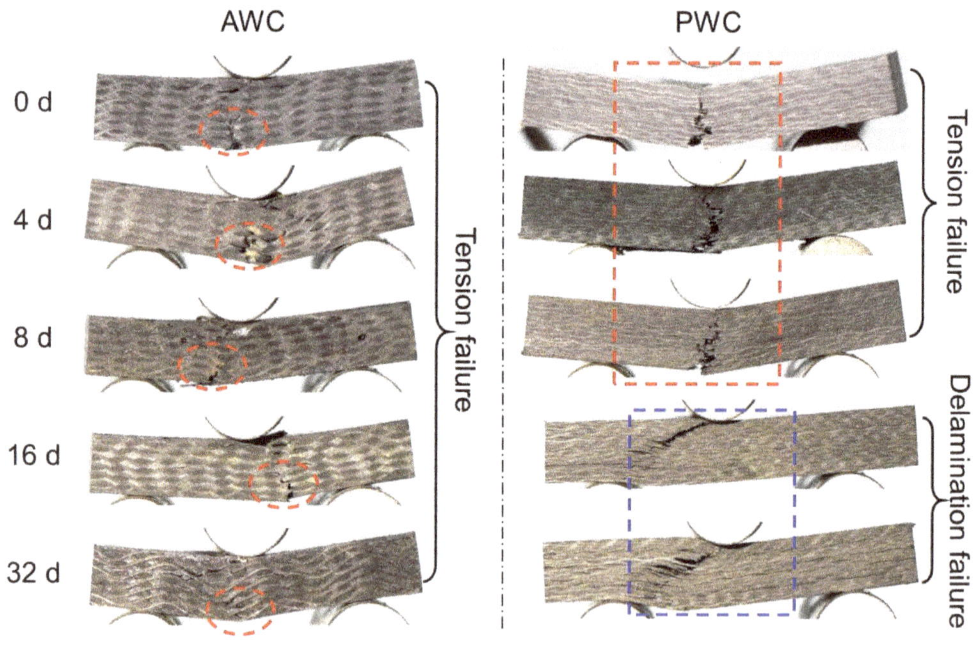

Figure 17. Failure morphologies of AWC and PWC with different ageing time.

The failure modes of the AWC did not change much before and after ageing; the upper side of the specimen was under compression, leading to the initiation of interface cracks. The bottom side was under tension, and the yarn's breakage under the tension led to the final failure of the specimen.

The PWC shows different failure modes after long-term exposure. At the early stage of ageing ($\leq$8 d), the failure is dominated by the fiber breakage at the bottom. At this stage, the PWC has good interface adhesion. The fiber tension failure occurred before the delamination failure. At this time, the short beam strength was determined by the tensile properties of the yarn, which cannot reflect the real interlaminar properties. The strength retention rate showed a slight decline consequently. With the increase of ageing time ($\geq$16 d), the interlaminar cracks propagated, forming continuous interlaminar cracks. The cracks were prone to extending under the shear loading, causing the delamination failure of the specimen.

5. Conclusions

This paper investigates the thermo-oxidative stability of 2-D plain woven composites and 2.5-D angle-interlock woven composites using both experimental and numerical approaches. The specimens have been exposed to 180 °C in an air-circulating oven for 4, 8, 16, and 32 days. The combined effect of matrix degradation and interface cracks leads to the reduction of properties in polymer composites after ageing.

ATR-FTIR spectroscopy and thermal analysis were conducted to illustrate the degradation of the resin matrix during the ageing process. The combined effect of thermolysis and oxidation led to chain scission and molecular rearrangement. The newly formed oxidized layer had higher thermal stability compared to the inner core. The glass transition temperature of epoxy resin decreased due to thermolysis after long-term exposure.

The distribution of interface cracks is closely related to the reinforced structure of the composites, and a mesoscale finite element method has been established to illustrate the structure effect in terms of matrix shrinkage and crack evolution. The AWC sample showed larger shrinkage displacement as well as internal stress compared to that of the PWC due to its regular arrangement and a larger yarn to yarn space. Continuous interlaminar cracks were restrained in the AWC sample due to the existence of reinforced yarns toward the thickness direction, which had a positive effect on the improvement of the interlaminar properties of the material after thermal oxidative ageing.

Finally, the stability of interlaminar properties was experimentally estimated by short beam shear tests. The retention rate of interlaminar shear strength was 81.1% and 74.9% for AWC and PWC, respectively, after ageing for 32 days, and the results prove that the 2.5-D angle-interlock woven composites have better thermo-oxidative stability after long-term thermal exposure due to better structural integrity.

Author Contributions: Conceptualization, B.T.; Data curation, T.H.; Funding acquisition, M.C.; Investigation, T.H.; Methodology, B.T.; Supervision, J.Y.; Validation, J.Y.; Writing—original draft, M.C.; Writing—review & editing, X.G. All authors have read and agreed to the published version of the manuscript.

Funding: This work was supported by the National Natural Science Foundation of China (No. 12102144 and 52101385); the Zhejiang Provincial Natural Science Foundation of China (No. LY22E020013); the Jiaxing Public Welfare Technology Application Research Project (No. 2021AD-10012), the Open Project Program of Key Laboratory of Yarn Materials Forming and Composite Processing Technology of Zhejiang Province (No. MTC-2021-04 and MTC-2020-22); Natural Science basic Research Program of Shaanxi Province [grant numbers 2021JQ-659]; Research Fund for the Doctoral Program of Xi'an Polytechnic University [grant numbers 107020527].

Institutional Review Board Statement: Not applicable.

Informed Consent Statement: Not applicable.

Data Availability Statement: The data presented in this study are available on request from the corresponding author.

Conflicts of Interest: The authors declare that they have no known competing financial interest or personal relationship that could have appeared to influence the work reported in this paper.

References

1. Shukla, S.; Behera, B.K. Auxetic fibrous structures and their composites: A review. *Compos. Struct.* **2022**, *290*, 115530. [CrossRef]
2. Lau, K.; Hung, P.; Zhu, M.; Hui, D. Properties of natural fibre composites for structural engineering applications. *Compos. Part B-Eng.* **2018**, *136*, 222–233. [CrossRef]
3. Hiremath, N.; Young, S.; Ghossein, H.; Penumadu, D.; Vaidya, U.; Theodore, M. Low cost textile-grade carbon-fiber epoxy composites for automotive and wind energy applications. *Compos. Part B-Eng.* **2020**, *198*, 108156. [CrossRef]
4. Abdelrahman, M.S.; Nassar, S.H.; Mashaly, H.; Mahmoud, S.; Maamoun, D.; El-Sakhawy, M.; Khattab, T.A.; Kamel, S. Studies of Polylactic Acid and Metal Oxide Nanoparticles-Based Composites for Multifunctional Textile Prints. *Coatings* **2020**, *10*, 58. [CrossRef]
5. Khattab, T.A.; Haggag, K.M.; Elnagdi, M.H.; Abdelrahman, A.A.; Aly, S.A. Microwave-Assisted Synthesis of Arylazoaminopyrazoles as Disperse Dyes for Textile Printing. *Z. Anorg. Allg. Chem.* **2016**, *642*, 766–772. [CrossRef]
6. Stoppa, M.; Chiolerio, A. Wearable Electronics and Smart Textiles: A Critical Review. *Sensors* **2014**, *14*, 11957–11992. [CrossRef]
7. Nurazzi, N.; Khalina, K.; Sapuan, S. Mechanical properties of sugar palm yarn/woven glass fiber reinforced unsaturated polyester composites: Effect of fiber loadings and alkaline treatment. *Polimery* **2019**, *64*, 665–675. [CrossRef]
8. Sapuan, S.M.; Aulia, H.S.; Ilyas, R.A.; Atiqah, A.; Dele-Afolabi, T.T.; Nurazzi, M.N.; Supian, A.B.M.; Atikah, M.S.N. Mechanical Properties of Longitudinal Basalt/Woven-Glass-Fiber-reinforced Unsaturated Polyester-Resin Hybrid Composites. *Polymers* **2020**, *12*, 2211. [CrossRef]
9. Aisyah, H.A.; Paridah, M.T.; Sapuan, S.M.; Ilyas, R.A.; Khalina, A.; Nurazzi, N.M.; Lee, S.H.; Lee, C.H. A Comprehensive Review on Advanced Sustainable Woven Natural Fibre Polymer Composites. *Polymers* **2021**, *13*, 471. [CrossRef]
10. Yildirim, F.; Kabakci, E.; Sas, H.S.; Eskizeybek, V. Multi-walled carbon nanotube grafted 3D spacer multi-scale composites for electromagnetic interference shielding. *Polym. Compos.* **2022**, *43*, 5690–5703. [CrossRef]
11. Zhang, K.; Zheng, L.; Pei, R.; Chen, L.; Xu, F. Light-weight, high-gain antenna with broad temperature adaptability based on multifunctional 3D woven spacer Kevlar/polyimide composites. *Compos. Commun.* **2022**, *30*, 101061. [CrossRef]
12. Zhao, J.; Ma, P. Comparative Study on the Tribological Properties of PTFE Coated Kevlar Fabric and Hybrid PTFE/Kevlar Fabric. *Fibers Polym.* **2022**, *23*, 1111–1118. [CrossRef]
13. Zhang, K.; Zheng, L.; Pei, R.; Chen, L.; Xu, F. High Temperature Aerospace Applications Epoxy Resin Transfer Molding Resins for High Temperature Aerospace Applications. *Sampe J.* **2020**, *56*, 38–43.
14. Aisyah, H.A.; Paridah, M.T.; Sapuan, S.M.; Khalina, A.; Berkalp, O.B.; Lee, S.H.; Lee, C.H.; Nurazzi, N.M.; Ramli, N.; Wahab, M.S.; et al. Thermal Properties of Woven Kenaf/Carbon Fibre-Reinforced Epoxy Hybrid Composite Panels. *Int. J. Polym. Sci.* **2019**, *2019*, 5258621. [CrossRef]
15. Haris, N.I.N.; Ilyas, R.A.; Hassan, M.Z.; Sapuan, S.M.; Afdzaluddin, A.; Jamaludin, K.R.; Zaki, S.A.; Ramlie, F. Dynamic Mechanical Properties and Thermal Properties of Longitudinal Basalt/Woven Glass Fiber Reinforced Unsaturated Polyester Hybrid Composites. *Polymers* **2021**, *13*, 3343. [CrossRef]
16. Buch, X.; Shanahan, M.E.R. Thermal and thermo-oxidative ageing of an epoxy adhesive. *Polym. Degrad. Stab.* **2000**, *68*, 403–411. [CrossRef]
17. Buch, X.; Shanahan, M.E.R. Migration of cross-linking agents to the surface during ageing of a structural epoxy adhesive. *Int. J. Adhes. Adhes.* **2003**, *23*, 261–267. [CrossRef]
18. Deng, J.; Song, Y.; Xu, Z.; Nie, Y.; Lan, Z. Thermal Aging Effects on the Mechanical Behavior of Glass-Fiber-Reinforced Polyphenylene Sulfide Composites. *Polymers* **2022**, *14*, 1275. [CrossRef]
19. Gigliotti, M.; Minervino, M.; Lafarie-Frenot, M.C. Assessment of thermo-oxidative induced chemical strain by inverse analysis of shrinkage profiles in unidirectional composites. *Compos. Struct.* **2016**, *157*, 320–336. [CrossRef]
20. Gigliotti, M.; Minervino, M.; Lafarie-Frenot, M.C. Thermo-oxidative induced shrinkage in Organic Matrix Composites for High Temperature Applications: Effect of fibre arrangement and oxygen pressure. *Compos. Struct.* **2016**, *146*, 176–186. [CrossRef]
21. Zhang, M.; Sun, B.; Gu, B. Experimental and numerical analyses of matrix shrinkage and compressive behavior of 3-D braided composite under thermo-oxidative ageing conditions. *Compos. Struct.* **2018**, *204*, 320–332. [CrossRef]
22. Kucher, M.K.; Yakovleva, O.S.; Chyzhyk, O.O. Thermal Expansion and Shrinkage of Unidirectional Composites at Elevated Temperatures. *Strength Mater.* **2020**, *52*, 790–797. [CrossRef]
23. Lévêque, D.; Schieffer, A.; Mavel, A.; Maire, J. Analysis of how thermal aging affects the long-term mechanical behavior and strength of polymer-matrix composites. *Compos. Sci. Technol.* **2005**, *65*, 395–401. [CrossRef]
24. Pochiraju, K.; Tandon, G.P. Interaction between cracking and oxidation in organic matrix composites. *J. Compos. Mater.* **2005**, *39*, 1371–1389. [CrossRef]
25. Liang, J.; Pochiraju, K.V. Oxidation-induced damage evolution in a unidirectional polymer matrix composite. *J. Compos. Mater.* **2015**, *49*, 1393–1406. [CrossRef]

26. Sebaey, T.A. Effect of Exposure Temperature on the Crashworthiness of Carbon/Epoxy Composite Rectangular Tubes Under Quasi-Static Compression. *Polymers* **2020**, *12*, 2028. [CrossRef]
27. Schoeppner, G.A.; Tandon, G.P.; Ripberger, E.R. Anisotropic oxidation and weight loss in PMR-15 composites. *Compos. Part A-Appl. Sci. Manuf.* **2007**, *38*, 890–904. [CrossRef]
28. Yang, L.; Yan, Y.; Ma, J.; Liu, B. Effects of inter-fiber spacing and thermal residual stress on transverse failure of fiber-reinforced polymer-matrix composites. *Comput. Mater. Sci.* **2013**, *68*, 255–262. [CrossRef]
29. Gigliotti, M.; Olivier, L.; Vu, D.Q.; Grandidier, J.; Lafarie-Frenot, M.C. Local shrinkage and stress induced by thermo-oxidation in composite materials at high temperatures. *J. Mech. Phys. Solids* **2011**, *59*, 696–712. [CrossRef]
30. Tandon, G.P.; Ragland, W.R. Influence of laminate lay-up on oxidation and damage growth: Isothermal aging. *Compos. Part A-Appl. Sci. Manuf.* **2011**, *42*, 1127–1137. [CrossRef]
31. el Mourid, A.; Ganesan, R.; Brochu, M.; Crochon, T.; Léévesque, M. Anisotropic oxidation due to aging in a triaxially braided composite and its influence on tensile failure. *Compos. Part B-Eng.* **2015**, *76*, 1–12. [CrossRef]
32. Ke, Y.; Sun, B.; Gu, B.; Zhang, W. Damage initiation and propagation mechanisms of 3-D angle-interlock woven composites under thermo-oxidative aging. *Compos. Struct.* **2021**, *259*, 113462. [CrossRef]
33. Wu, Y.; Cao, W.; Guo, J.; Xun, L.; Sun, B.; Gu, B. Near-fiber nanomechanical mapping and impact failure mechanism of 3D braided composites subjected to thermo-oxidative environment. *Compos. Sci. Technol.* **2021**, *216*, 109052. [CrossRef]
34. Xun, L.; Wu, Y.; Huang, S.; Sun, B.; Gu, B.; Hu, M. Degradation of torsional behaviors of 3-D braided thin-walled tubes after atmospheric thermal ageing. *Thin-Walled Struct.* **2022**, *170*, 108555. [CrossRef]
35. Wu, Y.; Xun, L.; Huang, S.; Ren, C.; Sun, B.; Gu, B. Crack spatial distributions and dynamic thermomechanical properties of 3D braided composites during thermal oxygen ageing. *Compos. Part A-Appl. Sci. Manuf.* **2021**, *144*, 106355. [CrossRef]
36. Fan, W.; Li, J.; Zheng, Y.; Guo, D. The effect of reinforced structure on thermo-oxidative stability of polymer-matrix composites. *J. Ind. Text.* **2016**, *46*, 237–255. [CrossRef]
37. Cao, M.; Wang, H.; Gu, B.; Sun, B. Impact damage and compression behaviours of three-dimensional angle-interlock woven composites after thermo-oxidation degradation. *J. Compos. Mater.* **2018**, *52*, 2085–2101. [CrossRef]
38. Huang, Z.M. A unified micromechanical model for the mechanical properties of two constituent composite materials. Part I: Elastic behavior. *J. Thermoplast. Compos. Mater.* **2000**, *13*, 252–271. [CrossRef]
39. Benzeggagh, M.L.; Kenane, M. Measurement of mixed-mode delamination fracture toughness of unidirectional glass/epoxy composites with mixed-mode bending apparatus. *Compos. Sci. Technol.* **1996**, *56*, 439–449. [CrossRef]
40. Xiao, G.Z.; Delamar, M.; Shanahan, M.E.R. Irreversible interactions between water and DGEBA/DDA epoxy resin during hydrothermal aging. *J. Appl. Polym. Sci.* **1997**, *65*, 449–458. [CrossRef]
41. Ohno, S.; Lee, M.; Lin, K.Y.; Ohuchi, F.S. Thermal degradation of IM7/BMI5260 composite materials: Characterization by X-ray photoelectron spectroscopy. *Mater. Sci. Eng. a-Struct. Mater. Prop. Microstruct. Process.* **2000**, *293*, 88–94. [CrossRef]
42. Ma, S.; He, Y.; Hui, L.; Xu, L. Effects of hygrothermal and thermal aging on the low-velocity impact properties of carbon fiber composites. *Adv. Compos. Mater.* **2020**, *29*, 55–72. [CrossRef]
43. Fan, W.; Li, J.; Zheng, Y.; Liu, T.; Tian, X.; Sun, R. Influence of Thermo-Oxidative Aging on Vibration Damping Characteristics of Conventional and Graphene-Based Carbon Fiber Fabric Composites. *Polym. Compos.* **2016**, *37*, 2871–2883. [CrossRef]
44. Audouin, L.; Langlois, V.; Verdu, J.; de Bruijn, J.C.M. Role of oxygen diffusion in polymer aging—Kinetic and mechanical aspects. *J. Mater. Sci.* **1994**, *29*, 569–583. [CrossRef]
45. Rivers, G.; Cronin, D. Influence of moisture and thermal cycling on delamination flaws in transparent armor materials: Thermoplastic polyurethane bonded glass-polycarbonate laminates. *Mater. Des.* **2019**, *182*, 108026. [CrossRef]
46. Fan, W.; Guo, D.; Li, J.; Zhou, Y.; Gong, L.; Xue, L.; Li, J.; Yuan, L.; Sun, R.; Meng, J. Effect of thermo-oxidative aging on compressive behavior of carbon fiber polymer matrix composites. *Text. Res. J.* **2018**, *88*, 510–519. [CrossRef]
47. Petkov, V.I.; Joffe, R.; Fernberg, P. Thermal oxidative aging of satin weave and thin-ply polyimide composites. *Polym. Compos.* **2022**, *43*, 2615–2627. [CrossRef]

polymers

MDPI

Article

Prediction of Long-Term Tensile Properties of Glass Fiber Reinforced Composites under Acid-Base and Salt Environments

Jihua Zhu [1], Yangjian Deng [1], Piyu Chen [2], Gang Wang [3,*], Hongguang Min [3,*] and Wujun Fang [3]

1 Guangdong Province Key Laboratory of Durability for Marine Civil Engineering, College of Civil and Transportation Engineering, Shenzhen University, Shenzhen 518060, China; zhujh@szu.edu.cn (J.Z.); yuangkin@126.com (Y.D.)
2 School of Civil and Environmental Engineering, Harbin Institute of Technology, Shenzhen 518055, China; 19b954012@stu.hit.edu.cn
3 Central Research Institute of Building and Construction (Shenzhen) Co., Ltd., MCC Group, Shenzhen 518055, China; fangwujun@cribc.com
* Correspondence: wgxh-dd@163.com (G.W.); minhongguang@163.com (H.M.); Tel.: +86-755-26521199 (G.W.)

Abstract: This study investigates the effects of deionized water, seawater, and solutions with various concentrations (5% and 10% by mass) of HCl and NaOH on the physical and mechanical properties of glass fiber reinforced polymers (GFRPs) through aging tests at 20 °C, 50 °C, and 80 °C. The tensile properties of GFRP were assessed by tensile testing at room temperature, and the strain during the tensile process was observed using digital image correlation. Additionally, the degradation mechanism was analyzed using scanning electron microscopy, and long-term tensile properties were predicted based on the Arrhenius model. The results indicated that the tensile strength of the GFRP decreased by 22%, 71%, and 87% after 56 d of exposure to 5% NaOH solutions at 20 °C, 50 °C, and 80 °C, respectively. The alkaline solutions had a more severe effect on the GFRP than deionized water, seawater, and acidic solutions. The experimental values and Arrhenius model predictions were found to be in good agreement with each other.

Keywords: GFRP; environmental degradation; tensile properties; DIC; long-term prediction

Citation: Zhu, J.; Deng, Y.; Chen, P.; Wang, G.; Min, H.; Fang, W. Prediction of Long-Term Tensile Properties of Glass Fiber Reinforced Composites under Acid-Base and Salt Environments. *Polymers* **2022**, *14*, 3031. https://doi.org/10.3390/polym14153031

Academic Editors: Rushdan Ahmad Ilyas, Salit Mohd Sapuan, Emin Bayraktar, Shukur Abu Hassan, Nabil Hayeemasae and Khubab Shaker

Received: 4 July 2022
Accepted: 25 July 2022
Published: 26 July 2022

1. Introduction

Pultruded glass fiber reinforced polymers (GFRPs), with high specific strength, non-magnetic properties, and ease of forming, have the added advantage of economy over carbon and basalt fiber-reinforced composites, and they are widely used in aerospace, automotive and marine, medical, and civil engineering fields [1,2]. However, when compared with steel, pultruded GFRPs are less ductile and the failure mode of the material is usually brittle. When the GFRP is applied to critical structures, such as offshore drilling platforms, wastewater treatment plants, power plants, storage tanks, and pipelines in aggressive environments, it is often subjected to aggressive media such as seawater, high temperatures and humidity, as well as acidic and alkaline environments [3–6]. Under long-term effects of the aggressive media, the GFRP is prone to aging that leads to the degradation of the GFRP's structural performance and causes significant social and economic loss. To better promote the application of the GFRP, it is necessary to obtain performance degradation data for the GFRP in actual service and predict its long-term performance using accelerated life characterization methods with the aim of revealing the degradation mechanism of the GFRP and its degradation law. Studies have shown that water-based solution immersion is one of the conditions that causes the greatest degradation in GFRPs [7] because water absorption by GFRP causes hydrolysis and plasticization of the resin matrix, weakening the fiber-matrix interface [8,9].

Several studies have been conducted in recent years to investigate the durability of GFRP composites in highly corrosive environments. Shao and Kouadio [10] found that the polyester matrix GFRP did not show signs of water-absorption saturation after 260 d of immersion in water at 23 °C, despite a 17.5% decrease in tensile strength. At 70 °C, its water absorption increased and then decreased, even reaching negative values, with a 56.3% decrease in tensile strength; however, the Young's modulus of the GFRP remained almost unchanged at the two temperatures. Grammatikos et al. [11] reported similar findings and concluded that the composite increases moisture absorption and water diffusion owing to the increase in temperature, and the entire process of water diffusion does not reach saturation. Based on the Levenberg-Marquardt algorithm, Xin and Liu [12] concluded that there are three dominant factors for the change in mass due to moisture absorption: diffusion, polymer relaxation, and composite damage.

Sousa et al. [13,14] further compared the effects of salt solution and tap water on the deterioration properties of polyurethane GFRP and epoxy resin GFRP lap joints. They found that the residual strength of the GFRP joints was higher in salt water than in tap water, which is consistent with the test results [15,16]. Furthermore, a significant post-curing phenomenon was found; the maximum reduction in the ultimate load of the GFRP was 27% during the aging period of 730 d in the 20 °C water immersion tests, but at the end of the aging period, the ultimate load reached, or even slightly exceeded, the initial ultimate load. The post-curing phenomenon has also been identified in numerous studies [11,17–19] and is attributed to the fact that polymer systems containing styrene polyesters and vinyl esters undergo rapid local homo-polymerization during the curing process, but the composite does not achieve complete curing in a short time, which leads to further curing during the aging period.

Feng Peng et al. [18] comparatively studied the effects of a sulfuric acid solution with pH = 5 and a mass fraction of 30% on the epoxy resin GFRP at a temperature of 60 °C and found that a high-concentration sulfuric acid solution had severe corrosion effects on the GFRP. After 90 d of immersion, the bending properties of the GFRP decreased by 34% and 73%, respectively, and the hardness decreased by 6.5% and 55.6%, respectively. Kanerva et al. [20] studied the properties of vinyl ester in a sulfuric acid solution at a temperature of 90 °C, pressure of 1.5 MPa, and concentration of 50 g/l. They found that the bending stiffness of vinyl ester decreased by 22% after one year of acid leaching, but bending strength increased by 109%. Bazli et al. [21] studied and compared the degradation of the compressive properties of vinyl ester GFRP with different cross-sectional shapes under different erosion environments and found that the O-shaped cross-section was more prone to erosion than the I- and C-shaped cross-sections. Additionally, alkali was found to cause the most serious deterioration in GFRPs. After 147 d of exposure to an alkali solution with pH = 13.6 at 20 °C, compressive strength decreased by as much as 46%. Xue et al. [22] further found that the compressive strength of the GFRP decreased faster than the tensile strength in the alkali solution. In addition, Hota et al. [19] found that the interlaminar shear stress decay of a vinyl ester GFRP plate in an alkaline environment was equivalent to that of a sample cut from the plate, indicating that there was no size effect during degradation.

To further study the deterioration law of pultruded GFRP profiles in the acid-base and salt environments, the effects of different chemical media (deionized water, artificial seawater, hydrochloric acid solution, and sodium hydroxide solution of different concentrations) at 20 °C, 50 °C, and 80 °C on the deterioration performance of the pultruded GFRP profiles were experimentally studied. Subsequently, the quality and appearance changes, tensile properties, and micro morphology of the cross section of the GFRP were analyzed. The global tensile-strain field in the fracture process was analyzed using the digital image correlation (DIC) technique. Finally, the long-term tensile properties of the pultruded GFRP profiles were predicted using the Arrhenius model.

2. Materials and Methods

2.1. Materials

A GFRP plate pultruded by an epoxy resin and E-glass fiber produced by Mushi Composite Materials Technology (Suzhou) Co., Ltd. (Suzhou, China) was selected as the raw material. The glass fiber was orthogonally braided and its mass fraction was 60%. To avoid stress concentration of the specimen during the tensile test, owing to the discontinuous notch formed by manual and mechanical cutting of the GFRP plate, the GFRP specimens used in the test were cut from the GFRP plate parallel to the fiber direction using laser-cutting technology. The specimen was cut into an I-shape, referring to *GB/T 1447-2005 Fiber-reinforced plastics composites determination of tensile properties*, with a nominal thickness of approximately 4 mm and length of 180 mm, as shown in Figure 1.

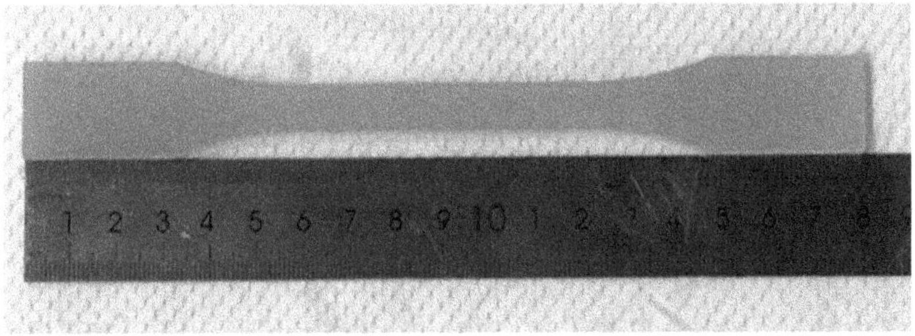

Figure 1. The specimen.

2.2. *Aging Test*

The aging experiment explored and compared the degradation effects of four chemical media, namely deionized water (DW), artificial seawater (SW), sodium hydroxide solution (SH), and hydrochloric acid solution (HA), at different temperatures and ages on the GFRP in order to evaluate the law and mechanism of deterioration. In this test, acid and alkali solutions with mass fractions of 5% and 10% were used to further explore the degradation law of the GFRP under different concentrations of acid and alkali solutions. For the composition standard of artificial seawater, refer to *GB/T 3857-2017 Test method for chemical medium resistance of glass fiber reinforced thermosetting plastics*. The acid solution was prepared using concentrated hydrochloric acid at a concentration of 33% by mass, and the alkaline solution was prepared using analytically pure sodium hydroxide. Table 1 presents the test protocol for the deterioration tests. The GFRP specimens were soaked in six types of chemical media at three temperature gradients (20 °C, 50 °C, and 80 °C) with age settings of 28 d and 56 d. The specimens were numbered according to the following rules: the part before "-" indicates the immersion solution; 05 and 10 indicate the 5% and 10% mass percentage concentrations, respectively; the part after "-" indicates the immersion temperature. For example: 05SH-80, implies that the specimen was immersed in a 5% sodium hydroxide solution at an immersion temperature of 80 °C. The specimens were clamped and immersed in an acid-resistant, alkali-resistant, and high-temperature resistant airtight plastic box, as shown in Figure 2. Eighteen sets of test conditions were set, with 5 parallel specimens per set of conditions for a total of 180 specimens (=5 parallel specimens × 6 types of chemical media × 3 test temperatures × 2 ages). All specimens with an exposure temperature of 20 °C were placed in a cabinet at a temperature setting of 20 ± 0.5 °C, whereas the specimens with 50 °C and 80 °C conditions were placed in a water bath for heating.

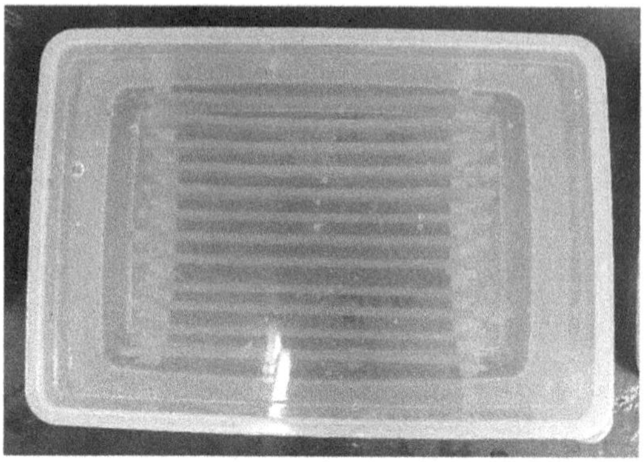

Figure 2. GFRP specimen immersion test.

Table 1. Experimental scheme.

Groups	Conditioning Environment	Immersion Medium	TEMP (°C)	Age (d)
1	DW-20	Deionized water	20	28, 56
2	DW-50		50	
3	DW-80		80	
4	SW-20	Sea water	20	28, 56
5	SW-50		50	
6	SW-80		80	
7	05HA-20	5% HCl solution	20	28, 56
8	05HA-50		50	
9	05HA-80		80	
10	10HA-20	10% HCl solution	20	28, 56
11	10HA-50		50	
12	10HA-80		80	
13	05SH-20	5% NaOH solution	20	28, 56
14	05SH-50		50	
15	05SH-80		80	
16	10SH-20	10% NaOH solution	20	28, 56
17	10SH-50		50	
18	10SH-80		80	

2.3. Mass Changes

The specimens were cleaned with running water, dried before and after the deterioration test, removed, and cooled to room temperature. The mass of the specimen was then measured (the mass of the specimen before and after exposure was noted as m_0 and m_t, respectively, accurate to ±1 mg). The rate of mass change (M) was calculated using Equation (1).

$$M = \frac{m_t - m_0}{m_0} \times 100\% \tag{1}$$

2.4. Mechanical Testing

After drawing auxiliary lines for each tested piece, a mechanical properties test was performed. The test apparatus were selected from the PWS-50 electro-hydraulic servo dynamic tester, and the tensile test at room temperature was performed at a constant speed of 0.2 mm/min with reference to *GB/T 1447-2005 Fiber-reinforced plastics composites determination of tensile* properties. Figures 3 and 4 show the test specimen and test setup with fixtures for the tensile test, respectively. Strain gauges were attached to both sides of the middle parallel section of each specimen parallel to the direction of tension and aligned centrally to obtain more accurate strain data in the initial stages of tension. An extensometer was used to indirectly measure strain over the entire scale.

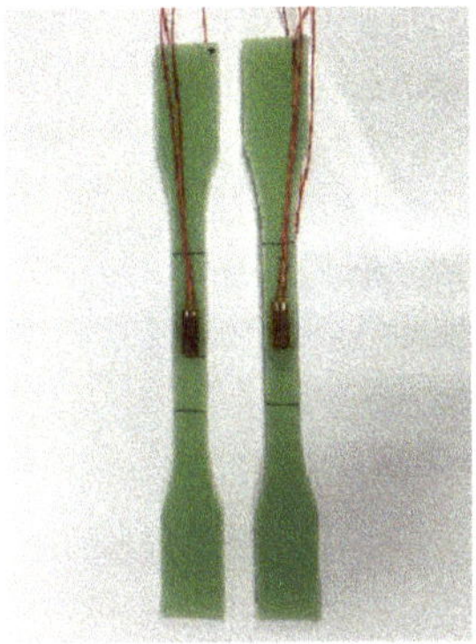

Figure 3. Specimens.

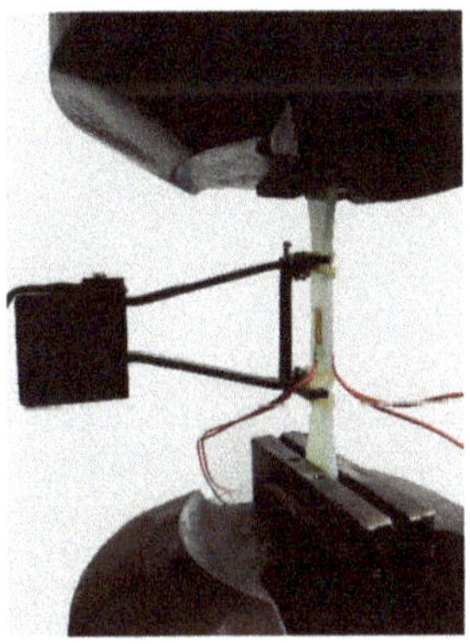

Figure 4. Test setup.

2.5. Digital Image Correlation

DIC is an optical metrology technique, based on digital image processing and numerical computation, which provides direct full-field displacement and strain with sub-pixel accuracy by comparing digital images of the test-object surface acquired before and after deformation [23]. To obtain a more visual and comprehensive understanding of the damage process in the GFRP during tension and deformation of specimens in various regions, one of the parallel specimens in each group was subjected to tensile testing using the DIC equipment. The DIC camera used was a Pointgrey-12.3 M, with a 4096 × 3000 resolution and a lens focal length of 50 mm; the acquisition interval was set to 400 ms. Before the test was performed, the specimens were marked with a scattering of black dots on a white background to allow the computer to better distinguish the movement of the scattering and to accurately analyze the displacement of the specimen.

2.6. Scanning Electron Microscopy

A Phenom Pure scanning electron microscope (SEM) was used to obtain a reasonable explanation of the test phenomena from a microscopic perspective. As the subject of this test was a non-conductive material requiring a carbon or metal coating, the SEM specimen was first sprayed with a 9–12 nm platinum metal coating using sputter coating equipment and then placed in a sample cup to capture microscopic images.

3. Results and Discussion

3.1. Appearance

All specimens were bright green before the test; however, after the test, the appearance of the specimens changed, as shown in Figure 5. For the specimens tested at 20 °C, the appearance remained largely unchanged; for the specimens tested at 50 °C, the color faded from bright to light green; and for the specimens tested at 80 °C, the color degradation was more significant, with the surface resin matrix dissolving and essentially turning white. Additionally, the appearance of the specimens at different ages under the same conditions was slightly different, with the specimens showing a more pronounced color change at 56 d than at 28 d; the appearance of the specimens immersed in different chemical media under the same test conditions was also slightly different, with the specimens immersed in the sodium hydroxide solution showing the most pronounced change in appearance. The microstructural properties and chemical reactions of the samples are discussed further in this paper.

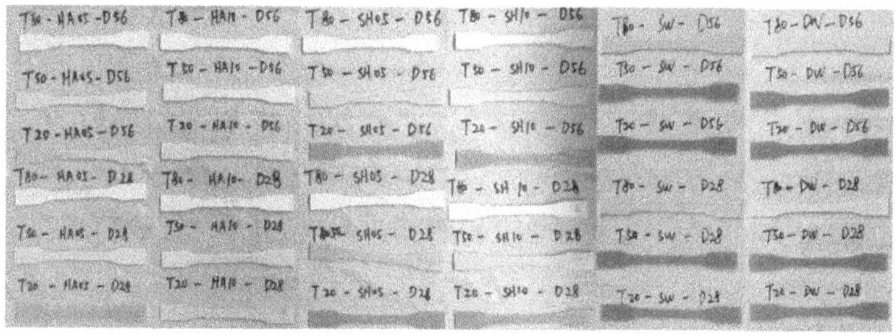

Figure 5. Appearance of specimens after immersion.

3.2. Mass Changes

Figure 6 shows the variation in the mass of the GFRP specimens over time for different hygrothermal environments. In general, mass absorption or loss under different degradation conditions increases with time; however, the rate of change can decrease or even tend to saturate. The change in mass is determined by a combination of the following: (i) water uptake, (ii) leaching of small molecular weight polymers, and (iii) leaching of dissolved hydrolysis products [4,5]. For specimens in the same chemical environment and exhibiting a consistent temperature effect, the degree of mass change increased with increasing immersion temperature. The masses increased by 0.09%, 0.34%, and 2.16% after 56 d of immersion in deionized water at 20 °C, 50 °C, and 80 °C, respectively. This was possibly because of the accelerated chemical reaction rate, owing to the increase in temperature, and a higher void pressure, resulting from the increased gas volume within the GFRP. This increase in pressure favors the extension of microcracks, thus increasing the free volume, within the specimen, that can be filled by the surrounding solution, such that the dendritic matrix absorbs more moisture [24].

In addition, it can be observed that the difference in mass change rates in the specimens, between the cases of deionized water and artificial seawater environments at 20 °C and 50 °C, is not significant; however, at 80 °C, the deionized water specimens absorb much more moisture than the artificial seawater environment specimens (the mass changes for DW-80-56 and SW-80-56 were 2.16 and 0.95%, respectively). This effect has been reported in earlier studies [13,25] and can be attributed to the cross-linking behavior of such polymers, which act as semi-permeable membranes. In the case of corrosive environments, such as acid and alkaline solutions, the specimens mainly showed mass loss (except for 05SH-80 and 10SH-80), contrary to the results [18,21]. This is because the dominant effect of mass change in the above environments is the decomposition of part of the polymer in the dendrimer matrix, its leaching into the solution [26], and dissolution of the resin on the surface of the specimen [6]. In contrast, for specimens immersed in an alkaline environment at 80 °C, an increase in mass occurred (2.77% and 3.49% after 56 d in 05SH-80 and 10SH-80 solutions, respectively), probably owing to the increase in cavities within the laminate at higher temperatures resulting in increased water uptake [27]. The overall mass increase was observed when the weight-gain due to water absorption exceeded the mass loss due to leaching of the small molecular weight segment polymer, with the surface coating of the specimen being dissolved.

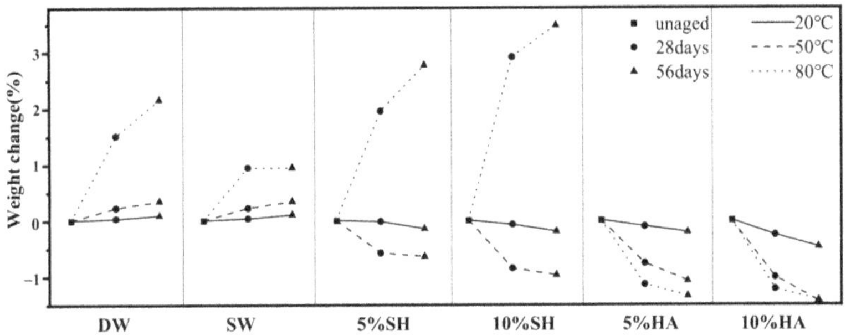

Figure 6. Change in specimen mass under various conditions.

3.3. Tensile Properties

3.3.1. Stress-Strain Curves and Failure Modes

Regardless of the exposure environment, all tested specimens showed greater degradation with increases in test temperature. As shown in Figure 7, the stress-strain relationship of various specimens exposed to 20 °C was similar, indicating that the elastic modulus of the tested GFRP specimen hardly decayed at 20 °C. With the increase in immersion temperature, the stress-strain curves of the tested specimens deviate from the control specimen and the maximum stress and strain decrease, indicating that the deterioration of the specimens is aggravated by the increase in temperature, especially for the specimen immersed in the 80 °C alkaline environment. Almost all the tested specimens exhibited failure near the clamping end (as shown in Figure 8a), which was mainly due to the stress concentration caused by the variable section near the clamping end of the specimen. However, the specimens immersed in acid at 50 °C and 80 °C exhibited delamination failure (Figure 8b). This is mainly because the binder resin is much worse than the fiber under high-temperature acid-leaching conditions, thus interlayer performance is weakened. Moreover, splitting occurs when the load reaches the limit [28].

3.3.2. Tensile Strength and Modulus

The test data under all conditions are shown in Table 2. Figure 9 shows the average strength retention and error bars of the samples after 28 d and 56 d under different exposure conditions. The three exposure temperatures of 20 °C, 50 °C, and 80 °C are indicated by solid, scribed, and dotted lines, respectively, and different symbols are used for different exposure periods. Error bars indicate the magnitude of the difference between parallel samples when calculating the average strength.

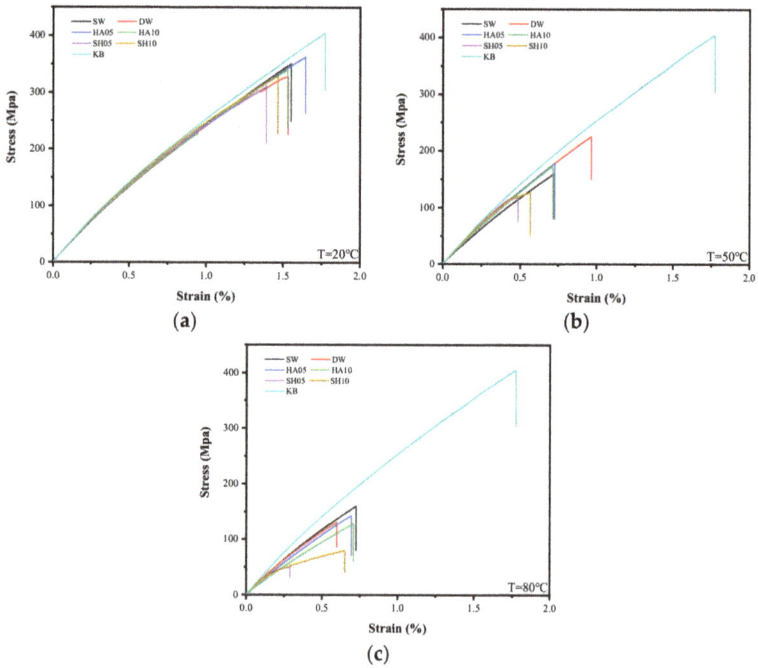

Figure 7. Stress-strain curves of specimens at (**a**) 20 °C, (**b**) 50 °C, and (**c**) 80 °C.

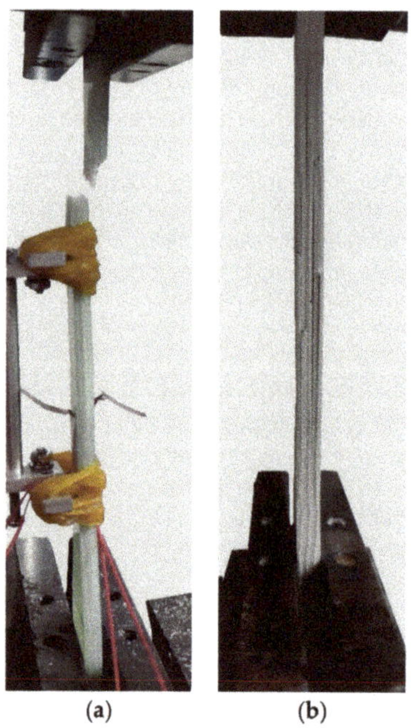

Figure 8. Tensile failure modes: (**a**) fiber fracture, and (**b**) delamination.

Table 2. Summary of mechanical test results.

Exp.	T (°C)	Age (Week)	Tensile Strength Avg (MPa)	Tensile Strength COV (%)	Tensile Strength Retention (%)	Modulus of Elasticity Avg (GPa)	Modulus of Elasticity COV (%)	Modulus of Elasticity Retention (%)
REF	20	-	407.9	5.3	100	30.6	0.0	100
DW	20	4	384.8	0.8	94	29.9	2.8	98
		8	351.6	1.6	86	29.1	5.2	95
	50	4	223.4	6.6	55	28.8	1.2	94
		8	217.5	3.2	53	27.6	2.8	90
	80	4	141.8	1.6	35	25.5	0.2	83
		8	136.4	2.0	33	25.0	1.2	82
SW	20	4	383.4	3.8	94	29.3	1.7	96
		8	367.0	2.2	90	27.8	0.4	91
	50	4	243.2	3.9	60	28.5	0.4	93
		8	226.0	3.3	55	26.8	2.5	88
	80	4	167.4	6.8	41	26.1	1.7	85
		8	166.5	4.6	41	25.2	1.8	82
5% SH	20	4	352.2	2.8	86	29.8	2.4	98
		8	317.8	2.7	13	29.4	1.1	96
	50	4	155.3	4.6	38	29.8	5.1	98
		8	119.4	3.3	29	28.4	0.9	93
	80	4	63.8	8.6	16	20.5	3.1	67
		8	51.4	5.0	13	17.2	3.9	56
10% SH	20	4	359.9	0.8	88	29.2	0.5	96
		8	333.3	4.0	82	29.2	0.6	96
	50	4	157.7	7.4	39	29.0	2.6	95
		8	124.7	1.1	31	28.3	0.5	92
	80	4	98.4	4.0	24	20.1	4.9	66
		8	80.0	9.4	20	17.6	3.6	57
5% HA	20	4	365.9	4.2	90	28.9	0.7	94
		8	349.1	1.3	86	28.4	2.4	93
	50	4	295.2	2.2	72	29.1	4.8	95
		8	274.3	3.1	67	27.1	1.1	89
	80	4	212.7	3.8	52	22.5	0.4	74
		8	194.2	3.8	48	22.4	0.6	73
10% HA	20	4	368.5	1.5	90	29.4	1.9	96
		8	348.4	1.8	85	28.9	1.8	94
	50	4	320.9	4.0	79	28.6	2.3	93
		8	297.4	1.8	73	26.8	0.7	88
	80	4	204.5	3.0	50	22.1	1.8	72
		8	194.7	3.1	48	20.8	0.6	68

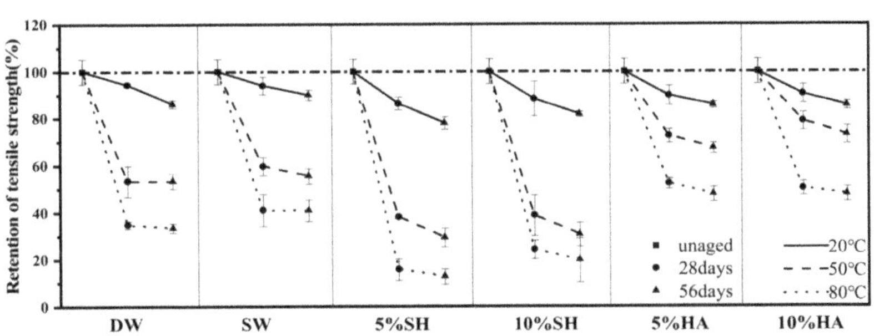

Figure 9. Variation of residual tensile strength as a function of immersion time and temperatures in various conditions.

As implied in Figure 9, regardless of the environment, temperature has a crucial impact on the deterioration of the specimens. Taking the sample immersed in artificial seawater as an example, after 56 d of exposure to 20 °C, the strength decreased slightly and the residual strength was 90.0%; however, when the temperature was raised to 50 °C, the deterioration was aggravated and the residual strength was 59.6%; at 80 °C, the degradation was further aggravated and the residual strength was only 40.8%. A hydrolysis reaction occurs when water comes in contact with the resin matrix. This reduces the compactness of the matrix, increases the number of pores, and reduces adhesion with the fibers, which may be the main reason for the worsening of the tensile properties of the GFRP [29]. The hydrolysis reaction between the resin matrix and water is shown in Equation (2). Hydrolysis of the resin polymer opens the ester bond and generates the corresponding carboxylic acid and free hydroxyl ions. An increase in temperature can accelerate the hydrolysis reaction, generate a higher void pressure conducive to microcrack propagation in the composite, and promote the entry of more water. However, it also affects the microstructure of the matrix, causing plasticization of the GFRP matrix [24,30] and weakening of the tensile properties of the GFRP. In addition, tensile strength does not decrease linearly with time, though most of the strength loss is generally observed after 28 d of exposure (except for DW-20), and the higher the exposure temperature, the more evident this phenomenon becomes. Taking 10SH-80 as an example, the residual tensile strength after 28 d immersion was only 24.1%, whereas the residual tensile strength after 56 d immersion was still 19.6%. It may be owing to the fact that the hydrolysis reaction of the composite is rapid, thus the rate of tensile strength loss is initially high before reducing due to the decrease in degradable reactants with the increase in degradation time. Consistent with the conclusions of many studies [7], the alkaline environment caused the most serious erosion of the GFRP. Specimens in 05SH-80-56 and 10SH-80-56 lost nearly all of their strengths, and the strength retention rates were only 12.6% and 19.6%, respectively. The tensile strength of the GFRP in deionized water and artificial seawater was lower than that of GFRP in acid. This may be because there are fewer free hydroxyl ions in the acidic environment, which inhibits the reaction between the glass fiber and hydroxyl ions to a certain extent, as shown in Equation (3). Sindhu et al. [30] suggested that fiber-matrix interface performance would be improved by immersion in an acid solution, thus improving tensile strength.

$$\text{R-COO-R}' + \text{H-OH} \rightarrow \text{R-COOH} + \text{R}' + \text{OH}^- \tag{2}$$

$$\text{-Si-O-Si-} + \text{OH}^- \rightarrow \text{Si-OH} + \text{Si-O-} \tag{3}$$

After 56 d of immersion in a deionized water environment at 20 °C, 50 °C, and 80 °C, the residual strengths were 86.2%, 53.3%, and 33.4%, respectively, compared to 90.0%, 55.4%, and 40.8%, respectively, for the seawater environment. Deionized water was found to reduce the tensile strength of the GFRP slightly more than artificial seawater. Compared to deionized water, the surface of the GFRP specimen is covered with a thin layer of salt in artificial seawater, which affects the exchange of matter between the interior and exterior of the specimen and, consequently, has less effect on the degradation of its tensile properties [13]. The effect of different concentrations of the sodium hydroxide solution and hydrochloric acid on the deterioration of the GFRP was also investigated. Unlike [31] and [32], it was found that a 5% sodium hydroxide solution deteriorated the GFRP more severely than the 10% solution, with the former having GFRP residual strengths of 77.9%, 29.3%, and 12.6% and the latter having the corresponding strengths of 81.7%, 30.6%, and 19.6% after 56 d of immersion at 20 °C, 50 °C, and 80 °C, respectively. One possible explanation is that, in the alkaline solution with higher concentrations of hydroxide, more hydrolysis products of the matrix and glass fibers were hoarded on the surface of the specimen at the beginning of the exposure, blocking the channels of the solution and inhibiting the matrix and glass fibers from undergoing hydrolysis reactions. In addition, it was observed that the strength degradation of the specimens immersed in 5% hydrochloric acid was slightly higher than the strength degradation of those immersed in 10% hydrochloric acid, apparently because fewer hydroxide ions inhibited the degradation of the fibers, as shown in Equation (2).

Figure 10 shows the mean modulus of elasticity retention and error bars for the specimens after 28 d and 56 d under different conditions. The pattern of stiffness change was generally consistent with that of strength change. The 20 °C and 50 °C environments had less of an effect on the stiffness of the GFRP, with the remaining stiffnesses after 56 d of exposure for DW-50, SW-50, 05SH-50, 10SH-50, 05HA-50, and 10HA-50 being 89.9%, 87.2%, 92.5%, 92.0%, 88.3%, and 87.2%, respectively. However, when the exposure temperature was increased to 80 °C, the GFRP had severe deterioration, and the stiffness residuals after 56 d of immersion in the above environments were 81.5%, 82.0%, 56.0%, 57.2%, 73.1%, and 67.6%, respectively, which is very different from GFRP grids [33]. The Young's modulus of FRP composites is mainly determined by the modulus of elasticity of the fibers and resin matrix as well as the corresponding volume fraction [32], and it can be deduced that the glass fibers had significant deterioration at 80 °C. Overall, when comparing various exposure environments, alkali had the most significant effect on the stiffness of the GFRP, followed by acid, deionized water, and seawater.

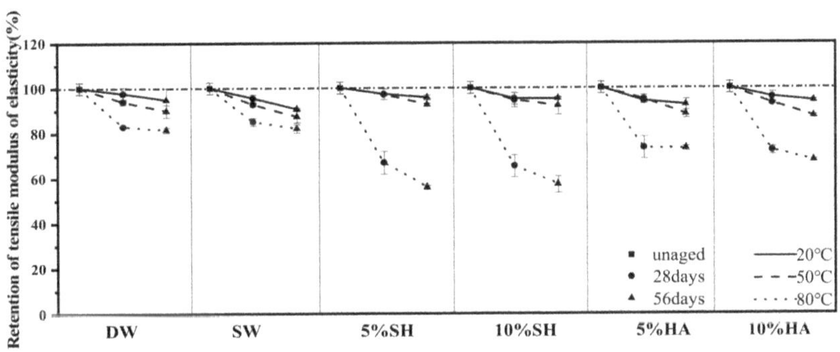

Figure 10. Variation of the residual tensile elastic modulus as a function of immersion time and temperatures in various conditions.

3.4. DIC Analysis

Figure 11 shows the tensile stress field distribution and fracture location of the specimens after 56 d of exposure to artificial seawater, 5% hydrochloric acid, and a 5% hydroxide nano solution at 20 °C, 50 °C, and 80 °C. The first two figures from the left show the stress field distribution in the middle of the tensile test and when reaching the ultimate load, respectively, while the picture on the right side shows the fracture position of the specimen.

In Figure 11, the failure section of the specimens occurs at the maximum strain, which is in accordance with the maximum tensile strain theory. Furthermore, with the exception of specimens 05SH-50-56 and 05SH-80-56, the 'maximum strain' region for all specimens in the exposed condition was close to the clamping end on both sides, with fracture and failure occurring in those locations, which are consistent with Section 3.3.1. In contrast, the 05SH-50-56 and 05SH-80-56 specimens had failure sections in the middle parallel section because the specimens deteriorated to a greater extent in this environment than would be expected from the stress concentration phenomenon. Interestingly, the 'maximum strain' region in the specimen is not fixed at a certain place but may shift with the test process and eventually break at the maximum tensile strain, as shown in Figure 11a,f,i. This phenomenon is most evident in 05SH-50-56, where the specimen was the first to reach maximum strain and develop a crack near the lower fixture during tension; however, this was quickly followed by the development of a new crack above the previous, where it fractured completely.

Additionally, the strain fields of the specimens in different environments (with the exception of 05SH-50-56 and 05SH-80-56) were relatively smooth and uniform. This correlates with the degree of material deterioration, and in relation to Section 3.3, the strain field for specimens with higher strength retention is smoother and more homogeneous. In contrast, the strain field distributions for specimens exposed to 05SH-50-56 and 05SH-80-56, which were the most severely weakened in terms of tensile properties, were patchy and discontinuous. This is due to the fact that pultruded GFRP profiles are generally relatively homogeneous. The GFRP which had not been severely eroded also showed relatively homogeneous deformation during tension. However, the distribution of defects within the GFRP, such as micropores, was not completely homogeneous, resulting in a considerable deterioration of the defective localities compared to other locations in an aggressive environment, which would be the first to induce greater strains and subsequently lead to a patchy distribution of the stress field.

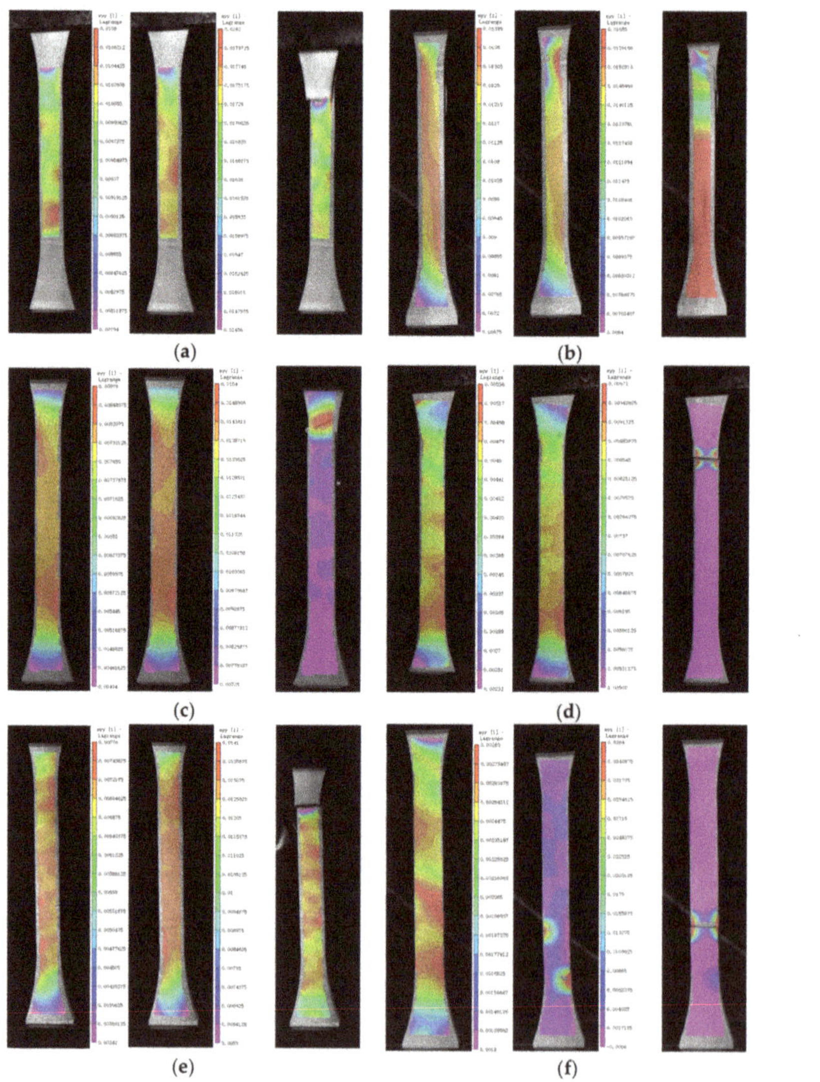

Figure 11. *Cont.*

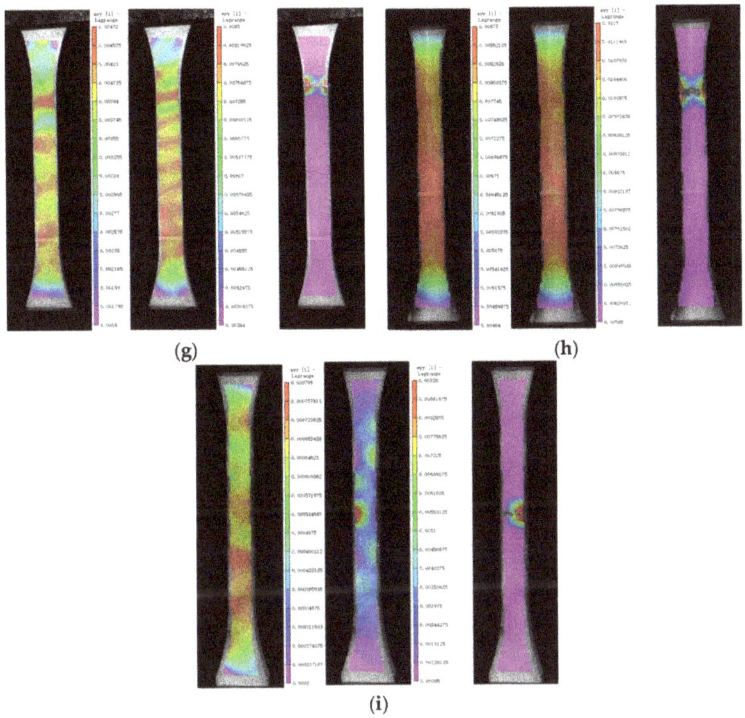

Figure 11. Tensile strain distribution exposed to SW, 5% NaOH, and 5% HCl for 56 d at 20 °C, 50 °C, and 80 °C, respectively. (**a**) SW-20-56, (**b**) 05HA-20-56, (**c**) 05SH-20-56, (**d**) SW-50-56, (**e**) 05HA-50-56, (**f**) 05SH-50-56, (**g**) SW-80-56, (**h**) 05HA-80-56, (**i**) 05SH-80-56.

3.5. SEM Analysis

Scanning electron micrographs of the fracture surfaces of the failed tensile specimens are shown in Figure 12. Comparing the microscopic morphology of the fracture surfaces after 56 d of immersion in SW, 05HA, and 05SH at 20 °C, 50 °C, and 80 °C, it was observed that there was no significant difference in the fracture morphology of the specimens exposed to different chemical environments at 20 °C. The fiber and resin sections were flat and compact, indicating that soaking at 20 °C had a good bonding effect between the fibers and resin, with no significant deterioration. When the ambient temperature was increased to 50 °C, gaps at the resin-glass fiber interface could be observed more clearly, with jagged fracture planes between the fibers. In particular, a large number of holes where the fibers were pulled out existed in the specimen section for the alkali environment. This indicates that, owing to the increased ambient temperature, the bonding between the fibers and resin matrix was substantially reduced, weakening the ability of the fibers to work perfectly with each other. When the immersion temperature was further increased to 80 °C, the bonding of the glass fibers to the epoxy resin in the specimen sections for all three chemical environments was further degraded (especially in the alkaline environment) and the nearly bare fibers, as well as the deboned strips of resin, were clearly visible, indicating that a large amount of resin matrix was withdrawn from the GFRP after 56 d of immersion at 80 °C. These observations support the discussions in Section 3.3.2 (Tensile strength and modulus).

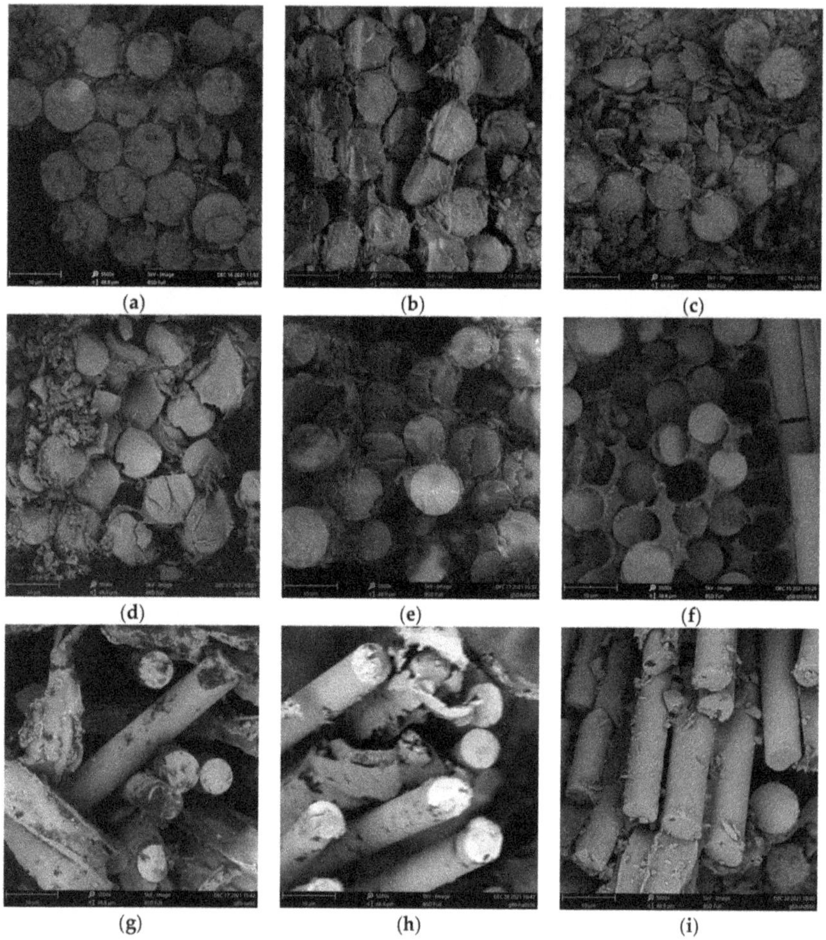

Figure 12. SEM images of fractured sections of specimens. (**a**) SW-20-56, (**b**) 05HA-20-56, (**c**) 05SH-20-56, (**d**) SW-50-56, (**e**) 05HA-50-56, (**f**) 05SH-50-56, (**g**) SW-80-56, (**h**) 05HA-80-56, (**i**) 05SH-80-56.

Figure 13 illustrates the microscopic morphology of the peeled layer surface of a specimen exhibiting the delamination failure mode. Figure 13a,b clearly show the neatly defined grooves left by the peeled fibers, whereas Figure 13c shows the smooth fiber surface after the resin has been peeled. During the stretching process, the fibers on the surface of the sandwich that comprise the GFRP sheet peel off from the resin between the bonded sandwich, resulting in delamination. The delamination only occurred in specimens exposed to higher temperatures in acidic and seawater environments and at the later stage of the tensile test. A plausible explanation for this is that the resin bond between the interlayers of the specimens in these environments was severely degraded beyond the tensile capacity of the single interlayers, with splitting damage occurring before the specimens were damaged; for specimens in deionized water and alkaline environments, the degradation in tensile properties of the specimens was even more severe, with the tensile strength limit being reached before the specimens became delaminated. In addition, a comparison in Figure 13 shows that the higher the immersion temperature, the smoother the grooves left after the fibers are peeled, indicating more severe resin debonding. This explains the macroscopic phenomenon in Section 3.3.1 that delamination damage is more likely to occur at 80 °C than at 50 °C in the same chemical environment.

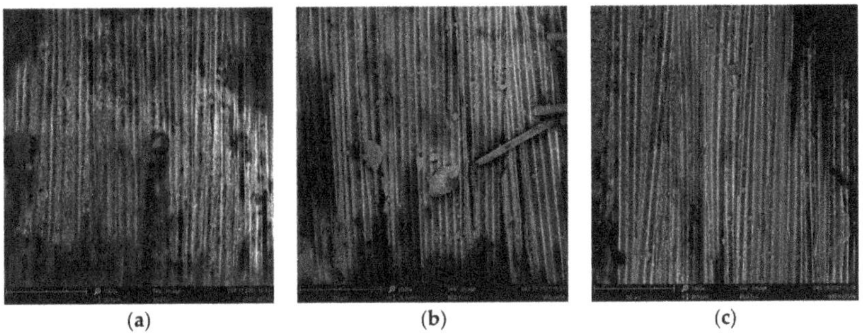

Figure 13. SEM images of the specimens' strip surfaces. (a) 05HA-50-56, (b) SW-80-56, (c) 05HA-80-56.

3.6. Prediction of the Long-Term Behavior and Service Life of GFRP Tensile Strength

Based on the results of the accelerated aging tests, the proposed deterioration model based on the Arrhenius theory [34] was used to predict the long-term tensile properties of the GFRPs. The following assumptions were made using the model: (1) Only one mode of chemical degradation can dominate the deterioration process, and the mode cannot vary with time or temperature. (2) The FRP must deteriorate in an aqueous solution but not in a dry environment [35]. The model considers temperature effects in accelerated aging tests and provides an accurate prediction for materials exposed to temperatures below their glass transition temperatures [36]. In this model, the relationship between performance retention Y and deterioration time t is given by Equation (4).

$$Y = (100 - Y_{\infty}) \exp\left(-\frac{t}{\tau}\right) + Y_{\infty} \quad (4)$$

where Y represents the retention of mechanical properties (%), t is the exposure time, τ is the regression fitting parameter, and Y_{∞} is the retention of mechanical properties (%) for an infinitely long exposure time. Shenzhen, China, was chosen as the target site for the predictions.

In summary, the prediction procedure for each specimen in different chemical exposure environments consisted of four steps. Step 1 involved fitting the test results using Equation (4) to determine the regression fit parameters τ and Y_{∞} at different temperatures. The fitting results are presented in Table 3.

Step 2 was for determining $\frac{E_a}{R}$ (i.e., the slope of the Arrhenius curve) of the samples for each exposure environment, where E_a is the activation energy, and R is the universal gas constant. To this end, a linear fit of the logarithm (ln(t)) of the time taken for the sample retention (%) to reach a certain value (i.e., 95%, 90%, and 85% in this study) was plotted against 1000/K at different exposure temperatures (where K represents the absolute temperature). ln(t) at different temperatures can be determined from Equation (4) and the fitting parameters obtained in step 1. Figure 14 shows the Arrhenius plots and $\frac{E_a}{R}$ values for each sample type. The fitted straight lines for each type are nearly parallel, and the regression lines have R^2 values of at least 0.8106 (all above 0.80), indicating that the accelerated deterioration test is valid and that the model can be used to predict degradation in the tensile strength of the GFRP. The mean values of these slopes represent $\frac{E_a}{R}$ values, with different $\frac{E_a}{R}$ values indicating different degradation rates and possibly different degradation mechanisms.

Table 3. Regression coefficients in Equation (4) and time shift factors (TSF) for the GFRP specimens.

Conditioning	TEMP (°C)	τ	Y_∞	Ea/R	TSF (Shenzhen-22.3 °C)
DW	20	125.93	61.53	6054	0.85
	50	8.13	53.26	6054	5.79
	80	7.18	33.42	6054	28.44
SW	20	70.11	81.75	7222	0.83
	50	12.41	54.89	7222	8.13
	80	4.96	40.81	7222	54.25
5% NaOH	20	58.05	64.30	5111	0.87
	50	14.35	27.81	5111	4.41
	80	8.43	12.49	5111	16.89
10% NaOH	20	47.44	73.58	5015	0.88
	50	13.82	29.33	5015	4.28
	80	9.92	19.33	5015	16.02
5% HCl	20	30.71	82.79	4226	0.89
	50	16.61	66.06	4226	3.41
	80	11.88	47.14	4226	10.50
10% HCl	20	41.37	80.33	4665	0.88
	50	21.39	70.76	4665	3.87
	80	9.23	47.61	4665	13.19

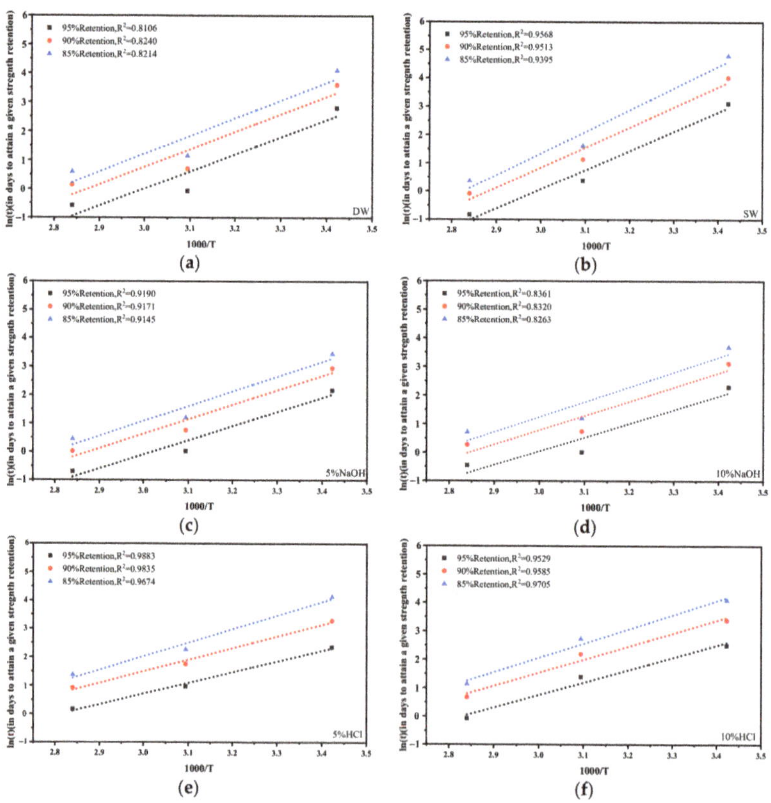

Figure 14. Arrhenius plots: specimens exposed to (**a**) DW, (**b**) SW, (**c**) 5% NaOH, (**d**) 10% NaOH, (**e**) 5% HCl, and (**f**) 10% HCl.

Step 3 was to calculate the time shift factor (TSF) for different exposure conditions using Equation (5).

$$TSF = \exp\left[\frac{E_a}{R}\left(\frac{1}{T_0} - \frac{1}{T_1}\right)\right] \quad (5)$$

where T_0 is the lowest temperature (the annual average temperature in Shenzhen, China) and T_1 is the highest temperature (the exposure temperature in this study). Table 3 lists the TSFs of the six exposure environments.

In the final step, the TSF was multiplied by the corresponding exposure time at different temperatures to obtain the conversion time, and Equation (4) was used to plot the conversion time against the corresponding tensile strength retention to predict the long-term tensile performance of the GFRP, as shown in Figure 15. R^2 of the fitted line is at least 0.94, indicating that the Arrhenius model can accurately predict the deterioration pattern of the GFRP. The predicted results are in agreement with the experimental results, and it is evident that most of the performance loss occurs at early stages, regardless of the exposure environment. Tensile properties were predicted to decrease by almost 40% after only 89 d in a 5% sodium hydroxide environment at 22.3 °C, whereas GFRP showed the best resistance to weak-acidic environments, maintaining 49.4% tensile properties after 500 d in a 5% hydrochloric acid environment at 22.3 °C. Because the test environment is a long-time immersion in a solution, the above prediction results may be conservative compared to those of actual working conditions.

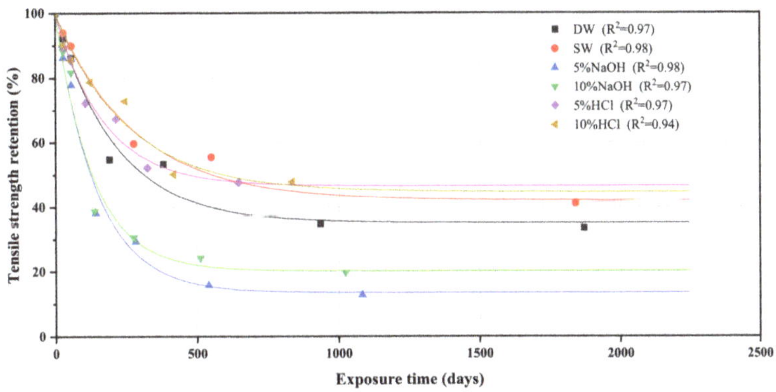

Figure 15. Predicted relationship between residual tensile strength and service life at the temperature of 22.3 °C for the GFRP exposed to various conditions.

4. Conclusions

The change in mass of GFRP increases with increasing ambient temperature and immersion time; however, its rate of mass change decreases with time gradually. In deionized water, artificial seawater, and alkaline solutions at 80 °C, the mass change of the GFRP was mainly characterized by an increase in mass; however, in acidic and alkaline solutions at 20 °C and 50 °C, it was mainly characterized by a mass loss.

The tensile strength and stiffness degradation of the GFRP were positively correlated with mass change. The residual tensile strength decreased with increasing ambient temperature and immersion time, but the rate of decrease gradually became slower with time. Compared to deionized water, artificial seawater, and acid, the alkaline environment had the greatest effect on the degradation of GFRP properties, with only 12.60% and 19.61% of the residual tensile strength after 56 d of exposure to 05SH-80 and 10SH-80, respectively, with degradation of tensile strength being more severe than that of tensile stiffness for GFRP.

The SEM results show that fiber-matrix debonding is most significant when the GFRP is exposed to alkaline environments, whereas the interlayer fibers of the GFRP are completely exposed in acidic and seawater environments at higher temperatures.

It can be deduced from the DIC image analysis that, at 50 °C and 80 °C, the GFRP corrodes uniformly when exposed to deionized water and acids but unevenly when exposed to alkaline solutions. The location of maximum strain may shift during tension, but failure nearly always occurs at the maximum strain.

The Arrhenius model was validated using experimental data, demonstrating that the model has good applicability. Compared to artificial seawater and alkaline environments, the GFRP is better at resisting long-term erosion in acidic environments. The tensile residual strength of the GFRP was predicted, using the Arrhenius model, to be 46.48% and 44.62% after exposure to 5% and 10% hydrochloric acid, respectively, for 2000 d in Shenzhen. The above test results and long-term performance prediction are of reference significance for the application of GFRP in Shenzhen.

Author Contributions: Conceptualization, all authors; methodology, J.Z., Y.D., P.C., G.W. and H.M; validation, Y.D. and P.C.; resources, J.Z.; data curation, Y.D. and P.C.; writing—original draft preparation, J.Z. and Y.D.; writing—review and editing, G.W., H.M. and P.C.; visualization, Y.D.; supervision, W.F.; project administration, W.F.; funding acquisition, G.W. and H.M. All authors have read and agreed to the published version of the manuscript.

Funding: This work was funded by the Key-Area Research and Development Program of Guangdong Province (grant number 2019B111107002) and the Major Projects of Central Research Institute of Building and Construction Co., Ltd., MCC Group (grant number JAA2018Kj01).

Institutional Review Board Statement: This study does not involve ethical research, so this statement is excluded.

Informed Consent Statement: Written informed consent has been obtained from the patient(s) to publish this paper.

Data Availability Statement: The data supporting this study are available from the corresponding author upon reasonable request.

Acknowledgments: We would like to recognize numerous co-workers, students, and research facility associates for giving specialized assistance on instrument use.

Conflicts of Interest: The authors declare no conflict of interest.

References

1. Rajak, D.K.; Pagar, D.D.; Menezes, P.L.; Linul, E. Fiber-Reinforced Polymer Composites: Manufacturing, Properties, and Applications. *Polymers* **2019**, *11*, 1667. [CrossRef] [PubMed]
2. Hollaway, L.C. Polymer composites in construction: A brief history. *Proc. Inst. Civ. Eng. Eng. Comput. Mech.* **2009**, *162*, 107–118. [CrossRef]
3. Lu, C.; Yang, Y.; He, L. Mechanical and durability properties of GFRP bars exposed to aggressive solution environments. *Sci. Eng. Compos. Mater.* **2021**, *28*, 11–23. [CrossRef]
4. Wang, M.; Xu, X.; Ji, J.; Yang, Y.; Shen, J.; Ye, M. The hygrothermal aging process and mechanism of the novolac epoxy resin. *Compos. Part B Eng.* **2016**, *107*, 1–8. [CrossRef]
5. Liao, K.; Schultheisz, C.R.; Hunston, D.L. Effects of environmental aging on the properties of pultruded GFRP. *Compos. Part B Eng.* **1999**, *30*, 485–493. [CrossRef]
6. Nishizaki, I.; Meiarashi, S. Long-term deterioration of GFRP in water and moist environment. *J. Compos. Constr.* **2002**, *6*, 21–27. [CrossRef]
7. Liu, T.; Liu, X.; Feng, P. A comprehensive review on mechanical properties of pultruded FRP composites subjected to long-term environmental effects. *Compos. Part B Eng.* **2020**, *191*, 107958. [CrossRef]
8. Ameli, A.; Datla, N.V.; Papini, M.; Spelt, J.K. Hygrothermal Properties of Highly Toughened Epoxy Adhesives. *J. Adhes.* **2010**, *86*, 698–725. [CrossRef]
9. Mubashar, A.; Ashcroft, I.A.; Critchlow, G.W.; Crocombe, A.D. Moisture absorption-desorption effects in adhesive joints. *Int. J. Adhes. Adhes.* **2009**, *29*, 751–760. [CrossRef]
10. Shao, Y.; Kouadio, S. Durability of Fiberglass Composite Sheet Piles in Water. *J. Compos. Constr.* **2002**, *6*, 280–287. [CrossRef]

11. Grammatikos, S.A.; Evernden, M.; Mitchels, J.; Zafari, B.; Mottram, J.T.; Papanicolaou, G.C. On the response to hygrothermal aging of pultruded FRPs used in the civil engineering sector. *Mater. Des.* **2016**, *96*, 283–295. [CrossRef]
12. Xin, H.; Liu, Y.; Mosallam, A.; Zhang, Y. Moisture diffusion and hygrothermal aging of pultruded glass fiber reinforced polymer laminates in bridge application. *Compos. Part B Eng.* **2016**, *100*, 197–207. [CrossRef]
13. Sousa, J.M.; Correia, J.R.; Cabral-Fonseca, S. Durability of an epoxy adhesive used in civil structural applications. *Constr. Build. Mater.* **2018**, *161*, 618–633. [CrossRef]
14. Sousa, J.M.; Correia, J.R.; Gonilha, J.; Cabral-Fonseca, S.; Firmo, J.P.; Keller, T. Durability of adhesively bonded joints between pultruded GFRP adherends under hygrothermal and natural ageing. *Compos. Part B Eng.* **2019**, *158*, 475–488. [CrossRef]
15. Cabral-Fonseca, S.; Correia, J.R.; Rodrigues, M.P.; Branco, F.A. Artificial Accelerated Ageing of GFRP Pultruded Profiles Made of Polyester and Vinylester Resins: Characterisation of Physical-Chemical and Mechanical Damage. *Strain* **2012**, *48*, 162–173. [CrossRef]
16. Xin, H.; Liu, Y.; Mosallam, A.; Zhang, Y.; Wang, C. Hygrothermal aging effects on flexural behavior of pultruded glass fiber reinforced polymer laminates in bridge applications. *Constr. Build. Mater.* **2016**, *127*, 237–247. [CrossRef]
17. Karbhari, V.M.; Zhang, S. E-glass/vinylester composites in aqueous environments-I: Experimental results. *Appl. Compos. Mater.* **2003**, *10*, 19–48. [CrossRef]
18. Feng, P.; Wang, J.; Wang, Y.; Loughery, D.; Niu, D. Effects of corrosive environments on properties of pultruded GFRP plates. *Compos. Part B Eng.* **2014**, *67*, 427–433. [CrossRef]
19. Hota, G.; Barker, W.; Manalo, A. Degradation mechanism of glass fiber/vinylester-based composite materials under accelerated and natural aging. *Constr. Build. Mater.* **2020**, *256*, 119462. [CrossRef]
20. Kanerva, M.; Jokinen, J.; Sarlin, E.; Pärnänen, T.; Lindgren, M.; Järventausta, M.; Vuorinen, J. Lower stiffness of GFRP after sulfuric acid-solution aging is due to degradation of fibre-matrix interfaces? *Compos. Struct.* **2019**, *212*, 524–534. [CrossRef]
21. Bazli, M.; Ashrafi, H.; Oskouei, A.V. Effect of harsh environments on mechanical properties of GFRP pultruded profiles. *Compos. Part B Eng.* **2016**, *99*, 203–215. [CrossRef]
22. Xue, W.; Li, Y.; Fu, K.; Hu, X.; Li, Y. Accelerated ageing test and durability prediction model for mechanical properties of GFRP connectors in precast concrete sandwich panels. *Constr. Build. Mater.* **2020**, *237*, 117632. [CrossRef]
23. Pan, B.; Qian, K.; Xie, H.; Asundi, A. Two-dimensional digital image correlation for in-plane displacement and strain measurement: A review. *Meas. Sci. Technol.* **2009**, *20*, 062001. [CrossRef]
24. Fergani, H.; Di Benedetti, M.; Miàs Oller, C.; Lynsdale, C.; Guadagnini, M. Durability and degradation mechanisms of GFRP reinforcement subjected to severe environments and sustained stress. *Constr. Build. Mater.* **2018**, *170*, 637–648. [CrossRef]
25. Sousa, J.M.; Garrido, M.; Correia, J.R.; Cabral-Fonseca, S. Hygrothermal ageing of pultruded GFRP profiles: Comparative study of unsaturated polyester and vinyl ester resin matrices. *Compos. Part A Appl. Sci. Manuf.* **2021**, *140*, 106193. [CrossRef]
26. Soykok, I.F.; Sayman, O.; Pasinli, A. Effects of hot water aging on failure behavior of mechanically fastened glass fiber/epoxy composite joints. *Compos. Part B Eng.* **2013**, *54*, 59–70. [CrossRef]
27. Correia, J.R.; Cabral-Fonseca, S.; Branco, F.A.; Ferreira, J.G.; Eusébio, M.I.; Rodrigues, M.P. Durability of pultruded glass-fiber-reinforced polyester profiles for structural applications. *Mech. Compos. Mater.* **2006**, *42*, 325–338. [CrossRef]
28. Xin, H.; Liu, Y.; Mosallam, A.S.; He, J.; Du, A. Evaluation on material behaviors of pultruded glass fiber reinforced polymer (GFRP) laminates. *Compos. Struct.* **2017**, *182*, 283–300. [CrossRef]
29. Huang, G.; Sun, H. Effect of water absorption on the mechanical properties of glass/polyester composites. *Mater. Des.* **2007**, *28*, 1647–1650. [CrossRef]
30. Chen, Y.; Davalos, J.F.; Ray, I.; Kim, H.Y. Accelerated aging tests for evaluations of durability performance of FRP reinforcing bars for concrete structures. *Compos. Struct.* **2007**, *78*, 101–111. [CrossRef]
31. Stamenović, M.; Putić, S.; Rakin, M.; Medjo, B.; Čikara, D. Effect of alkaline and acidic solutions on the tensile properties of glass-polyester pipes. *Mater. Des.* **2011**, *32*, 2456–2461. [CrossRef]
32. Amaro, A.M.; Reis, P.N.B.; Neto, M.A.; Louro, C. Effects of alkaline and acid solutions on glass/epoxy composites. *Polym. Degrad. Stabil.* **2013**, *98*, 853–862. [CrossRef]
33. D'Antino, T.; Pisani, M.A.; Poggi, C. Effect of the environment on the performance of GFRP reinforcing bars. *Compos. Part B Eng.* **2018**, *141*, 123–136. [CrossRef]
34. Wu, G.; Dong, Z.Q.; Wang, X.; Zhu, Y.; Wu, Z.S. Prediction of Long-Term Performance and Durability of BFRP Bars under the Combined Effect of Sustained Load and Corrosive Solutions. *J. Compos. Constr.* **2015**, *19*, 04014058. [CrossRef]
35. Bank, L.C.; Gentry, T.R.; Thompson, B.P.; Russell, J.S. A model specification for FRP composites for civil engineering structures. *Constr. Build. Mater.* **2003**, *17*, 405–437. [CrossRef]
36. Naya, S.; Meneses, A.; Tarrio-Saavedra, J.; Artiaga, R.; Lopez-Beceiro, J.; Gracia-Fernandez, C. New method for estimating shift factors in time-temperature superposition models. *J. Therm. Anal. Calorim.* **2013**, *113*, 453–460. [CrossRef]

MDPI
St. Alban-Anlage 66
4052 Basel
Switzerland
Tel. +41 61 683 77 34
Fax +41 61 302 89 18
www.mdpi.com

Polymers Editorial Office
E-mail: polymers@mdpi.com
www.mdpi.com/journal/polymers

www.ingramcontent.com/pod-product-compliance
Lightning Source LLC
Chambersburg PA
CBHW041018270926
51850CB00024B/1475
* 9 7 8 3 0 3 6 5 8 4 0 9 6 *